EUROPEAN DIRECTORY OF RESEARCH CENTERS IN THE FISHERIES SECTOR

European Directory of Research Centers in the Fisheries Sector

Second edition

Edited by

Commission of the European Communities

SPRINGER-SCIENCE+BUSINESS MEDIA, B.V.

FOR THE COMMISSION OF THE EUROPEAN COMMUNITIES

Library of Congress Cataloging-in-Publication Data

European directory of research centers in the fisheries sector / edited by Commission of the European Communities. -- 2nd ed.
p. cm.
Includes index.

1. Fisheries--Research--Europe--Directories. 2. Aquaculture--Research--Europe--Directories. I. Commission of the European Communities.
SH332.2.E85 1994
338.3'727'007204--dc20 93-47118

ISBN 978-94-010-4405-9 ISBN 978-94-011-0940-6 (eBook)
DOI 10.1007/978-94-011-0940-6

Publication arrangements by
Commission of the European Communities,
Dissemination of Scientific and Technical Knowledge Unit,
Directorate-General Telecommunications,
Information Market and Exploitation of Research, Luxembourg

EUR 15138 EN

LEGAL NOTICE
Neither the Commission of the European Communities nor any person acting on behalf of the Commission is responsible for the use which might be made of the following information.

Printed on acid-free paper

Table of contents

Preface

The future of the Common Fisheries Policy depends on progress in the relevant areas of research. This applies to the whole range of management decisions, where precise, reliable and complete data are essential to inform those who must decide on the pursuit of existing activities, especially in the area of maritime fisheries, and the development of promising new activities such as aquaculture. Every day the Director-General of DG XIV requires more and more information to prepare decisions which will affect the future of all those in the Community who are dependent on fishing and aquaculture.

There is thus a high level of direct demand from DG XIV. Over and above this immediate and specific requirement for short- and medium-term applications, research affects the competitivity of the Community. This is one area which favours the collaboration across frontiers of all those who seek to advance knowledge.

But although DG XIV is uniquely placed to appreciate the importance of research into fisheries and aquaculture, there is no question of succumbing to the temptation to directly control the scope of research or its conduct. The notion of subsidiarity can best be understood by examining the existing structures in the Member States. The Commission must act first and foremost as a catalyst, by promoting the circulation of information and the coordination of research programmes.

For this reason I take particular pleasure in writing the preface to this directory. It illustrates how the Commission can provide an inducement, by encouraging closer contacts between the different laboratories and research organizations which exist in the Member States. May all those who wish to be better assimilated into the Community research network find in this directory the information which will lead them to those of like mind.

J. ALMEIDA SERRA
Director General

Foreword to the second edition

The publication of the first edition of this directory was enthusiastically welcomed by the scientific community, private industry and those working in the administration of research in the field of fisheries and aquaculture.

It was realised from the outset that the first edition could not be complete. Therefore it was decided to publish a second edition very soon after the first to which laboratories which expressed the wish to be included could be added and corrections made to factual errors which were notified.

Fisheries and aquaculture research is becoming increasingly inter-national, going beyond the borders of the European Community. For this reason it was felt appropriate to include in the second edition research institutes working in this field from Austria, Finland, Iceland, Norway and Sweden.

We would like to thank all those who contributed to this revised edition and express the hope that it will contribute to successful international collaboration in fisheries science.

State of Research

AUSTRIA

Fishery research work in Austria is done by university departments, federal research centres and research organisations. Fish breeding is regulated by the Länder and organized within certain associations.

The Association for Austrian Trout Breeders undertakes advertising for Austrian trouts. These advertising campaigns are carried out in cooperation with the feeding stuff industry. The aim of this association is to promote sales of its products, to represent the association and to provide training for its members.

Carp producers are organized within two associations: the Pond Managers Association of Styria and the Association of Fishery Managers of Lower Austria and Vienna.

From an economic point of view aquaculture is of minor importance. Trout ponds are to be found especially in the alpine and prealpine regions of Austria. Carp production is centred in the northern parts of Austria ("Waldviertel") as well as in the south of the country (Styria, Burgenland).

The task of the so called County Fishery Association is to provide appropriate fishery management. This means:

* Promotion of fisheries, including the training of fishermen.
* Advising public authorities and all other public and private bodies or persons dealing with fishery and water management, by the provision of expert opinions.
* Surveillance and statistical evaluation of stocking and catches.
* Participation in water protection.

List of Institutes from AUSTRIA

1. Biologische Station Illmitz
2. Bundesanstalt für Wassergüte
3. Bundesanstalt für Fischereiwirtschaft
4. Forschungsstelle Wilhelminenburg
5. Institut für Seenforschung Kärnten
6. Ökologische Station Walderviertel
7. Österreichische Akademie der Wissenschaften
8. Österreische Akademie der Wissenschaften
9. Universität für Bodenkultur
10. Universität Graz
11. Universität Innsbruck
12. Universität Salzburg
13. Universität Wien
14. Veterinäermedizinische Universität Wien

1. Biologische Station Illmitz

Name : Biologische Station Illmitz
Address : Biologische Station
AUS - 7142 Illmitz
Head : Prof. Dr. Alois Herzig
Tel. : 02175.2328.29 **Fax**: 02175.2328.10

Type of institute
Research centre

Research activities
Fisheries management
Fishing analysis and stock assessment on running waters

Facilities
Research vessels
Chemistry laboratory
Microbiological laboratory
Microscope laboratory

Staff: Scientists: 15 Others: 18

Research budget (ECU/year): 58.200

Description
Autecology and synecology of economically unused cyprinids

2. Bundesanstalt für Wassergüte

Name : Bundesanstalt für Wassergüte
Address : Schiffmühlenstraße, 120
AUS - 1120 Wien
Head : Prof. Dr. Werner Kohl
Tel. : 1.23.45.91 **Fax**: 1.23.45.91.15

Type of institute
Research centre

Research activities
Applied limnological research
Ecotoxicological analysis and assessments

Facilities
Histological, analytical, microbiological, biological and ecotoxicological laboratories
Fish breeding installation
Microscopes, computers, testing equipment
Electric fishing equipment
Library

Staff: Scientists: 13 Others: 37

Research budget (ECU/year): 350.000

Description
Water quality analysis of running waters and lakes (chemical, biological, bacteriological).
Analysis of fish mortality.
Emission control.
Applied limnological research.
Ecotoxicological analysis of substances in all trophical levels.

3. Bundesanstalt für Fischereiwirtschaft

Name	:	Bundesanstalt für Fischereiwirtschaft
Address	:	Scharfling
		AUS - 5310 Mondsee
Head	:	Dr. Albert Jagsch
Tel.	:	06232.3847 **Fax**: 06232.3847.33

Type of institute
Research centre

Research activities
Human influences on running waters and lakes
Optimizing of fish production
Maintenance of endangered species
Health of fish

Facilities
Water chemistry laboratory
Research vessels
Aquaria room
Fish breeding facilities

Staff: Scientists: 6 Others: 23

Research budget (ECU/year): 36.400

Description
Consequences of electric fishing on the survival-rate and growth of fish.
Biological analysis of fish regarding acidification of waters.
Analysis of frozen-conservation and artificial insemination of native and endangered species.
Fishing utilization of flooded gravel pits.
Parasitological analysis of *Barbus Fluviatilis* and *Leuciscus Cephalus*.

4. Forschungsstelle Wilhelminenburg

Name : Institut für angewandte Ökoethologie
Address : Glasergasse, 20
AUS - 1090 Wien
Head : Dr. Hans Peter Kollar
Tel. : 0222.310.98.97 **Fax**: 0222.310.98.97

Type of institute
Research centre

Research activities
Hydroelectric installation (Greifenstein)
Settlement of cormorants
Cultivation of the landscape

Facilities
Research vessel
Water laboratory equipment

Staff: Scientists: 8 Others: 5

Description
As a result of the building of the Greifenstein generator more water is running through the river valley. The consequences of improved irrigation are examined.

5. Institut für Seenforschung Kärnten

Name : Institut für Seenforschung Kärnten
Address : Flatschacherstraße, 70
AUS - 9020 Klagenfurt
Head : Prof. Dr. Hans Sampl
Tel. : 0463.33190.243 **Fax**: 0463.33190.213

Type of institute
Research centre

Research activities
Fisheries management
Stock assessment
Analysis of productivity

Facilities
Vessels
Electric aggregates
Echo sounder
Test fishing nets

Staff: Scientists: 4 Others: 3

Research budget (ECU/year): 14.000

Description
Fish growth and composition of species in Carinthian lakes.
Problems of endangered species.
Rehabilitation of water and lake water priority.

6. Ökologische Station Walderviertel

Name : Ökologische Station Walderviertel
Address : Gebharts, 33
AUS - 3943 Schrems
Head : Dr. Günther Schlott
Tel. : 02853.78207 **Fax**: 02853.78463

Type of institute
Research centre

Research activities
Lake fishery management
Consulting

Facilities
Vessel
Water laboratory
Microscopes

Staff: Scientists: 2 Others: 3

Description
Feeding project Waldviertel: carps in natural feeding with low water density.
Assessment of the fish fauna in the Waldviertel.
Protection against otters in fish ponds.
Waldviertel anglers paradise-Southern Bohemia.

7. Österreichische Akademie der Wissenschaften

Name	:	Institut für Limnologie-Abt. Biologische Station Lunz
Address	:	Seehof, 4
		AUS - 3293 Lunz am See
Head	:	Prof. Dr. Gernot Bretschko
Tel.	:	07486.517 **Fax**: 07486.33031

Type of institute
Research centre

Research activities
Limnology of running waters
Ecology of detritus sediments

Facilities
Granulometry
Freescore Technic

Staff: Scientists: 8 Others: 10

Description
Ritrodat Lunz: ecology of detritus streams with low ordinal number, particularly the metabolism of allochthonal organic substances.
Dynamic ecothon-analysis of the Danube.

8. Österreische Akademie der Wissenschaften

Name	:	Institut für Limnologie-Abteilung Mondsee
Address	:	Gaisberg
		AUS - 5310 Mondsee
Head	:	Prof. Dr. Arnold Nauwerck
Tel.	:	06232.3125.12 **Fax**: 06232.3578

Type of institute
Research centre

Research activities
Variability of structure and adjustment of aquatic systems
Research of lakes
Fish biology
Paleolimnology

Facilities
Fishing vessels
Electron microscope
Auto-analyser
Carbon-analyser
Visual-analyser

Staff: Scientists: 8 Others: 8

Research budget (ECU/year): 85.000

Description
Dynamics of interactions between organisms and their habitat.
Long-term changes of ecosystems and organisms.
The investigation of natural and human influences on changes in aquatic systems.
Development of basic data for the recognition of changes and their consequences.

9. Universität für Bodenkultur

Name : Institut für Hydrobiologie, Fischereiwirtschaft und Aquakultur
Address : Feistmantelstraße, 4
AUS - 1180 Wien
Head : Prof. Dr. Matthias Jungwirth
Tel. : 0222.47564.11 **Fax**: 0222.31.95.670

Type of institute
University Department

Research activities
Ecology of running waters
Water pollution control
Environmental compatibility tests
Revitalization
Quality assessments

Facilities
Research vessels
Electric fishing equipment
Aquacultural laboratory
Quality laboratory

Staff: Scientists: 30 Others: 10

Research budget (ECU/year): 728.000

Description
Fundamental and applied research on running waters.

10. Universität Graz

Name	:	Institut für Zoologie
Address	:	Universitätsplatz, 2 AUS - 8010 Graz
Head	:	Prof. Dr. Heinrich Römer
Tel.	:	0316.380.5595 **Fax**: 0316.381255

Type of institute
University Department

Research activities
Theses and doctoral theses concerning fisheries
Eutrophic problems
Influence of the cultivation of the soil

Facilities
Fishing equipment
Microscopes

Staff: Scientists: 20 Others: 7

Description
Analysis of productivity and problems of endangered species.
Growth assessment in domestic lakes.
Rehabilitation of water.

11. Universität Innsbruck

Name : Institut für Zoologie, Abteilung Ökophysiologie
Address : Universitätsstraße, 4
AUS - 6020 Innsbruck
Head : Prof. Dr. Wolfgang Wieser
Tel. : 0512.218-5303 **Fax**: 0512.218.5358

Type of institute
University Department

Research activities
Physiology
Toxicology
Pathology

Facilities
Laboratory equipment
Microscopes
Photometry

Staff: Scientists: 5 Others: 5

Research budget (ECU/year): 70.000

Description
Ecotoxicological project: Consequences of extreme environmental factors for fish in alpine lakes.
Consequences of communal and industrial effluent for fish.
Biochemical analysis on fish.
Nitrate toxicity.

12. Universität Salzburg

Name : Institut für Zoologie
Address : Hellbrunnerstraße, 34
AUS - 5020 Salzburg

Head : Prof. Dr. Hans Adam
Tel. : 0662.8044.5600 **Fax**: 0662.8044.5698

Type of institute
University Department

Research activities
Ecology of running water
Reproduction
Ecology of fish
Environmental toxicology
Water assessment with protozoon

Facilities
Electron microscope
Atom absorption spectometry
Electric aggregates

Staff: Scientists: 4 Others: 3

Description
Reproductional biology of fish: morphological, biochemical and cryotechnical analysis of gametes of native fishes.
Circulation and population dynamics of mollusc and mussels.
Ecomorphology of fish and protozoon as bioindicator.
Ecology of small alpine waters.

13. Universität Wien

Name : Institut für Zoologie - Abteilung Limnologie
Address : Althanstraße, 19
AUS - 1090 Wien
Head : Prof. Dr. Heinz Löffler
Tel. : 0222.31336.1225 **Fax**: 0222.31336.700

Type of institute
University Department

Research activities
Ecology of flood waters
Nature conservation
Functional role of fish in aquatic systems

Facilities
Electron microscope
Atom-absorption
Echography
Laboratory for genetics of population
Hydrochemical laboratory

Staff: Scientists: 6 Others: 3

Description
Ecology of fish particularly regarding large river systems.
Analysis of metabolism of fish.
Limnological field site analysis.

14. Veterinäermedizinische Universität Wien

Name	:	Institut für Fisch- und Bienenkunde
Address	:	Linke Bahngasse 11
		AUS - 1030 Wien
Head	:	Prof. Dr. Walter Grünberg
Tel.	:	0222.71155.551

Type of institute
University Department

Research activities
Ecotoxicology
Aquatic toxicology
Immunology of aquatic organisms

Facilities
Vessel
Echo sounder
Specialised analytical apparatus
Adequate laboratory equipment

Staff: Scientists: 5 Others: 7

Description
Prophylaxis, diagnosis and treatment of diseases of domestic fishes and other organisms kept in aquaculture.

State of Research

BELGIUM

The Belgian coast is very short (67 km) and has only three fishing ports (Nieuwpoort, Oostende annd Zeebrugge), with a total of 200 ships. The Belgian fleet is specialised in fishing for flat-fish.

Research on Fisheries in Belgium is carried out in the research centres of the Ministry of Agriculture, the Ministry of National Health and Environment, and also relevant departments of the different Belgian universities.

Research priorities were defined as:
* research focused on high net output and research allowing a valorisation of the products.
* socio-economical and biological research in fisheries.

Most of this research is conducted at the Ostend Station (Rijksstation voor Zeevisserij), which depends on the Ministry of Agriculture. The research programme of the station can be considered as being the national programme in R&D for fisheries. It is conducting research in three fields: marine biology, fishing techniques, and the technology of fishing products.

The Station participates within the framework of marine research in the scientific research programmes of several international organisations. At the domestic level, there are collaboration programmes with universities, other institutes or research stations and other government departments, where the Station is involved.

Meetings between the representatives of the Station, representatives of Ministry, and representatives of the industry take place to comment on the research activities. Specific questions may also be brought before the Professional Fishing Council, which is made up of representatives from industry, fishermen and fisheries specialists.

No university training exists in fisheries, which makes it difficult to recruit scientific staff for marine research. Scientists come from the biological, agronomical and veterinary faculties and have to be formed at the research institutes.

Transfer of technology towards developing countries is done through different training courses, and twice yearly, practical courses are organised by the "International Training and Contact" body.

Aquaculture should be more developed in Belgium, as the country, like most European countries, depends highly on imports to satisfy its needs. However, the main rivers (Scheldt, Meuse) are highly polluted by industrial and urban discharges, which make them unsuitable for aquaculture. Aquaculture is only done, on a small scale, in the

Ardennes and near the nuclear power plant near Namur - as yet, the whole Belgian potential is not used.

Belgian aquaculture is still in its development phase, yet a maximum will be quickly reached because of the lack of proper water resources.

However the research activities in aquaculture which are being carried out by the Research Station, different universities and companies are relevant.

List of Institutes from BELGIUM

1. Artemia Systems
2. EUROGENTEC
3. IZWO - Institute for Marine Scientific Research
4. MUMM - Math Models of North Sea and Scheldt Estuary
5. Musée Royal de l'Afrique Centrale / Koninklijk Museum voor Midden-Afrika
6. Rijksstation voor Zeevisserij
7. Univ. Bruxelles - ULB - Lab. Chimie Bio-Organique
8. Univ. Gent - RUG - Artemia Center
9. Univ. Gent - RUG - Fac. Agricultural Sciences
10. Univ. Gent - RUG - Fac. Environment
11. Univ. Gent - RUG - Lab. Ecologie der Dieren
12. Univ. Gent - RUG - Lab. Medical Biochemistry
13. Univ. Gent - RUG - Marine Biologie
14. Univ. Leuven - KUL - Zool. Inst. - Lab. Ecologie Aquacultuur
15. Univ. Louvain la Neuve - UCL - Biochimie de la Nutrition
16. Univ. Louvain la Neuve - UCL - Fish Technology Group
17. Univ. Louvain la Neuve - UCL - Genetique CGEN
18. Univ. Louvain la Neuve - UCL - Lab. Pisciculture - M. Huet
19. Univ. Namur - ND de la Paix - Ecologie Eaux Douces

1. Artemia Systems

Name : ARTEMIA Systems
R&D Dept.
Address : Wiedauwkaai 79
B - 9000 Gent 5
Head : M. Coomans / P. Léger
Tel. : 091.534142 **Fax**: 091.536893 **Telex**: 11475 ARTSYS

Type of institute
Processing Industry/agro-food/diets producer
Engineering/consulting firm
Research centre
Distribution

Research activities
Aquaculture
Reproduction
Early stages of reared species
Growth, nutrition, feed composition, live food production for aquaculture
Disease, immunology, stress

Facilities
Rearing tank
Field site specialised analytical apparatus

Staff: Scientists: 10 Others: 7

Research budget (ECU/year): 518.000

Description
Research projects:
Development of a complete larval feed product range for supplying to fish, shrimp and mollusc hatcheries reducing the dependence on live food organisms. Studies of microbiological aspects in aquaculture hatcheries which scope the possibility of biomedication through the food chain.

2. EUROGENTEC

Name : Eurogentec S.A.
Société Européenne de Biotechnologie S.A.
Address : Parc de Recherche de la Cense Rouge
Rue Bois Saint Jean 14
B - 4102 Seraing

Head : A. Renard
Tel. : 41.660150 **Fax**: 41.655103 **Telex**: 41.562968

Type of institute
Research centre
Veterinary Pharmaceuticals Producer

Research activities
Aquaculture
Physiology
Growth, nutrition, feed composition, live food production for aquaculture
Disease, immunology, stress
Aquaculture economics

Facilities
Field site, electron microscope, specialized analytical apparatus
Experimental unit for vaccine trials
Biotechnology research laboratory
Fermentation unit

Staff: Scientists: 20 Others: 5

Research budget (ECU/year): 615.000

Description
Research activities:
Recombinant vaccines against VHS, IHN and SVC.
Smoltification of Salmon.
Growth hormone, protection of fish.
Cold Renstance.
Drug delivery into fish.
Oral delivery of proteins.

3. IZWO - Institute for Marine Scientific Research

Name : IZWO-Instituut voor Zeewetenschappelijk Onderzoek (vzw)
Institute for Marine Scientific Research
Address : Victorialaan 3
B - 8400 Oostende
Head : Dr. ir. E. Jaspers
Tel. : 32(0)59.321045 **Fax**: 32(0)59.321135/321813

Type of institute
Non-profit organization

Research activities
Coordination of marine research projects effectuated by laboratories of Flemish Universities and Governmental Scientific Institutes;
Carrying out marine research projects (all disciplines) upon demand from Governmental and private organizations.

Facilities
Offices, field laboratory, marine scientific library

Staff: Scientists: 1 Others: 2

Description
IZWO is a non-profit organization that coordinates the marine research projects of the various Flemish universities. It organizes symposia, workshops etc. in marine sciences, and an interuniversitary postgraduate programme in 'Marine Ecology'. IZWO holds a substantial scientific library devoted to marine sciences. Annually it publishes the "IZWO Collected Reprints" available on exchange basis upon request.

4. MUMM - Math Models of North Sea and Scheldt Estuary

Name	:	MUMM-Management Unit of the Mathematical Models of the North Sea and Scheldt Estuary
Address	:	Gulledelle 100 (AEE) B - 1200 Bruxelles
Head	:	G. Pichot
Tel.	:	32.2.7732123 **Fax**: 32.2.7706972 **Telex**: 65752 MUMM B

Type of institute
Government Department/regional or local authority
Engineering/consulting firm
Research centre

Research activities
Fisheries management
Environmental problems and their impact on resources

Facilities
Data bases, extensive libraries
Specialised computer hardware

Specialised software
Research vessel "Belgica", potentially open to external users.

Staff: Scientists: 15 Others: 10

Research budget (ECU/year): 697.300

Description
Research projects:
Modelling of the ice-seas in the Antarctic area. Mathematical simulation of the phytoplancton (Phaeocystis) spring bloom in nutriment enriched coastal waters. Development of software aiming at the simulation of oil and other substances disposed of at sea. Models to be used in case of marine pollution incidents. Hydrodynamical models (currents, tides, dispersion). Ecological modelling cycle of pollutants.

5. Musée Royal de l'Afrique Centrale / Koninklijk Museum voor Midden-Afrika

Name : Musée Royal de l'Afrique Centrale/Koninklijk Museum voor Midden-Afrika
Laboratoire d'Ichtyologie
Address : Leuvensesteenweg 18
B - 3080 Tervuren
Head : D. Thys Van Den Audenaerde / G.G. Teugels
Tel. : 32.2.7695229 **Fax**: 32.2.7670242

Type of institute
Research centre
Institute for professional training/Higher Education Institute

Research activities
Fisheries management
- Multi-species biological models
- Multi-disciplinary approach to fisheries management
- Environmental problems and their impact on resources

Aquaculture
- Genetics and conservation of gametes
- Early stages of reared species

Facilities
Data bases, extensive libraries

Staff: Scientists: 6 Others: 3

Description
The Laboratory is specialised in different aspects of especially African freshwater fish. It studies, in collaboration with national and international organizations, problems in taxonomy (classical and recent techniques); aspects of fishculture (strain-identification ; early development of cultured species; caryology; enzymatic variation); effects of insecticides, oil drillings, etc. on fresh-and brackish water fishfauna. On going projects cover West-Africa, from Senegal to Nigeria (collaboration with ORSTOM), Cameroon, Congo Brazzaville and Burundi. It has an important reference collection of fish specimens and an important library on various aspects related to African fish, fisheries and fishculture.

6. Rijksstation voor Zeevisserij

Name : Ministerie van Landbouw
Rijksstation voor Zeevisserij
Agricultural Research Centre
Address : Ankerstraat 1
B - 8400 Oostende
Head : Dr. P. Hovart / Dr. ir. W. Vyncke
Tel. : 32(0)59 320388 **Fax**: 32(0)59 330629

Type of institute
Research centre

Research activities
Fisheries management
Stock assessment
Environmental problems and their impact on resources
Fishing techniques
Upgrading of fishery products
Improvement of techniques for handling, storing, processing, packaging fish
Methods for grading and quality assessment of fish

Facilities
Research vessel
Specialised computer hardware
Specialised software

Staff: Scientists: 12 Others: 25

7. Univ. Bruxelles - ULB - Lab Chimie Bio-Organique

Name : Université Libre de Bruxelles (ULB)
Faculté des Siences / Lab. de Chimie Bio-Organique
Address : CP 160
Av. Fr. Roosevelt, 50
B - 1050 Bruxelles
Head : J.C. Braekman
Tel. : 32.(0)2.6422961

Type of institute
University Department

Research activities
Environmental problems (pollution due to algea blooms) and their consequences on resources, mainly edible molluscs

Facilities
Specialized analytical apparatus
Mass spectrometry
Nuclear, magnetic resonance spectroscopy
HPLC
Gas chromatography
Laboratory of aquatic chemistry
UV and IR spectroscopy
GC:MS instrumentation

Staff: Scientists: 15

Research budget (ECU/year): 62.000

Description
Research projects:
Development of HPLC analytical procedures to quantify the PSP and DSP toxins in consumable shellfish in contaminated waters.

8. Univ. Gent - RUG - Artemia Center

Name : RUG - Rijksuniversiteit Gent
Faculty of Agricultural Sciences
Laboratory of Aquaculture and Artemia Reference Center
Address : Rozier 44
B - 9000 Gent

Head : Prof. P. Sorgeloos
Tel. : 32(0)91.643754 **Fax**: 32(0)91.644193 **Telex**: 12754 B

Type of institute
Research centre
University Department/Higher Education Institute

Research activities
Aquaculture
Early stages of reared species
Growth, nutrition, feed composition, live food production for aquaculture
Disease, immunology, stress
Aquaculture structures, systems, techniques

Facilities
Different rearing installations for fish and shellfish larvae and their food (algae, Brachionus, Artemia)
Analytical apparatus for lipid classes and fatty acids
Library and Artemia data base

Staff: Scientists: 16 Others: 14

Research budget (ECU/year): 348.600

Description
1. Techniques and feeds for intensive Artemia culturing (international coop.)
2. Production, enrichment and use of Brachionus in marine fish larvae culture.
3. Optimization of larviculture methods through improved nutrition feeding strategies and zootechnical aspects (international coop.)
4. Detection of nutritional requirements and role of lipid classes and vitamins in larviculture of fish, shrimp, prawn and bivalves.
5. Larviculture medication through bio-encapsulation of chemotherapeutics and vaccines in the live food (international coop.)
6. Microbiology of marine fish hatchery systems in Europe (international coop.)
7. Development and testing of artificial diets as replacement/supplementation products for algae, rotifers, and Artemia.
8. Brood stock nutrition and improvement of egg quality in marine fish, shrimp and prawn (international coop.)

9. Univ. Gent - RUG - Fac. Agricultural Sciences

Name : Rijksuniversiteit Gent-RUG
Faculteit van de Landbouwwetenschappen
Address : Coupure Links 653
B - 9000 Gent
Head : C. Pelerents
Tel. : 32.91.236961 **Fax**: 32.91.232867

Type of institute
University Department/Higher Education Institute
Government Department/regional or local authority
Research centre

Research activities
Fisheries management
Stock assessment
Environmental problems and their impact on resources
Aquaculture
Aquaculture/environment interaction

Staff: Scientists: 2 Others: 1

10. Univ. Gent - RUG - Fac. Environment

Name : Rijksuniversiteit Gent - RUG
Department of Applied Ecology and Environmental Biology
Laboratory for Biological Research in Aquatic Pollution
Address : J. Plateaustraat 22
B - 9000 Gent
Head : G. Persoon/N. De Pauw
Tel. : 32.(0)9.264.37.65 **Fax**: 32.(0)9.264.41.99

Type of institute
University Department/Higher Education Institute

Research activities
Fisheries management
Environmental problems and their impact on resources
Ecotoxicological testing
Bioremediation/Bioaugmentation/High rate algal ponds

Aquaculture
  Aquaculture/environment interaction
  Environmental impact assessment
  Algoculture

Facilities
Micro- and Mesocosms

Staff: Scientists: 12 Others: 6

11. Univ. Gent - RUG - Lab. Ecologie der Dieren

Name	:	Rijksuniversiteit Gent - RUG Laboratorium voor Ecologie der Dieren Zoogeografie en Natuurbehoud
Address	:	KL Ledeganckstraat 35 B - 9000 Gent
Head	:	Prof. H. Dumont
Tel.	:	32.91.227821

Type of institute
University Department/Higher Education Institute
Fish/Crustacean/mollusc farming

Research activities
Fisheries management
  Stock assessment
  Environmental problems and their impact on resources
Aquaculture
  Aquaculture/environment interaction
  Aquaculture structures, systems, techniques

Facilities
Rearing tank
Mesocosm
Data bases, extensive libraries
Specialised software
Field site, electron microscope, specialized analytical apparatus

Staff: Scientists: 5 Others: 4

Research budget (ECU/year): 116.200

12. Univ. Gent - RUG - Lab. Medical Biochemistry

Name : Rijksuniversiteit Gent - RUG
Fac Pharm. & Sciences
Lab. voor Medische Biochemie
Address : Harelbekestraat 72
B - 9000 Gent
Head : Prof. A.P. De Leenheer / Dr. H.J. Nelis
Tel. : 32.91.218951 **Fax**: 32.91.217902

Type of institute
University Department/Higher Education Institute
Research centre

Research activities
Aquaculture
Growth, nutrition, feed composition, live food production for aquaculture
Upgrading of fishery products
Consumer response and demand studies

Facilities
Analytical equipment: HPLC, GC-Ms

Staff: Scientists: 2 Others: 2

Research budget (ECU/year): 11.600

Description
Multidisciplinary research on Artemia, coordinated by the Artemia Reference Center.
Analysis of Artemia fish and shellfish larvae.
Determination of pigments, vitamins, hormones and therapeutics.
Analytical support of multidisciplinary projects.

13. Univ. Gent - RUG - Marine Biologie

Name : Universiteit Gent
Zoology Institute
Marine Biology Section
Address : K.L. Ledeganckstraat 35
B - 9000 Gent
Head : Prof. Dr. M. Vincx/Dr. O. Hamerlynck
Tel. : 32.91.64.52.10 **Fax**: 32.91.64.53.44

Type of institute
University Department/Higher Education Institute

Research activities
Fisheries management
Environmental problems and their impact on resources
Parasites of marine fish
Ecology of fish

Facilities
Scanning Electron Microscope
Specialised analytical equipment (HPLC, Coulter Size Analyser)

Staff: Scientists: 12 Others: 5

Research budget (ECU/year): 500.000

Description
Research projects:
Community ecology of fish and epibenthic animals, trophic role of fishes in coastal waters and estuaries, predator-prey relations energy fluxes from meiobenthos and hyperbenthos to higher trophic levels. Daily growth ring studies. Fishes of east-African mangrove systems and seagrass beds. Interspecies relations, general marine ecology.

14. Univ. Leuven - KUL - Zool. Inst. - Lab. Ecologie Aquacultuur

Name	:	KUL-Katholieke Universiteit Leuven Zoologisch Instituut Laboratory for Ecology and Aquaculture
Address	:	Naamsestraat 59 B - 3000 Leuven
Head	:	F. Ollevier
Tel.	:	016.283966 **Fax**: 016.284575

Type of institute
University Department

Research activities
Fisheries management
Stock assessment
Environmental problems and their impact on resources
Aquaculture
Genetics and conservation of gametes

Growth, nutrition, feed composition, live food production for aquaculture
Disease, immunology, stress

Facilities
Field site, electron microscope, specialized analytical apparatus
Rearing tanks

Staff: Scientists: 8 Others: 3

Research budget (ECU/year): 500.000

Description
Research projects:
Studies of sex-determination, growth, pathology, survival and feeding in the field of farmed fishes (eel and seabass) Regulation of fish gen expression. Reproduction and fingerling production of Clarias gariepinus, the African catfish. Characterization of the microbiological environment in hatcheries for Mediterranean marine fish: influence on larval health and development of prophylaxis. Impact of w-3 fatty acids and vitamins on growth, body composition and the immune system in aquaculture organisms. Studies upon the natural eel population in western Europe.

15. Univ. Louvain-la-Neuve - UCL - Biochimie de la Nutrition

Name : Université Catholique de Louvain (UCL)
Unité de Biochimie de la Nutrition
Address : Place Croix du Sud, 3 - Sc 15 D2
B - 1348 Louvain-la-Neuve
Head : M. Vanbelle
Tel. : 32.10.473785 **Fax**: 32.10.474745 **Telex**: 59037 B

Type of institute
Research centre
University Department/Higher Education Institute

Research activities
Aquaculture
Growth, nutrition, feed composition, live food production for aquaculture
Digestibility test, requirements

Facilities
Research vessel
Mesocosm
Controlled environmental chamber

Specialised software
Field site, electron microscope, specialized analytical apparatus
Rearing tanks
Software
Digestibility tanks
Ichthyoacoustic chamber (Unique in Europe!)
Aquaria

Staff: Scientists: 2

Research budget (ECU/year): 23.200

Description
Our laboratory is mainly concerned with fish nutrition. We investigate
1. The amino acids requirements of the Atlantic Salmon (Salmo Salar) and we determine the validity of the dose/response method used for these investigations.
2. The fatty acids requirements of the goldfish (Carassius Auratus) and we study the interactions between nutrition and reproduction.
3. The PALATABILITY of Azolla spp. - an aquatic fern - and we determine its nutritional interest on TILAPIA RENDALLI. We are also concerned with the biology of HAPLOCHROMIS SP. from IHEMA and MUHAZI LAKES (Rwanda).

16. Univ. Louvain-la-Neuve - UCL - Fish Technology Group

Name	:	Université Catholique de Louvain (UCL) Génie génétique, 'Fish Technology Group'
Address	:	Place Croix du Sud, 3/5 B - 1348 Louvain-la-Neuve
Head	:	P. Querinjean
Tel.	:	32.10.473504 **Fax**: 32.10.472999

Type of institute
University Department/Higher Education Institute
Fish/Crustacean/Molluscm Farming
Research centre
Fishing

Research activities
Fisheries management
Multi-disciplinary approach to fisheries management
Aquaculture
Early stages of reared species
Disease, immunology, stress

Facilities
Rearing tank
Specialised software

Staff: Scientists: 6 Others: 10

Description
This group is a new one designed to approach some problems of the follow-up of the fisheries by immunological methodologies.

17. Univ. Louvain-la-Neuve - UCL - Genetique CGEN

Name : Université Catholique de Louvain (UCL)
Unité de génétique (CGEN)
Address : Place Croix du Sud, 2
B - 1348 Louvain-la-Neuve
Head : M. Bourgois
Tel. : 32.10.473667 **Fax**: 32.10.474745 **Telex**: 59037

Type of institute
Research centre
University Department/Higher Education Institute

Research activities
Aquaculture
- Reproduction
- Genetics and conservation of gametes
- Early stages of reared species
- Growth, nutrition, feed composition, live food production for aquaculture

Facilities
Research vessel
Rearing tank

Staff: Scientists: 1

18. Univ. Louvain-la-Neuve - UCL - Lab. Pisciculture - M. Huet

Name : Laboratoire de Pisciculture M. Huet
Université de Louvain-la-Neuve (UCL)

Address : Place Croix du Sud, 3
B - 1348 Louvain-la-Neuve
Head : J.C. Micha
Tel. : 32.10.472530 **Fax**: 32.10.474745 **Telex**: 59307 B

Type of institute
Research centre
University Department/Higher Education Institute
Fish farming

Research activities
Aquaculture
Early stages of reared species
Growth, nutrition, feed composition

Facilities
Rearing tank
Field site, specialized analytical apparatus
Aquarium

Staff: Scientists: 3 Others: 1

Research budget (ECU/year): 100.000

Description
Nutritional demand in amino-acids for rearing fingerlings of atlantic salmon.
Development of agro-piscicultural system in tropical marschlands (Rwanda, Mexico).
Consumption of aquatic floating fern, Azolla spp, by fish (Tilapia spp) in tanks and ponds.

19. Univ. Namur - ND de la Paix - Ecologie Eaux Douces

Name : Facultes Universitaires ND de la Paix
Unité d'Écologie des Eaux Douces
Address : Rue de Bruxelles 61
B - 5000 Namur
Head : J.C. Micha
Tel. : 32.81.724111 **Fax**: 32.81.230391 **Telex**: 59222 B

Type of institute
University Department/Higher Education Institute
Fish farming/Fish biology

Research activities

Fisheries management

Stock assessment

Environmental problems and their impact on resources

Aquaculture

Reproduction

Early stages of reared species

Live food production for aquaculture

Facilities

Rearing tank

Specialised software

Field site, electron microscope, specialized analytical apparatus

Aquarium

Staff: Scientists: 7 Others: 3

Research budget (ECU/year): 250.000

Description

Stock assessment of Cyprinidae (Rutilus rutilus) in the river Meuse and impact of pollution. Restoration of the atlantic salmon in the river Meuse. Stock assessment of Cichlidae in lakes (Rwanda) and rivers (Mexico). Artificial spawning and intensive fingerlings production of Cyprinidae (Gobio gobio, Leuciscus idus, Carassuis auratus). Mass rearing of freshwater rotiferous (Brachionus Calyciflorus).

State of Research

DENMARK

Fisheries research is mainly carried out at the three government institutes. The Danish Institute for Fisheries and Marine Research (DIFMAR), the Technological Laboratory of the Danish Ministry of Fisheries, and the Danish Inland Fishery Laboratory.

DIFMAR is the principal fishery biology institute. It takes active part in the work of ICES, and covers most of the Danish advisory function in the EEC concerning fisheries biology and fishery management.

The Technological Laboratory covers research in fish products.

The Danish Inland Fishery Laboratory covers fishery resarch and stock enhancement in fresh water.

The Danish Institute for Fisheries Technology and Aquaculture (DIFTA) is a non-governmental institute, partly sponsored by the Ministry of Industry and by the Ministry of Fisheries.

The Danish Institute of Fisheries Economic Research, which is a research and development part of the University of South Jutland, works on Fisheries Economics.

Other laboratories - private, non profit or universities - do research in fisheries, but their role is difficult to evaluate.

Most large companies have a co-called research department, more appropriately a product development department. These departments often work together with technological research institutes, which are generally self owning non-profit making institutes sponsored by the Ministry of Industry. They act as a basic link between sector institutes and universities conducting pure scientific research on the one hand, and industry on the other.

Also consultancy companies make a link between pure research and industry. They do not, however, possess any of the special testing equipment associated with fishing techniques research. Their work is concentrated on computer based economic and technical evaluation plus the development of specialised computer software.

Some universities incorporate lectures on fishing techniques within biology engineering courses.

List of Institutes from DENMARK

1. BioProcess
2. Bornholms Fiskerilaboratorium
3. CARNITECH
4. Cowi Consult A/S Lyngby
5. Danish Institute for Fisheries and Marine Research - DIFMAR
6. Danish Maritime Institute
7. Danish Trout Culture Research Station
8. Den kgl. Veterinær- og Landbohoejskole
9. DIFTA - Danish Institute of Fisheries and Aquaculture
10. DTH - Danmarks Tekniske Hoejskole - AI and Vision Group - EMI
11. Esbjerg Teknikum
12. European Association of Fisheries Economists
13. Fish Biology Research Center
14. Fiskeriministeriets Forsøegslaboratorium - FF - Lyngby
15. Højmarklaboratoriet
16. Inland Fisheries Laboratory - IFLAB
17. JAM A/S
18. Jysk Teknologisk Institut
19. Levnedsmiddelstyrelsen
20. LUMETECH A/S - Carlsberg A/S
21. MATCON
22. Nordsømuseet
23. Rauf og Sorensen A/S Shipmate
24. Univ. Ålborg - Institut for Produktion
25. Univ. Ålborg - Jord Vand og Miløteknik
26. Univ. København - Lab. Marinebiologi
27. Univ. København - Lab. Zoofysiologi
28. Univ. Sydjylland - Fiskeriøkonomisk Inst. - DIFER

1. BioProcess

Name : BioProcess
Address : Nordsøcentret
Postboks 104
DK - 9850 Hirtshals
Head : N.H. Norsker
Tel. : 45.98.944266 **Fax**: 45.98.944833

Type of institute
Engineering/consulting firm
Equipment manufacturer

Research activities
Aquaculture
Early stages of reared species
Growth, nutrition, feed composition, live food production for aquaculture

Facilities
Specialised computer hardware

Staff: Scientists: 1 Others: 1

Research budget (ECU/year): 110.000

Description
Pilot plant for production of live feed. BioProcess is at present engaged in 2 R and D projects:
a) developing a continuous micro algae cultivation plant for full scale hatchery operation.
b) developing copepod cultivation techniques including research in physiology of the reared copepod species.

2. Bornholms Fiskerilaboratorium

Name : Bornholms Fiskerilaboratorium
Address : Østre Flak 2
DK - 3730 Nexø
Tel. : 45.56.493553 **Fax**: 45.56.493435 **Telex**: 48127 delnx

Type of institute
Research centre
Engineering-consulting firm
Fishing

Research activities

Fisheries management
- Multi-disciplinary approach to fisheries management
- Environmental problems and their impact on resources

Aquaculture

Upgrading of fishery products
- Improvement of techniques for handling, storing, processing, packaging fish
- Methods for grading and quality assessment of fish
- Novel processing methods for by-products, discards, and process wastes
- Consumer response and demand studies

Facilities

Data bases, extensive libraries
Specialised software
General laboratory for fisheries technology and biology investigations

Staff: Scientists: 2 Others: 4

Research budget (ECU/year): 648.700

Description

Bornholm Fisheries Laboratory is a local development centre for fisheries. BFL main activities is concerned with optimization of raw materials from the sea, turning low value raw materials into value added products. Projects for reducing discard of intestines from the cod fishery. Projects concerned with utilization of intestines of cods: catchtreatment, conservation techniques, yield, product evaluation, feasibility analyses. Projects for improving processing of flounder with the aim of increasing value for the fisherman. Projects for developing new ready meal products. Projects on fisheries technology in India, Thailand, United Arab Emirates, Pacific Ocean. Projects on assisting Baltic fishermen in their to change to tropical fishery. Project on quality control on Baltic fish. Project on sewage treatment in the fishing industry.

3. CARNITECH

Name : CARNITECH
Address : Industriparken Sørup
DK - 9530 Støvring
Head : Poul Erik Korsgaard

Tel. : 45.98.373577 **Fax**: 45.98373790 **Telex**: 69817 DK

Type of institute

Engineering/consulting firm
Equipment manufacturer

Research activities

Upgrading of fishery products

Improvement of techniques for handling, storing, processing, packaging fish
Novel processing methods for by-products, discards, and process wastes

Staff: Scientists: 4 Others: 4

4. COWI Consult A/S Lyngby

Name : COWI Consult
Consulting Engineers and Planners A/S
Address : Parallelvej 19
DK - 2800 Lyngby
Head : S.P. Christensen / A. Randløv
Tel. : 45.45.972211 **Fax**: 45.45972212 **Telex**: 33 580 cowl dk

Type of institute

Consulting firm

Research activities

Fisheries management

Multi-disciplinary approach to fisheries management
Environmental problems and their impact on resources

Facilities

Data bases, extensive libraries
Specialised computer hardware
Specialised software

Staff: Scientists: 12 Others: 4

Description

Studies of Danish fisheries sector in order to provide the basis for policy decisions.
Studies of fishing boats and fish processing plants, in Denmark and abroad.
Ecotoxical tests.
Ecological Mapping.
Methods for Marine Pollution Monitoring.

5. Danish Institute for Fisheries and Marine Research - DIFMAR

Name : Danish Institute for Fisheries and Marine Research - DIFMAR
Address : Charlottenlund Castle
DK - 2920 Charlottenlund
Head : J.Møller Christensen
Tel. : 45.33.96.33.00 **Fax**: 45.33.96.33.33 **Telex**: 19960 dfh DK

Type of institute
Research Institute under the Ministry of Fisheries

Research activities
Development of models for interaction between fish stocks and fisheries (multispecies, fleet structure etc.)
Studies of interaction between environment, fish stocks and fisheries
Studies of processes within the food chain structure, e.g. plankton - fish larvae, feeding biology and physiology and recruitment
Studies of effects of enhancement of marine species
Fish diseases
Development of hydro acoustic techniques

Facilities
Research vessel (2484 GRT), and two minor fishery vessels
Indoor-outdoor small and large scale experimental tank systems
Data bases, libraries, laboratories

Staff: Scientists: 35 Others: 70

Research budget (ECU/year): 5.000.000

Description
The objective of DIFMAR's research is to produce information which can be used for management of living marine resources.
DIFMAR carries out research, monitoring, and development, and acts as an advisory capacity in matters relating to fisheries, fisheries management, sea ranching, and the aquatic environment.
DIFMAR evaluates and monitors the status and interactions of fisheries and stocks for all fish species of importance to the Danish fishing industry.
Studies are also conducted on the effect on fish and fish stocks of changes in the marine environment.
Headquarters and two research departments in Charlottenlund.
One research department and the research vessel DANA in Hirtshals:

Address: North Sea Center, PO Box 101, DK 9850 Hirtshals,
Tel.: 45.33963200 Fax: 45.33963262 Telex: 67765 dfhh DK

6. Danish Maritime Institute

Name	:	Danish Maritime Institute
Address	:	Hjortekærsvej, 99 DK - 2800 Lyngby
Head	:	A. Hasle Nielsen, Director
Tel.	:	45.45.879325 **Fax**: 45.45.879333 **Telex**: 37223 DK

Type of institute
Research centre

Research activities
Fishing techniques
- Testing techniques/instruments
- Hydrodynamics of fishing gear
- Improvements in fishing gear and/or fishing vessels
- Overtrawlability of subsea structures

Facilities
240 m deep-water towing tank with wave generator.
24 m shallow water basin.
Cavitation tunnel.
Manoeuvring simulator with bridge instrumentation and perspective view by image generation.
Arbitrary motions of single or multiple bodies and their connections, floating or submerged, in real or fast time in 6 degrees of freedom.

Staff: Scientists: 40 Others: 45

Research budget (ECU/year): 1.500.000

Description
Research projects:
Design and development of various towed subsea bodies for observation of fish in the catch situation and for stock assessment.
Interaction of subsea wellhead protection with fishing gear.
Hydrodynamic design and model testing of numerous fishing vessels, trawlers, fisheries inspection vessels, etc.
Measurement of flow in and around large scale (1:3) trawl net models.

7. Danish Trout Culture Research Station

Name : Danish Trout Culture Research Station
Address : Brøns Møllevej 7
DK - 6780 Skærbæk
Head : Viggo Hørlyck
Tel. : 45.74.753234 **Fax**: 45.74.753682

Type of institute
Research centre
Engineering/consulting firm
Fish

Research activities
Aquaculture

Facilities
Rearing tank
Controlled environmental chamber
Full scale trout farm.

Staff: Scientists: 5 Others: 13

Research budget (ECU/year): 50.000

8. Den Kgl. Veterinær- og Landbohoejskole

Name : Den Kgl. Veterinær- og Landbohøjskole
Institut for veterinær mikrobiologi og Hygiejne
Address : Bülowsvej 13
DK - 1870 Frederiksberg C
Head : Dr. J. L. Larsen
Tel. : 45.31351788 **Fax**: 45.31373193 **Telex**: DVJBIB 15061 DK

Type of institute
Research centre
University Department/Higher Education Institute

Research activities
Fisheries management
Environmental problems and their impact on resources
Aquaculture
Disease, immunology, stress
Aquaculture/environment interaction

Facilities
Research vessel
Field site, electron microscope, specialized analytical apparatus

Staff: Scientists: 5 Others: 4

Research budget (ECU/year): 106.800

Description
Bacterial branch:
1. Taxonomial and secrological studies of fish-pathogenic vibrios.
2. Virulence factors of Aeromonas salmonicida.
3. Bacterial diseases in cultured fish.
4. Development and testing of vaccines (bacterial infections).
5. Plasmid and RFLP analysis of bacterial fish deseases.
6. Relationship between pollution and fish diseases.
7. Exchange of pathogens between fish in aquaculture and feral fish.
8. Disease problems in cultured turbot.

Parasite branch:
1. Humoral immune response of eels.
2. Sero-diagnosis of parasitic infections.
3. Treatment of parasitosis in fish farms.
4. Experimental selection of anthelmintic resistant parasites.
5. Parasites of wild fishes.
6. Eco-parasitological conditions in fish farms.

9. DIFTA - Danish Institute of Fisheries and Aquaculture

Name : DIFTA Danish Institute for Fisheries Technology and Aquaculture
Address : North Sea Centre
DK - 9850 Hirtshals
Head : A. Thygesen
Tel. : 45.98.94.43.00 **Fax**: 45.98.94.22.26 **Telex**: 67757 FTI DK

Type of institute
Engineering/consulting firm
Research centre

Research activities
Fisheries management
Multi-disciplinary approach to fisheries management

Fishing techniques
Aquaculture
Upgrading of fishery products

Facilities
Flume tank
Rearing tank
Data bases, extensive libraries
Specialised software
Field site
Pilot plant for fish meal processing
Specialized analytical apparatus
Computer aided trawl design and simulation equipment
Equipment for measuring trawl geometry and drag
Towed underwater vehicle
Chemical and microbiological laboratories
Recirculated, intensive aquaculture hatchery and production systems
Documentation centre

Staff: Scientists: 26 Others: 15

Research budget (ECU/year): 2.500.000

Description
Research projects:
- Behaviour of different fish species/sizes in order to design selective fishing gears.
- Problems involved in using codend mesh-size.
- Marine aquaculture.
- Sector and feasibility studies.
- Recirculated systems for aquaculture production.
- Intensive hatchery techniques for marine species.
- Plant based waste water treatment.
- Development of new technologies to reduce the environmental problems of the fish processing industry.

10. DTH - Danmarks Tekniske Hoejskole - AI and Vision Group - EMI

Name : Technical University of Denmark/AI and Vision Group-EMI
Address : BLDG 348/EMI
DK - 2800 Lyngby
Head : L.F. Pau
Tel. : 45.42881444 **Fax**: 45.42931634

Research activities
Fishing techniques
  Electronics/acoustics
Upgrading of fishery products
  Methods for grading and quality assessment of fish

Facilities
5 vision computers, special live scan, high scan rate camera.

Description
Research projects and contract research/development work for industry (Scandinavia, Iceland, Japan) in the area of use of computer vision and expert systems for fish inspection, fish sorting (length, species, colour) and alignment of fish for filleting.

11. Esbjerg Teknikum

Name : ESBJERG TEKNIKUM Ingeniørhøjskolen
Address : Niels Bohrsvej 8
DK - 6700 Esbjerg
Head : H. Karlby / H.V. Pedersen
Tel. : 45.75.127666 **Fax**: 45.75.453643

Type of institute
University Department/Higher Education Institute
Institute for professional training/Secondary school

Research activities
Fisheries management
  Multi-disciplinary approach to fisheries management
  Environmental problems and their impact on resources
Aquaculture
Upgrading of fishery products

Facilities
Data bases, extensive libraries
Specialised software
Microbiological/Fishprocessing laboratories

Staff: Scientists: 5 Others: 3

12. European Association of Fisheries Economists

Name : European Association of Fisheries Economists
Address : c/o DIFER
Niels Bohrsvej, 9
DK - 6705 Esbjerg
Head : Mr. Hans Frost

Type of institute
Engineering/consulting firm

Research activities
Fisheries management
Multi-disciplinary approach to fisheries management
Aquaculture
Aquacuture economics
Upgrading of fishery products
Consumer response and demand studies

Facilities
Data bases, extensive libraries
Specialised software

Staff: Scientists: 40

Description
Objects:
To promote co-operation in economic research in fisheries and aquaculture.
To assist in the dissemination of information about fisheries economics among members.
To undertake any activity deemed appropriate to further the understanding of fisheries economics.
To serve as a channel of communication with other interested bodies.

13. Fish Biology Research Center

Name : Fish Biology Research Center
Address : c/o DIFMAR
Head : Prof Morten Simonsen
Tel. : 45.31371779

Type of institute
Research centre
University Department/Higher Education Institute

Research activities
Fisheries management
Stock assessment
Environmental problems and their impact on resources
Aquaculture
Physiology
Genetics and conservation of gametes
Early stages of reared species
Growth, nutrition, feed composition, live food production for aquaculture.
Disease, immunology, stress.

Facilities
Field site, electron microscope, specialized analytical apparatus

Staff: Scientists: 18 Others: 10

Research budget (ECU/year): 622.000

Description
This centre is a framework institution established under the Danish "Biotechnological Research and Development programme 1987-90". It is meant to encourage collaboration between university institutes, other public research institutes, technological services institutes and research division in private companies.
Research projects:
Resistence to infection in rainbow trout.
Vaccine on site.
Genetic disposition for disease.
Digestion and growth in marine fish larvae.

14. Fiskeriministeriets Forsøegslaboratorium - FF - Lyngby

Name : Danmarks Tekniske Højskole
Fiskeriministeriets Forsøgslaboratorium (FF)
Address : Bygning 221, DTH
DK - 2800 Lyngby
Head : T.A. Børresen
Tel. : 45.42883322 **Fax**: 45.42884774 **Telex**: 37528 DTHDIA DK

Type of institute
Research centre
University Department/Higher Education Institute

Research activities
Upgrading of fishery products
Handling, storing, preservation and upgrading of fish and shellfish
Fishery biochemistry, microbiology and technology
Integrated quality assurance and HACCP
Marine enzymes and microorganisms
Upgrading of fishery products
Handling, storing, preservation and upgrading of fish and shellfish

Facilities
Specialized analytical apparatus as amino acid analyser,
HPLC, GC bactometer, electroforese equipment.
Pilot plant facilities for fish meal production
Pilot plant, freezing quipment, smoking and drying kiln, chilling and freezing stones, autoclave, meat-bone separator.

Staff: Scientists: 27 Others: 27

Description
FF has the following purpose: The laboratory shall carry out investigations and research into the handling, storage, preservation and other industrial utilization of fish. The information obtained shall be made available to the fisheries and the fishing industry. FF is located at the Technical University of Denmark, where it has served as a teaching centre for students of chemical engineering, who desire to specialize in fisheries technology and to further this technology by carrying out research and development work for the Danish fishing industry. FF's activities take place in laboratories and pilot plants of its own as well as on commercial fishing vessels, in fishing harbours and in fish processing plants ashore.

15. Højmarklaboratoriet

Name : Højmarklaboratoriet a/s
Address : Adelvej 11
DK - 6940 Lem st.
Head : Greta Jakobsen
Tel. : 45.97.343366 **Fax** 45 97 343435

Type of institute
Research and development centre
Private quality control company

Research activities
Upgrading of fishery products
Improvement of techniques for handling, storing, processing, packaging fish
Methods for grading and quality assessment of fish
Novel processing methods for by-products, discards, and process wastes

Facilities
Analytical laboratory for bacteriological and chemical analysis approved by DANARK, Danish Accreditation Scheme
Laboratory and pilot plant for biotechnology.

Staff: Scientists: 4 Others: 11

Description
Development of functional fishproteins.
Minimizing the outlet of pollution from fish industries.

16. Inland Fisheries Laboratory - IFLAB

Name	:	Inland Fisheries Laboratory - IFLAB
Address	:	Lysbrogade 52 DK - 8600 Silkeborg
Head	:	G. Rasmussen
Tel.	:	45.33963100 **Fax**: 45.33963150

Type of institute
Research Institute under the Ministry of Fisheries.

Research activities
- Changes in fish stocks during time
- Influence of fish gears on fish stocks
- Tagging experiments and development
- Interactions between stock enhancement and environment (biomanipulation)
- Aquaculture (fish diseases and fish growth modelling).

Facilities
Indoor-outdoor small and large scale experimental tank systems.
Laboratories, data bases, libraries.

Staff: Scientists: 10 Others: 9

Research budget (ECU/year): 1.228.000

Description
The objective of IFLAB's research is to produce information which can be used for management of the fish stocks in lakes, rivers and coastal areas.
IFLAB carries out research and development, and acts as an advisory capacity in matters relating to inland fisheries, fisheries management and fish farming.
IFLAB administrates the Danish Stocking Enhancement Project and works out stocking plans for eel and salmonids.
Headquarter in Silkeborg and one minor department in Charlottenlund at DIFMAR.
From 1995 all activities will take place in Silkeborg.

17. JAM A/S

Name : JAM A/S
Address : Alsvej 2
DK - 5800 Nyborg
Head : N. Johnstad-Møller
Tel. : 45.65.311617 **Fax**: 45.65.312325 **Telex**: 50471 DK

Type of institute
Equipment manufacturer

Research activities
Upgrading of fishery products
- Improvement of techniques for handling, storing, processing, packaging fish
- Methods for grading and quality assessment of fish
- Consumer response and demand studies

Facilities
Specialised computer hardware
Specialised software

Staff: Scientists: 2 Others: 20

18. Jysk Teknologisk Institut

Name : Jysk Teknologisk Institut
The Jutland Technological Institute
Address : Teknologiparken
DK - 8000 Århus C
Head : H. D. Smidt
Tel. : 45.86.142400 **Fax**: 45.86.147722 **Telex**: 68722 DK

Research activities

Upgrading of fishery products
  Improvement of techniques for handling, storing, processing, packaging fish
  Methods for grading and quality assessment of fish
  Novel processing methods for by-products, discards, and process wastes

Facilities

Controlled environmental chamber
Data bases, extensive libraries
Specialised computer hardware
Specialised software
Field site, electron microscope, specialized analytical apparatus

Staff: Scientists: 3 Others: 1

19. Levnedsmiddelstyrelsen

Name	:	Levnedsmiddelstyrelsen Centrallaboratoriet		
Address	:	Mørkhøj Bygade 19 DK 2860 Søborg		
Head	:	Klaus Holch		
Tel.	:	45.31696600	**Fax**: 45.31696126	**Telex**: 16298 DK

Type of institute

Government Department / Regional or local authority

Research activities

Upgrading of fishery products
  Improvement of techniques for handling, storing, processing, packaging fish

Staff: Scientists: 3 Others: 4

Description

Research projects:

Pollution in fish: dioxins, Mustard Gas, Heavy Metals, organochlorine compound residues, drug residues (aquaculture). Composition of fish products: Analysis of Carotenoid Pigments in Salmonids, Fatty acid composition of fish.

20. LUMETECH A/S - Carlsberg A/S

Name : LUMETECH A/S (Carlsberg A/S)
Address : Strandevejen 50
DK - 2900 Hellerup
Head : P. Haagensen/S. Reenberg
Tel. : 45.31185600

Type of institute
Processing Industry/ agro-food / diets producer
Equipment manufacturer
Engineering/consultancy firm
Research centre

Research activities
Upgrading of fishery products
- Improvement of techniques for handling, storing, processing, packaging fish
- Methods for grading and quality assessment of fish
- Novel processing methods for by-products, discards, and process wastes

Facilities
Data bases, extensive libraries
Specialised computer hardware
Specialised software

Staff: Scientists: 6 Others: 7

Research budget (ECU/year): 827.000

21. MATCON

Name : MATCON
Consulting Engineers and Planning Ltd
Address : Nordsøcentret
Postboks 104
DK - 9850 Hirtshals
Head : M. Hans-Peder Pedersen
Tel. : 45.98.944903 **Fax**: 45.98.944601

Type of institute
Engineering/consulting firm

Research activities
Fisheries management
  Multi-disciplinary approach to fisheries management
Aquaculture
  Aquaculture structures, systems, techniques
  Aquacuture economics
Upgrading of fishery products

Facilities
Data bases, extensive libraries
Specialised software

Staff: Scientists: 19 Others: 7

Research budget (ECU/year): 51.000

Description
The major research activities concern:
- the upgrading and rationalization of existing plants;
- the computerized production control;
- the development of new fish technology; quality management;
- the design of new high-technology fish and shellfish industries;
- business development.

22. Nordsømuseet

Name	:	Nordsømuseet
Address	:	Postboks 125 DK - 9850 Hirtshals
Head	:	Susanne Fibiger
Tel.	:	45.98944844

Type of institute
Non-profit private foundation, Public Museum and Aquarium

Research activities
Documentation of Danish fisheries and associated service and industry.

Facilities
Exhibition
Aquariums
Micrarium

Staff: Scientists: 3 Others: 15

Description
Nordsømuseet is a contemporary museum and aquarium. It carries out exhibitions dealing with the North Sea and activities related to the North Sea (biological, commercial).
The main subjects are: fishes, fisheries, fishing methods, environment and seals.
Non-profit foundation, part of the North Sea Centre.
Contemporary museum of Danish fisheries and aquarium with educational facilities, wet laboratory etc. Level: national.

23. Rauf og Sorensen A/S Shipmate

Name	:	R and S Shipmate A/S
Address	:	Østre Allé 6 DK - 9530 Støvring
Head	:	P. Elstrøm
Tel.	:	98.373499 **Fax**: 98.373807 **Telex**: 69838

Type of institute
Engineering/consulting firm
Equipment manufacturer

Research activities
Fishing techniques
Electronics/acoustics

24. Univ. Ålborg - Institut for Produktion

Name	:	Ålborg Universitet Institut for Produktion
Address	:	Aalborg Universitetscentret Fibigerstræde 13 DK - 9220 Ålborg
Head	:	B. Rønsholdt
Tel.	:	45.98158522 **Fax**: 45.98153030

Type of institute
University Department/Higher Education Institute

Research activities
Aquaculture
Aquaculture/environment interaction
Upgrading of fishery products
Improvement of techniques for handling, storing, processing, packaging fish
Methods for grading and quality assessment of fish

Facilities
Field site, electron microscope, specialized analytical apparatus
Research laboratory with standard equipment. Availability: negotiable.
Colorimeter. Availability: negotiable.

Staff: Scientists: 4 Others: 1

Research budget (ECU/year): 20.000

Description
Environmental load from Marine Fish Farms and Land-based salt water fish farms. Project Leaders: Jesper Horsted and Kim Diget Christensen. Duration: 3 years. Involved N° of scientists: 5. The project undertaken by the Danish Marine and Fisheries Research Institute for the Danish Ministry of Fisheries. Our part of the project was concerned with determination of the nitrogen and phosphorous content in farmed fish (Rainbow trout). Quality assessment of colour of fish flesh and fish products. Project Leader: Bent Ronsholdt. Duration: on-going. Involved n° scientists: 1. The main objective of this project is to enable the fish industry to use colour as quality parameter in a quality management concept. Project Leader: Jesper Raakjaer. Duration: 2 1/2 years. Involved n° scientists: 3. Quality design of reared fish species. Project leader: Jacob Faergemand. Duration: 2 1/2 years. Involved n° of scientists: 3. These are Ph.d projects.

25. Univ. Ålborg - Jord, Vand og Miløteknik

Name	:	Universitet Ålborg Institut for Jord, Vand og Miløteknik
Address	:	Sohngaardshomsvej 57 DK - 9000 Ålborg C
Head	:	J.O. Frier
Tel.	:	45.98142333 **Fax**: 45.98142555 **Telex**: 69523 DK

Type of institute
University Department

Research activities
Environmental problems

Facilities
Research facilities for full scale aquaculture of freshwater fish.

Staff: Scientists: 1 Others: 2

Research budget (ECU/year): 100.000

Description
We work with sublethal effect of external factor on fish populations. Currently we work with the effect of sublethal amounts of oxygen and food.

26. Univ. København - Lab. Marinebiologi

Name : Københavns Universitet
Marinebiologisk Laboratorium
Address : Strandpromenaden 5
DK - 3000 Helsingør
Head : T. Fenchel/Marianne Køie
Tel. : 45.49213344 **Fax**: 45.49261165

Type of institute
University Department/Higher Education Institute

Research activities
Fisheries management
- Environmental problems and their impact on resources

Aquaculture
- Physiology
- Early stages of reared species
- Growth, nutrition, feed composition, live food production for aquaculture

Facilities
Research vessel
Rearing tank
Controlled environmental chamber
Field site, electron microscope, specialized analytical apparatus

Staff: Scientists: 12 Others: 9

Description
The major research activities concern:
The parasites, especially helmints, of fishes, molluscs and crustaceans.
The life-cycle studies of marine fishes digeneans.

27. Univ. København - Lab. Zoofysiologi

Name : Københavns Universitet
Zoofysiologisk Laboratorium
Address : August Krogh Institutet
Universitetsparken 13
DK - 2100 København
Head : Per Rosenkilde
Tel. : 45.3532 1550 **Fax** 45 3532 1567

Type of institute
University Department/Higher Education Institute

Research activities
Fisheries management
Environmental problems and their impact on resources
Physiological effect on fish of heavy metal ions
Aquaculture
Physiology
Osmoregulation, growth, development and reproduction

Facilities
Controlled environmental chamber(s)
Field site, electron microscope, specialized analytical apparatus
12 cubic meters recirculated experimental systems for fresh and salt water, plus smaller units (controlled temperature and daylight).

Staff: Scientists: 2 Others: Varying

Description
The research is to a wide extend based upon projects for persons aiming at a career in aquaculture, fisheries or environmental marine research, administration or business. The main lines are:
a) physiological regulation of growth, development and reproduction, and
b) osmoregulation in euryhaline species.
The main purpose is attempts to define and secure wellness, i.e. defining conditions that allow optimal physiological function under experimental (and) rearing conditions.

28. Univ. Sydjylland - Fiskeriøkonomisk Inst. - DIFER

Name : SUC-Sydjysk Universitetscenter
Fiskeriøkonomisk Institut (DIFER)
Address : Niels Bohrsvej, 9
DK - 6705 Esbjerg
Head : Inst. Led. Hans Frost
Tel. : 45.79.141111 **Fax**: 45.79.141199

Type of institute
Research centre
University Department/Higher Education Institute

Research activities
Fisheries management
Multi-disciplinary approach to fisheries management
Aquaculture
Aquaculture economics
Upgrading of fishery products
Consumer response and demand studies

Facilities
Data bases, extensive libraries
Specialised software

Staff: Scientists: 4 Others: 1

Research budget (ECU/year): 152.300

State of Research

FINLAND

Finland has a long coastline in the Baltic Sea and about 187 000 lakes. The main part of the commercially exploited fish stocks are situated in the Baltic Sea. Despite the amount of lakes, the inland waters are only of minor importance for the Finnish fishing industry. The main reason for the under-utilization of the inland waters is uneven and small value of catch per unit of effort and thus most of the professional fishermen in the lake fisheries only get a rather small share of their income from fishing.

The fisheries research is mainly carried out at the governmental institutes, private enterprises and consulting firms. The number of big research units is small and that of small research units high.

The main governmental fisheries research institute, Finnish Game and Fisheries Research Institute (FGFRI) takes actively part in several of the ICES coordination programmes and it covers most of the Finnish advisory functions in the International Baltic Sea Fisheries Commission concerning fisheries biology and fisheries management. FGFRI also participates in several programmes coordinated by the Nordic Council of Ministers. FGFRI is also taken actively part in FAO coordinated work at European Inland Fishery Advisory Commission (EIFAC). FGFRI-Aquaculture Division ensure the supply of eggs and fry by breeding brood fish at its fish farms.

National Board of Waters and Environment and its water districts are responsible for water quality monitoring in fresh waters and coastal areas in the Baltic Sea as well as research on eutrophication and other ecological issues related to fish and fisheries.

Several universities incorporate higher education on fish biology, environmental ecology, stock assessment and many other issues related to fisheries and university departments and laboratories carry out basic as well as applied research in fisheries field.

Most of the private companies and consult firms have a research department or laboratory, which carry out fisheries research giving advice for local authorities in their region's specific problems in special circumstances.

List of Institutes from Finland

1. Åbo Akademi University - Husö Biological Station
2. Åbo Akademi University - Institute of Parasitology
3. Agricultural Research Centre - Animal Breeding
4. Engineering Agency Väylä Ltd.
5. Federation of Finnish Fisheries Associations
6. Finnish Environmental Research Group (Part of the European Environmental Research Group Ltd.)
7. Finnish Game and Fisheries Research Institute - Aquaculture Division
8. Finnish Game and Fisheries Research Institute - Fisheries Division
9. Finnish Game and Fisheries Research Institute - Merenkurkku Fisheries Research Station
10. Finnish Game and Fisheries Research Institute - Perämeri Research Station
11. Fish and Water Research Ltd.
12. Lappland Water Research Ltd.
13. National Board of Waters and Environment - Water and Environment Research Institute
14. National Veterinary and Food Institute
15. Soil and Water Ltd. - Consulting Engineers
16. Technical Research Centre of Finland - Food Research Laboratory
17. University of Helsinki - Department of Food Technology
18. University of Helsinki - Department of Limnology and Environmental Protection
19. University of Helsinki - Department of Zoology - Division of Physiology
20. University of Helsinki - Department of Zoology - Integrative Ecology Unit
21. University of Helsinki - Kilpisjärvi Biological Station
22. University of Helsinki - Lammi Biological Station
23. University of Joensuu - Karelian Institute - Section of Ecology
24. University of Jyväskylä - Department of Biology
25. University of Kuopio - Department of Applied Zoology
26. University of Oulu - Department of Zoology
27. University of Oulu - Department of Zoology - Bothnian Bay Research Station
28. University of Turku - Archipelago Sea Research Institute
29. University of Turku - Department of Biology - Laboratory of Ecology and Systematics
30. Vesi - Hydro Consulting Engineers Ltd.
31. Water and Environment District of Central Finland
32. Water and Environment District of Kokkola
33. Water and Environment District of Lappland
34. Water and Environment District of Turku
35. Water Protection Association of Savo - Karjala
36. Water Protection Association of Southwest Finland

37. Water Protection Association of the River Kokemäenjoki - Fisheries Department
38. Water Protection Society of Central Finland
39. Water Protection Society of Lake Saimaa
40. Water Research and Engineering PSV Ltd.

1. Åbo Akademi University - Husö Biological Station

Name : Åbo Akademi University - Husö Biological Station
Address : Åland Islands
FIN-22 220 Emkarby
Finland
Head : Dr. E. Bonsdorff
Tel. : 358.28.37221 **Fax**: 358.0.37244

Type of institute
University Department, Higher education

Research activities
Basic aquatic ecology
Teaching
Monitoring
Fish ecology
Long-term monitoring of coastal fish stocks
Community analysis
Monitoring of fish farms
Food consumption studies
Recruitment

Facilities
Modern laboratory
Aquarium facilities
Several small research vessels

Staff: Scientists: 1 Others: 5

Research budget (ECU/year): 40.000

Description
Small university laboratory and field station. About 20 persons working in May-September and less than 6 in October-April. Research focused on basic aquatic ecology and local monitoring.

2. Åbo Akademi University - Institute of Parasitology

Name : Åbo Akademi University - Institute of Parasitology
Address : BioCity
Tykistökatu 6
FIN-20 520 Turku
Finland
Head : Dr. G. Bylund
Tel. : 358.21.654.301 **Fax**: 358.21.654.748

Type of institute
University institute, research centre

Research activities
Aquaculture
Fish disease diagnostics
Immunology and vaccination of fish
Molecular methods in fish virology
Fisheries
Diseases in wild fish stocks
Diseases transmission wild / farmed fish
Relationships between fish diseases and pollution

Facilities
Laboratories for virology, bacteriology and parasitology
Histology and EM-laboratories
Aquarium rooms with quarantine facilities.

Staff: Scientists: 4 Others: 2

3. Agricultural Research Centre - Animal Breeding

Name : Agricultural Research Centre - Animal Breeding
Address : FIN-31600 Jokioinen
Finland
Head : Prof. A. Mäki-Tanila
Tel. : 358.16.1881 **Fax**: 358.16.188.618

Type of institute
State research centre

Research activities
Genetics in Aquaculture
Breeding programs

Facilities
Data banks and registers
Specialized software

Staff: Scientists: 1.5

Research budget (ECU/year): 44.000

Description
General aim is to apply genetic methods in improving the productivity of fish farming. Major emphasis is on rainbow trout in both material testing and development of breeding schemes.

4. Engineering Agency Väylä Ltd.

Name : Engineering Agency Väylä Ltd.
Address : Rasinkatu 10
Merimiehenkatu 38 A
FIN-01 360 Vantaa
FIN-80 120 Joensuu
Finland
Head : O. Timonen
Tel. : 358.0.874.3044 **Fax**: 358.0.874.2010

Type of institute
Engineering agency

Research activities
Geotechnics
Watercourse construction
Environmental monitoring
Fisheries

Facilities
Applied computer software

Staff: Scientists: 9 Others: 12

Research budget (ECU/year): 750.000

Description
Assessment of fish stocks in lakes. Environmental impact assessment and priority settings between various environmental control strategies. Environmental hazard assessment of watercourse constructions.

5. Federation of Finnish Fisheries Associations

Name : Federation of Finnish Fisheries Associations
Address : Köydenpunojankatu 7 B 23
FIN-00180 Finland
Head : M. Myllylä
Tel. : 358.0.640.126 **Fax**: 358.0.608.309

Type of institute
Advisory Organization of Fisheries

Research activities
Experimental activities concerning fisheries management, commercial fisheries and organizing fisheries.

Facilities
Office and library

Staff: Others: 15

Description
The Federation of Finnish Fisheries Associations is a central advisory and promoting organization of fisheries. Most of its members are owners of water areas and/or commercial fishermen. Federation has about 490 000 members. Federation publishes three magazines, which are an important forum for discussion and dialogue concerning commercial and recreational fisheries.

6. Finnish Environmental Research Group (Part of the European Environmental Research Group Ltd.)

Name : Finnish Environmental Research Group
Address : Tekniikantie 12
FIN-02150 Espoo
Finland
Head : K.-J. Lehtinen
Tel. : 358.0.435.42076 **Fax**: 358.0.455.4403

Type of institute
Research and consulting firm

Research activities
Fisheries management
Stock assessment

Environmental problems and their impact on resources
Fish physiology and biology
Reproduction
Growth
Early life stages
Parasites
Environmental impact assessment
Mesocosm studies

Facilities
Research station
Baltic Sea laboratory in the Archipelago of Turku
Mesocosm pools
Aquariums for exposing fish and invertebrates
Toxicity testing equipments

Staff: Scientists: 4 Others: 4

Research budget (ECU/year): 367.000

Description
The Finnish Environmental Research Group Ltd. (MFG-FIN) is an independent company and a part of the European Environmental Research Group consortium founded in Sweden 1984. The European Environmental Research Group has offices in Denmark, Finland, Germany and Sweden. Expertise is found in fields of environmental research and consulting in cooperation with universities and research centres. Research is focused on consulting for private and governmental organizations. Main research fields are environmental impact assessment, priority settings between various environmental control strategies for industrial development projects, environmental hazard assessment of chemicals and industrial effluents, ecotoxicological testing and research using both laboratory scale test systems and mesocosm scale model ecosystems. Ecological field monitoring covering both physical, chemical, biochemical and biological aspects form an important part of activities.

7. Finnish Game and Fisheries Research Institute - Aquaculture Division

Name : Finnish Game and Fisheries Research Institute - Aquaculture Division
Address : Merimiehenkatu 36 D
P. O. Box 202,
FIN-00151 Helsinki,
Finland
Head : K. Westman
Tel. : 358.0.624.211 **Fax**: 358.0.631.513 **Telex**: 19101236 VDX FIN

Type of institute
Research institute

Research activities
Aquaculture
  Production of eggs and fry
  Development of aquaculture methods for crayfish and signal crayfish
  Improvement of breeding stocks
  Development of rearing techniques
Assessment and reduction of environmental damage resulting from fish farming

Facilities
Central unit is located in Helsinki. In all 14 State fish farms with modern equipments for experimental fish farm studies in different parts of the country. Capacity of 4 m^3 for salmonid and whitefish eggs, 4 000 m^2 tanks for fingerlings and 73 000 m^2 ponds for broodfish. Natural rearing ponds with total area of 2 000 hectares

Staff: Scientists: 13 Others: 86

Research budget (ECU/year): 5.455.000

Description
In aquaculture, research is concentrated on production of high-quality eggs for fish and crayfish culture and fingerlings for stocking. The most important fish culture species are Baltic salmon, whitefish, pikeperch, grayling, the Saimaa landlocked salmon and the Saimaa char. State aquaculture and fish farms are rearing in all about 20 fish species, 60 different stocks and 2 crayfish species. Production is annually about 100 million whitefish, 25 million salmon, trout and char and 25 million pikeperch, grayling and other fish eggs and fry. State fish farms ensure the supply of eggs and fry by breeding brood fish at the fish farms. Stocking and farming experiments are performed to find new stocks adapted to the Finnish climate. Cooperation exists with Nordic, Japanese and Canadian institutes and scientists.

8. Finnish Game and Fisheries Research Institute - Fisheries Division

Name : Finnish Game and Fisheries Research Institute - Fisheries Division
Address : Merimiehenkatu 36 D
P. O. Box 202
FIN-00151 Helsinki
Finland
Head : P. Tuunainen
Tel. : 358 0 624 211 **Fax**: 358 0 631 513 **Telex**: 19101236 VDX FIN

Type of institute
Research institute

Research activities
Stocking and introductions
Fishery management
Fishing technology
Fish stock assessment
  Multi-species biological models
  Multidisciplinary approach to fisheries management
Statistics and economy
  Catch statistics
  Fisheries statistics and economy
Environmental research
  Environmental problems and their impact on resources

Facilities
Small research vessel
Data-bases
Extensive library
Field stations along the Baltic Sea coast

Staff: Scientists: 70 Others: 47

Research budget (ECU/year): 4.800.000

Description
Stocking and introductions are an essential part of fish stock management. Research on stocking is performed to ensure the success and economic feasibility of the programmes. The species mainly studied are coregonids and salmonids. In fishery management subjects studied are the effects of different gears and fishing methods on the state of the stocksand catch composition, the regional and temporal distribution of the catch and the effects of regulation on the fisheries and fishermen. In fishing technology main topics are to develop more economical and effective fishing methods. Fish stock assessment is used for regulation and development of fisheries at regional and international level. The research in statistics and economy is directed at causal relationships in the fisheries, the major aim being to clarify the role of social, economic and biological factors in determining the development of the fisheries. The environmental research monitors the changes in the state of the environment, studies their effects on fish and fishery and carries out research on ways to diminish the deleterious impact of the alterations. Changes due to human activities are mainly considered. International cooperation is maintained with the International Council for the Exploration of the Sea (ICES), International Baltic Sea Fisheries Commission (IBSFC) and North Atlantic Salmon Conservation Organization (NASCO). Institute also contributes to the work of the European Inland Fisheries Advisory Commission (EIFAC, FAO) as well as Nordic Council of Ministers.

9. Finnish Game and Fisheries Research Institute - Merenkurkku Fisheries Research Station

Name	:	Finnish Game and Fisheries Research Institute - Merenkurkku Fisheries Research Station
Address	:	Korsholmanpuistikko 16 FIN-65 100 Vaasa Finland
Head	:	R. Hudd
Tel.	:	358.61.175.073 **Fax**: 358.61.125.197

Type of institute
Fisheries research station

Research activities
Stock assessment
Fishing technology
Marketing
Environmental impact

Facilities
Laboratory
Packhouse
Small research vessel

Staff: Scientists: 6 Others: 4-5

Research budget (ECU/year): 500.000

Description
Main research activities are stock assessment of whitefish, semlt, burbot, perch, sea trout and salmon populations in the Gulf of Bothnia area and acidification studies in estuaries and near shore areas.

10. Finnish Game and Fisheries Research Institute - Perämeri Research Station

Name	:	Finnish Game and Fisheries Research Institute - Perämeri Research Station
Address	:	Simontie 9 FIN-95200 Simo Finland

Head : E. Jokikokko
Tel. : 358.0.698.876694 **Fax**: 358.0.698.876784

Type of institute
Research station

Research activities
Fisheries management
Stock assessment

Facilities

Staff: Scientists: 2 Others: 10

Research budget (ECU/year): 50.000

Description
Activities focused on fisheries in the Bothnian Bay sea area and rivers bordering Bothnian Bay. Main activities are monitoring and assessment of fish stocks. Species considered are wild and reared salmon stocks in rivers Simonjoki and Tornionjoki. Electrofishing and smolt trap nets are used to estimate the annual production. In the Bothnian Bay sea area target species are Baltic salmon, sea trout, whitefish and vendace.

11. Fish and Water Research Ltd.

Name : Fish and Water Research Ltd.
Address : Vattuniemenkatu 27
FIN-00210 Helsinki
Finland
Head : J. Niinimäki / K. Korhonen
Tel. : 358.0.692.7100 **Fax**: 358.0.692.7124

Type of institute
Consulting firm for biology and limnology

Research activities
Water research
Water treatment
Fishery management
Fish and shrimp farming
Economics and marketing

Facilities
Research vessel
Specialized analytical equipments
Computers, microscopes etc.

Staff: Scientists: 4 Others: 4

Research budget (ECU/year): 250.000

Description
Research and consulting studies on fishery management, water biology and limnology. In all about 30 surveys and feasibility studies as well as management assignments carried out annually.

12. Lappland Water Research Ltd.

Name : Lappland Water Research Ltd.
Address : P. O. Box 96
FIN-96 101 Rovaniemi
Finland
Head : O. Salo
Tel. : 358.60.347.401 **Fax**: 358.60.22.885

Type of institute
Consulting firm

Research activities
Fish stock assessment
Fisheries management

Facilities
Chemical and microbiological laboratory

Staff: Scientists: 2 Others: 6

13. National Board of Waters and the Environment - Water and Environment Research Institute

Name : National Board of Waters and the Environment
Water and Environment Research Institute
Address : P. O. Box 250
FIN-00101 Helsinki
Finland
Head : Prof. S. Mustonen
Tel. : 358.0.192.191 **Fax**: 358.0.192.9577 **Telex**: 126086 VYH sf

Type of institute
State environment research institute

Research activities
Water research
Nature conservation research
Chemical research
Technical research

Facilities
Chemical, biological and microbiological laboratories
Soil and waste laboratory

Staff: Scientists: 90-100 Others: 100

Research budget (ECU/year): 6.000.000

Description
Research related to fisheries sector:
Monitoring of water quality in freshwaters and in coastal areas;
Research on eutrophication and other limnological issues;
Ecotoxicology using fish as an indicator species.

14. National Veterinary and Food Institute

Name : National Veterinary and Food Institute
Address : Hämeentie 57
FIN-00580 Helsinki
Finland
Head : Prof. E. Nurmi
Tel. : 358.0.393.101 **Fax**: 358.0.393.1811

Type of institute
Diagnostic laboratory for animal diseases and food hygiene

Research activities
Aquaculture
- Diseases
- Vaccines
- Antibiotic residues

Wild fishes
- Diseases
- Antibiotic residues

Facilities
Modern central laboratory in Helsinki, regional laboratories in Oulu and Kuopio

Staff: Scientists: 5 Others: 5

Research budget (ECU/year): 175.725

Description
Bacterial branch: Bacterial diseases in cultured and wild fish. Antibiotic resistence of fish pathogen bacteria, antibiotic residues in farmed and wild fish, testing of vaccines (vibriosis, furunculosis). Virological branch: Virus isolations of cultured and wild fish.

15. Soil and Water Ltd. - Consulting Engineers

Name	:	Soil and Water Ltd. - Consulting Engineers
Address	:	Itälahdenkatu 2 FIN-00210 Helsinki Finland
Head	:	P. Piri / H. Leppänen
Tel.	:	358.0.682.661 **Fax**: 358.0.682.6600 **Telex**: 121963 swcon sf

Type of institute
Multidisciplinary consulting company including infrastructure and road network design. Water supply and sanitation, wastewater treatment plants, environmental technology etc.

Research activities
Aquatic ecology
Ecological capacity assessment
Limnology
Environmental impact assessment
Feasibility studies

Facilities
Library
Advanced computer hardware
Specialized software
Water laboratory

Staff: Scientists: 3 Others: 5

Description
Studies concerning impacts of construction and waste water on fisheries and aquatic environment, assessing compensation to fishermen and land owners, ecotoxicological tests, fisheries management plans, feasibility studies, limnological studies, tropical fisheries and aquatic ecology studies.

16. Technical Research Centre of Finland - Food Research Laboratory

Name	:	Technical Research Centre of Finland - Food Research Laboratory
Address	:	Biologinkuja 1 P.O. Box 203 FIN-02151 Espoo Finland
Head	:	Prof. J. Ahvenainen, Mr. T. Hattula (fish research)
Tel.	:	358.0.456.5160 **Fax**: 358.0.455.2103 **Telex**: 122972 vttha FIN

Type of institute
Research institute

Research activities
Fish as food
Development of quality measurement methods for fish and fish products
Sensory analysis of fish as an indicator of environmental pollution

Facilities
Research facilities and laboratory for food analysis
Various equipments for food processing

Staff: Scientists: 1 Others: 1

Research budget (ECU/year): 100.000 for fish research

Description
The fish research in the Technical Research Centre of Finland-Food Research Laboratory is mostly based on projects financed by the Ministry of Agriculture and Forestry, fishing organisations and industry. The principal objects have been Baltic herring and farmed salmon and trout. The area of research has varied from fishing conditions to fish products with special emphasis on the quality and development of quality measurement methods.

17. University of Helsinki - Department of Food Technology

Name	:	University of Helsinki-Department of Food Technology
Address	:	P. O. Box 27 (Viikki B) FIN-00014 University of Helsinki Helsinki Finland
Head	:	Dr. H. Salovaara
Tel.	:	358.0.70851 **Fax**: 358.0.7085212

Type of institute
Higher Education Institute, University Department

Research activities
Processing
Packaging
Sensory evaluation

Facilities
Pilot plant
Sensory laboratory

Staff: Scientists: 32 Others: 10

Research budget (ECU/year): 100.000

Description
Most of the research directed on food packaging, fermentation, physical properties of food and sensory science. Research on fish and fish products is occasional.

18. University of Helsinki - Department of Limnology and Environmental Protection

Name : University of Helsinki - Department of Limnology and Environmental Protection
Address : P. O. Box 27 (E-building)
FIN-00014 Univeristy of Helsinki
Finland
Head : Prof. P. Eloranta
Tel. : 358.0.708.5466 **Fax**: 358.0.708.5257

Type of institute
University Department. Higher Education and Research Institute

Research activities
Stock assessment
Multispecies interactions
Environmental biology
Food-web manipulations

Facilities
Research and teaching laboratories with standard equipments. Full scale facilities for fisheries research in the field.

Staff: Scientists: 12 Others: 7

Research budget (ECU/year): 130.000

Description
Department is dealing with fisheries research in all its major fields such as stock assessment and management, species interaction, food-web manipulation and mathematical modelling of fish stocks. Studies are made in inland waters and in the Baltic Sea. Department is the major higher level education institute in fisheries field in Finland.

19. University of Helsinki - Department of Zoology - Division of Physiology

Name : University of Helsinki - Department of Zoology
Division of Physiology
Address : P.O. Box 17 (Arkadiankatu 7)
FIN-00014 University of Helsinki
Finland

Head : Prof. K. Kaila
Tel. : 358.0.1911 **Fax**: 358.0.1917301

Type of institute
University Department, Higher education

Research activities
Fisheries management
Physiological effects of pulp mill effluents on fish
Environmental effects on respiration
Aquaculture
Respiratory physiology of fish

Facilities
Aquaria, fluorescence ratio imagine system, other specialized analytical equipments

Staff: Scientists: 5 Others: 1

Research budget (ECU/year): 30.000

Description
The research on fish physiology aims at explaining the mechanisms of respiratory adjustments in fish and how these are disturbed by environmental changes. Special emphasis is focused on gas transport.

20. University of Helsinki - Department of Zoology - Integrative Ecology Unit

Name : University of Helsinki - Department of Zoology
Integrative Ecology Unit
Address : Division of Ecology
P.O. Box 19
P. Rautatiekatu 13
FIN-00014 University of Helsinki
Finland
Head : Dr. E. Ranta
Tel. : 358.0.191.7420 **Fax**: 358.0.191.7492

Type of institute
Higher education, University Department

Research activities
Fish and fisheries biology

Facilities
University of Helsinki field stations with aquarium rooms and small research vessel are available.

Staff: Scientists: 5

Research budget (ECU/year): 80.000 (funded by Finnish Academy of Sciences)

Description
The main topics are behavioural ecology schooling, foraging, sexual selection, population ecology, theory of fish yield in lakes and predator prey interactions in crayfish aquaculture.

21. University of Helsinki - Kilpisjärvi Biological Station

Name : University of Helsinki - Kilpisjärvi Biological Station
Address : FIN-99 490 Kilpisjärvi
Finland
Head : Dr. A. Järvinen
Tel. : 358.0.77.712 **Fax**: 358.0.77.709

Type of institute
Biological and geographical research and education station

Research activities
Ecology
  Population biology
Monitoring
  Environmental changes

Facilities
Lodging for 30 persons
Small library
Laboratory with standard equipments

Staff: Scientists: 2 Others: 4-6

Description
Ecology of subarctic plant and animal populations. Long-term follow-up of environmental changes. Growth, food and feeding studies on whitefish populations in subarctic lakes.

22. University of Helsinki - Lammi Biological Station

Name : Lammi Biological Station
Address : FIN-16900 Lammi
Finland
Head : Dr. J. Syrjämäki
Tel. : 358.17.631.111 **Fax**: 358.17.631.1166

Type of institute
University Department

Research activities
Ecology
Hydrobiology

Facilities
Spectrophotometers, HPLC, ion chromatograph, carbon analyzer, auto-analyzer etc.

Staff: Scientists: 8 Others: 20

Research budget (ECU/year): 450.000

Description
Principal aim of the station is to act as a field base for instruction and research in biological, mainly ecological and related sciences.

23. University of Joensuu - Karelian Institute - Section of Ecology

Name : University of Joensuu - Karelian Institute - Section of Ecology
Address : P. O. Box 111
FIN-80101 Joensuu
Finland
Head : Dr. P. Huttunen
Tel. : 358.73.151.3500 **Fax**: 358.73.151.3449 **Telex**: 46183 joy FIN

Type of institute
University Department, Research centre

Research activities
Fish biology and ecology
Larval fish ecology
Food consumption and food resources

Multispecies stock assessment
Environmental parameters and their impact on resources
Ecotoxicology

Facilities
Research vessel
Experimental hatchery
Laboratory for fish and water analysis
Aquariums
Hydrobiological and fish biological sampling equipments

Staff: Scientists: 4 Others: 7

Research budget (ECU/year): 67.000

Description
Activities focused on research projects and consulting projects. Research projects include mechanisms of stock fluctuations in Coregonids, estimation of the food resources in planktivorous fishes, energy flow in lake ecosystems, food consumption and energy allocation in fish, multispecies stock assessment of fish in Lake Saimaa, effects of mass removal of Cyprinids on water quality, benthic invertebrates and zooplankton. Consulting project focus on monitoring the effects of waste waters and other environmental changes on fisheries and fishes.

24. University of Jyväskylä - Department of Biology

Name : University of Jyväskylä - Department of Biology
Address : P. O. Box 35
FIN-40 351 Jyväskylä
Finland
Head : Prof. M. Kulomaa
Tel. : 358.41.602.320 **Fax**: 358.41.602.321

Type of institute
University Department, Higher education

Research activities
Fish biology
- Environmental biology
- Physiological ecology
- Parasitology
- Immunology

Fisheries management in lakes and rivers
  Stock-recruitment relationships
  Stock adjustments
  Environmental problems

Facilities
SIMRAD acoustic system, biomonitoring of heart pulse and ventilation, experimental hall

Staff: Scientists: 25 Others: 24

Research budget (ECU/year): 2.500.000

Description
About 20 % of activities deals with fisheries. Annually 5 MSc thesis and one Lic.Phil. or PhD thesis on fish biology. Research groups on parasitology and immunology including crayfish plague and environmental, community and life cycle group. Fisheries management is focused on vendace fisheries and its stock-recruitment relationship as well as lake and river manipulation. Experimental fish biology deals with competition of early and young stages of dominant species in lakes and environmental ecophysiology.

25. University of Kuopio - Department of Applied Zoology

Name : University of Kuopio - Department of Applied Zoology
Address : P. O. Box 1627
FIN-70 211 Kuopio
Finland
Head : Prof. O. V. Lindqvist
Tel. : 358.71.163.122 **Fax**: 358.71.163.148 **Telex**: 42218 kuy FIN

Type of institute
Higher Education Institute / University Department

Research activities
Fish biology
  Ecophysiology
  Nutrition
  Genetics
  Pathology
Fisheries management and economics
  Inland fishing technology
  River restoration
  Fishery management and development

Fishery tourism
Crayfish fishery
Aquaculture
Fish and crayfish culture
Breeding
Transgenic fish
Tropical limnology and fishery in developing countries

Facilities
Data bases
Extensive libraries
Laboratories for fish and water analysis
Access to Biotechnology and Environmental laboratories
Aquaculture research unit
Field research equipments

Staff: Scientists: 12 Others: 4

Research budget (ECU/year): 40.000

Description
Fishery and aquaculture education on MSc and PhD levels includes fish and aquatic biology, production management, processing and marketing aquatic products as well as business management. Research activities aim broadly at fishery and aquaculture development with emphazis on small-scale, inland fisheries on rural areas. The department has wide national and international connections to the institutions of fishery, biotechnology and environmental sciences. Department works closely with the fishing and aquaculture industry on fishing gear development, aquaculture technology and genetics, national rainbow trout breeding programme, new species technology and waste technology. Department is strongly involved in the international development cooperation, tropical limnology and fishery research and it provides a MSc-Programme for students from developing countries.

26. University of Oulu - Department of Zoology

Name : University of Oulu - Department of Zoology
Address : FIN-20570 Oulu
Finland
Head : Prof. E. Pulliainen
Tel. : 358.81.553.1210 **Fax**: 358.81.553.1060

Type of institute
University Department, Higher Education and Research

Research activities
Ecology of fresh and brackish water fish
  Assessment of fish stocks, especially Coregonidae
  Control of parasites and diseases
Fisheries management
Aquaculture

Facilities
Ecological and physiological research facilities, laboratories with water temperature control

Staff: Scientists: 3 Others: About 4 undergraduate annually

Description
Several research projects in fish ecology, physiology and parasitology. Research activities based on students aiming at higher university degree and doctors aiming at a career in fishery biology in research, administration and environmental protection.

27. University of Oulu - Department of Zoology - Bothnian Bay Research Station

Name	:	University of Oulu - Department of Zoology - Bothnian Bay Research Station
Address	:	Linnanmaa FIN-90570 Oulu Finland
Head	:	S. Sulkava
Tel.	:	358.81.553.1211 **Fax**: 358.81.556.1278

Type of institute
University Department, Higher Education and Research

Research activities
Population biology
  Habitat selection
  Food selection
Restoration of rivers
Aquaculture
Fish physiology

Facilities
Field sampling and physiological sampling equipments. Aquariums.

Staff: Scientists: 4 Others: 2

Research budget (ECU/year): 30.000

Description
Population biology of vendace, habitat selection and food selection studies on fish in rapids. Studies on restoration of rivers for fisheries, controlling studies of various fish diseases at hatcheries and heavy metal and hormonal studies of fish physiology.

28. University of Turku - Archipelago Sea Research Institute

Name : University of Turku - Archipelago Sea Research Institute
Address : University of Turku
FIN-20500 Turku
Finland
Head : Dr. I. Vuorinen
Tel. : 358.21.633.5974 **Fax**: 358.21.633.6592

Type of institute
University research institute and field station

Research activities
Brackish water ecology
Ecology of fish
Geology
Sedimentology
Interdisciplinary research in the Archipelago Sea area

Facilities
Research vessel and small boats
CTD
Flowthrough fluorometer
Standard hydrographic and aquatic ecology sampling gears

Staff: Scientists: 1 Others: 9

Research budget (ECU/year): Included in the budget of the University of Turku

29. University of Turku - Department of Biology - Laboratory of Ecology and Systematics

Name	:	University of Turku - Department of Biology - Laboratory of Ecology and Systematics
Address	:	FIN-20500 Turku Finland
Head	:	Prof. E. Haukioja
Tel.	:	358.21.63351 **Fax**: 358.21.633.6550

Type of institute
University Department and a Laboratory, Higher Education Institute

Research activities
Reproduction of Baltic Sea herring
Regulation of vendace stocks
Fish stock assessment
Fish bioenergetics models
Fish population models

Staff: Scientists: 1 (+2) Others: 1

Research budget (ECU/year): Variable depending on external funding

Description
Research is based on external funding and the work of doctoral as well as graduate students. Main study areas are the Archipelago Sea and certain inland lakes.

30. Vesi-Hydro Consulting Engineers Ltd.

Name	:	Vesi-Hydro Consulting Engineers Ltd.
Address	:	Sentnerinkuja 1 FIN-00440 Helsinki Finland
Head	:	J. Korpela
Tel.	:	358.0.56.501 **Fax**: 358.0.565.0385

Type of institute
Consulting firm/Water and Environmental Research Department

Research activities
Fishing surveys
Limnological and hydrobiological research
Water quality research
Pollution assessments and forecasts
Damage and loss estimates

Facilities
Water laboratory
Hydrobiological laboratory
Fisheries biology and limnology field study facilities

Staff: Scientists: 8 Others: 10

Research budget (ECU/year): 290.000

Description
Vesi-Hydro is an independent private consulting engineering firm. The water and environmental department deals with environmental problems e.g. pollution and its impact on fisheries, aquatic life and water quality. The specialities of the department are fisheries biology and economy, limnology, hydrobiology, water quality and sediment studies as well as development of specialized computer software. Clients are municipalities, governmental authorities and industry in Finland and abroad.

31. Water and Environment District of Central Finland

Name : Water and Environment District of Central Finland
Address : PL 110
FIN-40 101 Jyväskylä
Finland
Head : M. Ranta
Tel. : 358.41.697.211 **Fax**: 358.41.614.273

Type of institute
Regional centre of water and environment research and administration

Research activities
Fisheries management
  Stock assessment
  Electric fishing
Environmental problems and their impact
Fishways
Restoration of rivers

Facilities
Modern water laboratory

Staff: Scientists: 2 Others: 2

Description
The main research activities are population ecology of burbot, brown trout and greyling and environmental problems of aquaculture.

32. Water and Environment District of Kokkola

Name : Water and Environment District of Kokkola
Address : Torikatu 40 B
FIN-67 100 Kokkola
Finland
Head : O. Hjelt
Tel. : 358.68.827.9111 **Fax**: 358.68.827.9237

Type of institute
Regional centre of water and environment research and administration

Research activities
Research and monitoring of water ways
Quality
Hydrology
Fisheries
Restoration of rivers and lakes
Environmental problems
Aquaculture

Facilities
Chemical laboratory
Aquariums
Data bases and water quality models software

Staff: Scientists: 5 Others: 14

Research budget (ECU/year): 440.000

Description
Basic population studies of fish stocks and fisheries. Fisheries economics in relation to water construction projects. Studies on tolerance of main fish species, lamprey and crayfish on changes of water quality. Aquaculture of lamprey larvae. Restoration of dredged rivers.

33. Water and Environment District of Lappland

Name : Water and Environment District of Lappland
Address : P. O. Box 8060
FIN-96 101 Rovaniemi
Finland
Head : Dr. K. Kinnunen
Tel. : 358.60.2941 **Fax**: 358.60.310.340 **Telex**: 126086 vyh FIN

Type of institute
Regional centre of water and environment research and administration

Research activities
Research and monitoring of water ways
Quality
Hydrology
Fisheries
Restoration of demaged rivers and lakes
Environmental problems

Facilities
Water laboratory
Electric fishing equipments

Staff: Scientists: 5 Others: 15

Research budget (ECU/year): 440.000

34. Water and Environment District of Turku

Name : Water and Environment District of Turku
Address : P. O. Box 47
FIN-20 801 Turku
Finland
Head : V. M. Tiainen
Tel. : 358.21.661.799 **Fax**: 358.21.661.730

Type of institute
Regional centre of water and environment research and administration

Research activities
Water protection
Monitoring of the quality of surface and ground waters

Water engineering
Environmental research
Baltic marine research
Regional environmental monitoring

Facilities
Chemical laboratory for water quality analyses. Regional field activities in hydrobiological and water quality sampling.

Staff: Scientists: 5 Others: 10

Research budget (ECU/year): 550.000

Description
Research department is about 28 % of the total staff. The entire staff is 70 persons.

35. Water Protection Association of Savo - Karjala

Name : Water Protection Association of Savo - Karjala
Address : Yrittäjäntie 24
FIN-70 150 Kuopio
Finland
Head : J. Ronkainen / J. Hartikainen
Tel. : 358.71.233.644 **Fax**: 358.71.233.404

Type of institute
Consulting firm

Research activities
Environmental problems and their impact to water and fishes. Aquatic pollution.

Facilities
Laboratory for water analysis and biological investigations
Specialized computer hardware

Staff: Scientists: 3 Others: 11

Research budget (ECU/year): 490.000

Description
Water Protection Association of Savo-Karjala is a consulting firm and the main activities are monitoring of water quality and investigations on eutrophication and pollution effects on water and fishes as well as collection of primary data on biological and water quality.

36. Water Protection Association of Southwest Finland

Name : Water Protection Association of Southwest Finland
Address : Telekatu 16
FIN-20 360 Turku
Finland
Head : Dr. K. Jumppanen
Tel. : 358.21.538.222 **Fax**: 358.21.381.838

Type of institute
Water research institute, consulting firm

Research activities
Monitoring of waste water load and its effects on a state and usability of water courses and coastal waters including fish and fisheries

Facilities
Water research laboratory and equipments for sampling and field studies

Staff: Scientists: 5 Others: 12

Research budget (ECU/year): 590.000

37. Water Protection Association of the River Kokemäenjoki - Fisheries Department

Name : Water protection Association of the River Kokemäenjoki - Fisheries Department
Address : P. O. Box 265
FIN-33 101 Tampere
Finland
Head : H. Westerling
Tel. : 358.31.246.1111 **Fax**: 358.31.246.1200

Type of institute
Engineering and consulting firm

Research activities
Fisheries management
- Fishery statistics
- Stock assessment
- Environmental problems and their impact on resources

Facilities
Analytical laboratory for chemical and microbiological analysis
Biological laboratory
Equipments for field studies
Specialized software

Staff: Scientists: 5 Others: 2

Research budget (ECU/year): 280.000

Description
Association and its department is a non-profit engineering and consulting firm for monitoring of eutrofication and pollution in the basin of the River Kokemäenjoki and adjacent sea-area on the west coast of Finland.

38. Water Protection Society of Central Finland

Name : Water Protection Society of Central Finland
Address : Jyväskylän kaupunki
PL 192
FIN-40101 Jyväskylä
Head : Kari Hokkanen
Tel. : 358.41.625.402 **Fax**: 358.41.625.068

Type of institute
Water protection society

Research activities
Communal and industrial waste water treatment and hydrographical research in lakes

Staff: Scientists: 1

Research budget (ECU/year): 3.000

39. Water Protection Society of Lake Saimaa

Name : Water Protection Society of Lake Saimaa
Address : Hietakallionkatu 2
P. O. Box 17
FIN-53 850 Lappeenranta
Finland
Head : P. Laine
Tel. : 358.53.26.650 **Fax**: 358.53.26.653

Type of institute
Water research, consulting firm

Research activities
Water research
Biomanipulation
Utilization and treatment projects for lake fisheries

Facilities
Research vessel

Staff: Scientists: 2 Others: 7

Research budget (ECU/year): 220.000

40. Water Research and Engineering PSV Ltd.

Name : Water Research and Engineering PSV Ltd.
Address : Kalevalankuja 8
P.O. Box 20
FIN-90 571 Oulu
Finland
Head : P. Antila
Tel. : 358.81.363.222 **Fax**: 358.81.363.772

Type of institute
Consulting firm

Research activities
Water research
Limnological surveys and planning
Fish biology and fisheries
Environmental monitoring and engineering

Facilities
Water research and analytical laboratory
Data processing division
Mobile sampling and measuring unit for various field conditions

Staff: Scientists: 14 (4 for biological section) Others: 20

Research budget (ECU/year): 1.200.000-1.600.000 (for biology 200.000)

Description
Continuous monitoring programs for fisheries and fish biology based on catch statistics, fish samples, electric fishing chemical analysis of hazardous compounds in fish etc. Case studies for licence applications of industry, construction, mining, aquaculture and waste waters.

State of Research

FRANCE

Une part importante de la recherche océanographique française est assurée par l'IFREMER (Institut Français de Recherche pour l'Exploitation de la Mer) créé en 1984 par fusion de l'Institut Scientifique et Technique des Pêches Maritimes et du Centre National d'Exploitation des Océans; il gère la majeure partie de la flotte océanographique (moyens lourds) et, par des contrats d'étude et de recherche, peut solliciter et inciter des partenaires extérieurs.

L'institut a de nombreuses implantations (centres, stations, délégations) sur le territoire métropolitain ainsi que dans les départements et territoires d'outre-mer. Il possède un certain nombre de filiales et a contribué à la création de Centres techniques (trois) chargés de relayer son action de transfert dans le domaine de la transformation des produits de la mer (y compris les algues).

IFREMER est un établissement public a caractère industriel et commercial placé sous les tutelles des Ministères de la Recherche et de l'Enseignement supérieur, de l'Equipement des Transports et du Tourisme, de l'Agriculture et de la Pêche.

Les liens entre les départements ministériels et la recherche sont assurés par une mission de la recherche pour les problèmes généraux, mais l'interlocuteur opérationnel pour les questions d'avis pour l'exploitation des ressources vivantes est la direction des Pêches Maritimes et des Cultures Marines du Ministère de l'Agriculture et de la Pêche.

L'organisation fonctionnelle de l'IFREMER repose sur une répartition entre quatre directions "opérationelles" correspondant à quatre grands domaines d'activité:
- Direction de l'Ingénierie, de la technologie et de l'Informatique
- Direction des Recherches Océaniques
- Direction des Ressources Vivantes
- Direction de l'Environnement Littoral

D'autres organismes contribuent aux recherches en océanographie mais l'IFREMER exerce un quasi monopole en matière de recherche concernant la pêche maritime et l'aquaculture marine sur le domaine national, et d'avis scientifique pour la gestion des ressources marines renouvelables, l'INSU (Institut National des Sciences de l'Univers) qui regroupe des équipes du CNRS (Centre National de la Recherche Scientifique) et celles des universités, a les équipes numériquement les plus importantes; l'ORSTOM (Institut Français de Recherche Scientifique pour le Développement en Co-opération) s'intéresse essentiellement aux zones tropicales et intertropicales.

La co-opération avec les autres organismes est souvent formalisée par des accords cadres qui peuvent donner naissance à des coopérations concrètes destinées à conforter les moyens et compétences disponibles dans des recherches à caractère fondamental. Des "unités mixtes" de recherche ont été créées, associant l'IFREMER à l'ORSTOM, au CNRS, à l'INRA et à l'INSU; ainsi par exemple la station mixte IFREMER-INSU sur les problèmes écologiques de l'aquaculture en marais et zones côtières; le laboratoire mixte l'IFREMER-ORSTOM sur la sclérochronologie des animaux marins; la station expérimentale mixte IFREMER-INRA (Institut National de la Recherche Agronomique).

Les chercheurs de l'IFREMER sont également sollicités par l'enseignement universitaire. S'il existe en France des formations dédiées à l'océanographie, la politique de l'IFREMER a cependant été d'engager régulièrement des spécialistes autres qu'océanographes (généticiens, informaticiens...) pour enrichir les équipes existantes par des compétences très spécifiques et du plus haut niveau.

Plusieurs instances sont attachées à l'IFREMER en ce qui concerne la liaison avec les utilisateurs.

* la première correspond au Conseil d'Administration dans lequel apparaissent les Ministères de tutelle et d'autres Ministères concernés (Environnement, Finances, Affaires Etrangères) ainsi que des personnalités du monde professionnel et des industries liées à la mer, le dispositif est complété par des membre élus;
* Il existe en outre divers organes consultatifs:
* Le Conseil scientifique éclaire l'IFREMER sur les questions de stratégie scientifique, il est essentiellement composé de scientifiques de différents organismes extérieurs comme de l'IFREMER et s'appuie sur des commissions spécialisées;
* deux comités différents sont destinés à renforcer les liens entre l'IFREMER et le corps social qui doit bénéficier de ces travaux:
* le Comité Technique et Industriel, s'occupe de la Direction de l'Ingénierie et de la Technologie
* le Comité des Ressources Vivantes, s'occupe de la Direction des Ressources Vivantes. Il est composé entre autres de représentants des différentes professions de la pêche et des cultures marines et des Ministère concernés.

Un gros effort de vulgarisation est conduit, à l'aide de revues, de manuels et d'autres supports didactiques souvent créés avec la collaboration de professionnels de la communication.

Les priorités actuelles sont l'amélioration de la qualité des milieux côtiers, la valorisation des produits marins, et la formation de spécialistes destinés aux domaines de l'exploitation et de la valorisation des ressources vivantes de la Mer.

List of Institutes from FRANCE

1. ADRIA
2. Agropolis-Halieutique et Aquaculture
3. C.N.R.S. U.R.A 1451 - Laboratoire d'Ecologie des Systèmes Fluviaux
4. CAEC - Centre Atlantique Etudes Cosmétologie
5. CEMAGREF - Centre National du Machinisme Agricole du Génie Rural des Eaux des Forêts
6. CEMAGREF - Centre National du Machinisme Agricole du Génie Rural des Eaux des Forêts
7. Centre Etudes de Projets
8. CNEVA - Lab. Central de Recherches Vétérinaires
9. CNEVA - Lab. Pathologie Animaux Aquatiques
10. Collège de France - Lab. Biologie Marine
11. COM - Centre d'Océanologie de Marseille URA no. 41 du CNRS
12. Conseil Supérieur de la Pêche
13. Digital Equipment Europe - European Technical Center
14. Dynecar - Dynamics of Caribbean Marine Ecosystems and Biology of Associated Species
15. Ecole Nationale Supérieure Chimie Paris-Lab. Bioorganique et Biotechnologie
16. ENSMP - Ecole Nat. Supér. des Mines de Paris-Centre de Géostatistique
17. ID. Mer - Lorient
18. IFREMER - Environnement et Aménagement Littoral
19. IFREMER - Génie Océanique - Service Navires Pêche Aquaculture
20. IFREMER - Service Economie Maritime
21. IFREMER - Ressources Vivantes - Lab. Recherche Aquacole
22. IFREMER - Ressources Vivantes - Lab. Algoculture
23. IFREMER - Ressources Vivantes - Lab. Aquaculture Tropicale
24. IFREMER - Ressources Vivantes - Lab. Biologie et Ecologie Invert. Marins
25. IFREMER - Ressources Vivantes - Lab. Défense et Résistance chez les Invertébrés Marins
26. IFREMER - Ressources Vivantes - Lab. L'Houmeau CREMA
27. IFREMER - Ressources Vivantes - Lab. Recherches Vivantes Méditerranéen
28. IFREMER - Ressources Vivantes - Lab. Aquaculture en Marais Noirmoutier
29. IFREMER - Ressources Vivantes - Lab. Recherche Aquacole Martinique
30. IFREMER - Ressources Vivantes - Lab. Côtier Ressources Aquacole Nouvelle Calédonie
31. IFREMER - Ressources Vivantes - Lab. Côtier Ressources Aquacole Trinité sur Mer
32. IFREMER - Ressources Vivantes - Lab. Côtier Ressources Aquacole Port en Bessin
33. IFREMER - Ressources Vivantes - Lab. Dynamique Syst. Production
34. IFREMER - Ressources Vivantes - Lab. Ecologie Halieutique
35. IFREMER - Ressources Vivantes - Lab. Mathematiques Appl. à l'Exploitation Ress. Halieutiques et Aquacoles
36. IFREMER - Ressources Vivantes - Lab. Sclérochronologie
37. IFREMER - Ressources Vivantes - Lab. Ressources Halieutiques Boulogne sur Mer

38. IFREMER - Ressources Vivantes - Lab. Côtier Ressources Halieutiques Brest
39. IFREMER - Ressources Vivantes - Lab. Côtier Ressources Halieutiques Guyane
40. IFREMER - Ressources Vivantes - Lab. Côtier Ressources Halieutiques La Réunion
41. IFREMER - Ressources Vivantes - Lab. Côtier Ressources Halieutiques La Rochelle
42. IFREMER - Ressources Vivantes - Lab. Côtier Ressources Halieutiques Lorient
43. IFREMER - Ressources Vivantes - Lab. Côtier Ressources Halieutiques Port en Bessin
44. IFREMER - Ressources Vivantes - Lab. Côtier Ressources Halieutiques Sète
45. IFREMER - Ressources Vivantes - Lab. Côtier Ressources Halieutiques St. Pierre et Miquelon
46. IFREMER - Ressources Vivantes - Lab. Côtier Ressources Halieutiques Tahiti
47. IFREMER - Ressources Vivantes - Lab. Biochimie et Molécules Marines
48. IFREMER - Ressources Vivantes - Lab. Etudes Technico-Réglémentaires
49. IFREMER - Ressources Vivantes - Lab. Génie Alimentaire
50. IFREMER - Ressources Vivantes - Lab. Qualité et Physico-Chimie
51. INRA - Dept. Hydrobiologie et Faune Sauvage
52. INRA - Laboratoire de Génétique des Poissons
53. INRA - Laboratoire de Physiologie des Poissons
54. INRA - Molecular Virology and Immunology Unit Fish Diseases Teams
55. INRA - Station d'Hydrobiologie de St. Pee sur Nivelle
56. INRA - Station d'Hydrobiologie Lacustre
57. INTECHMER
58. IUT - La Rochelle - Lab. Biologie Biochemie Marine
59. IUT - St. Nazaire - Lab. Génie des Procédés
60. Lab. Veterinaire Départemental - Arras
61. Lab. Veterinaire Départemental - Lons le Saunier
62. Museum Histoire Naturelle - Lab. Ichtyologie Générale et Appl.
63. Museum Histoire Naturelle - Lab. Physiologie Générale et Comparée
64. SANOFI - France Aquaculture
65. SEPIA Internationale
66. Univ. Blaise Pascal Clermont II - Equipe d'Hydrobiologie
67. Univ. Bordeaux I - Laboratoire d'Oceanographie Biologique d'Arcachon
68. Univ. Caen - Centre Régional Etudes Cotières - Lab. Luc sur Mer
69. Univ. Claude Bernard Lyon I - Institut Michel Pacha Station Maritime de Physiologie
70. Univ. de Nice - Sophia Antipolis - Lab. Physiologie Cellulaire et Comparée
71. Univ. Languedoc - Sciences et Techn. - Station Biol. Marine et Lagunaire
72. Univ. Limoges - Lab. Immunophysiologie
73. Univ. Lyon I - Claude Bernard - Lab. Physiologie Générale et Comparée
74. Univ. Lyon I - Lab. Biologie Animale et Ecologie U.R.A. CNRS 1451 - Ecologie des Eaux Douces et des Grands Fleuves
75. Univ. Lyon - Catholique - Inst. Reg. Rech. Appl. Aquaculture
76. Univ. Marseille III - Fac. Sciences et Techn. - St. Jerome - CERAM
77. Univ. Montpellier II - Dept. Aquaculture et d'Halieutique
78. Univ. Montpellier II - Lab. Biologie Animale
79. Univ. Montpellier II - Lab. Ichthyologie
80. Univ. Montpellier II - Station Méditerranéenne de l'Environnement Littoral

81. Univ. Nantes - Lab. Biologie Marine
82. Univ. Paris VI - P. et M. Curie - I.N.S.U - Observatoire Océanologique - Villefranche-sur-Mer
83. Univ. Paris-Sud XI - Lab. Physiologie Cellulaire et Métabolique Poissons
84. Univ. Perpignan - Lab. Biologie Marine
85. Univ. Rennes I - Equipes Biologie Moléculaire et Physiologie des Régulations de l'URA CNRS 256
86. Univ. Rennes I - Station de Biologie Marine - Ile de Bailleron
87. Univ. Strasbourg Louis Pasteur - Lab. Psychophysiologie
88. Univ. Toulouse - ENSAT Lab. Ichtyologie

1. ADRIA

Name	:	Association pour le Développement de la Recherche Appliquée aux Industries Agricoles et Alimentaires (ADRIA)
Address	:	6 rue de l'Université - BP 313 F - 29191 Quimper Cedex
Head	:	C. Bourgeois
Tel.	:	98906232 **Fax**: 98907328 **Telex**: 941122

Type of institute
Research centre
Institute for professional training/Secondary school

Research activities
Upgrading of fishery products
 Improvement of techniques for handling, storing, processing, packaging fish
 Methods for grading and quality assessment of fish
 Novel processing methods for by-products, discards, and process wastes

Staff: Scientists: 3 Others: 2

Description
Research Programmes:
Improvement of the fish freshness from the capture to the detail sale.
Setting up of detection techniques of irradiated crustaceans.
Setting up of method to measure histaminic rate through HPLC.
Conceiving a method to open the shell of molluscs by the use of infra red keys.
Decoloration of fish meat and pulp by the use of ozone.
Pilot workshop and related equipments to test the transfer of technology.
Outlets of surimi in prepared meals and delicatessen (FAR).
Foaming of delicatessen of fish products.

2. Agropolis - Halieutique et Aquaculture

Name	:	Agropolis Secteur Halieutique et Aquaculture
Address	:	2 place Pierre Viala F - 34060 Montpellier Cedex 01
Head	:	L. Malassis/J.P. Quignard
Tel.	:	67612526 **Fax**: 67612580 **Telex**: 490818 F

Type of institute
Government Department / Regional or local authority
Engineering/consulting firm
Research centre

Research activities
Fisheries management
- Stock assessment
- Multi-disciplinary approach to fisheries management
- Environmental problems and their impact on resources

Fishing techniques
- Testing techniques/instruments

Aquaculture
- Reproduction
- Early stages of reared species
- Growth, nutrition, feed composition, live food production for aquaculture
- Disease, immunology, stress
- Aquaculture/environment interaction
- Aquaculture structures, systems, techniques

Facilities
Research vessel
Rearing tank
Data bases, extensive libraries
Field site, electron microscope, specialized analytical apparatus
Calculation center CNUSC

Staff: Scientists: 80 Others: 50

3. C.N.R.S. U.R.A 1451 - Laboratoire d'Ecologie des Systemes Fluviaux

Name : Laboratoire d'Ecologie des Systèmes Fluviaux
U.R.A. CNRS 1451
Address : 1, rue Parmentier
F - 13200 Arles
Head : D. Pont
Tel. : (33)90.93.86.66 **Fax**: (33)90.96.07.99

Type of institute
Research Laboratory

Research activities

Fish communities and fisheries in large rivers: the Rhône catchment
- Evaluation of professional and recreational catches in large rivers
- Biological strategies of French fauna
- Analysis of fish community structure at the catchment scale (eco-ions)
- Longitudinal zonation in large impounded rivers
- Local patterns and young fish-habitat relationships

Facilities

Boat
Sampling apparatus (electro-fishing, nets, ...)
Scalimetric materials
Automatic sampler
Basic materials for physico-chemical measurements
Possibilities for training courses and lodging (sleeping-rooms, kitchen, bathroom, ...)

4. CAEC - Centre Atlantique Etudes Cosmétologie

Name : Centre Atlantique d'Etudes en Cosmétologie
Laboratoire de Cosmétologie et de Pharmacie Industrielle
Address : 68, bld Eugène Orieux
F - 44000 Nantes
Head : Prof. Y. De Roeck-Holtzhauer
Tel. : 33.40742322 **Fax**: 33.40743289

Type of institute

Research centre
Engineering/consulting firm
University Department/Higher Education Institute

Research activities

Upgrading of fishery products
- Improvement of techniques for handling, storing, processing, packaging fish
- Novel processing methods for by-products, discards, and process wastes

Facilities

Data bases, extensive libraries

Staff: Scientists: 5 Others: 2

Research budget (ECU/year): 3.000

Description
The "Centre Atlantique d'Etudes en Cosmetologie" is also carrying studies to improve the use of seaweeds in cosmetics, dietetics or pharmacy.

5. CEMAGREF - Centre National du Machinisme Agricole du Génie Rural des Eaux des Forêts

Name : Centre National du Machinisme Agricole du Génie Rural des Eaux des Forêts
(French Institute of Agricultural and Envrionmental Engineering Research)
Division Aquaculture et Píche
(Fisheries and Aquaculture Division)
Address : 50, Avenue de Verdun-BP 3
F - 33611 Gazinet Cedex
Head : Pierre Elie (Division Chief)
Tel. : (33)56.36.09.40 **Fax**: (33)56.36.75.11

Type of institute
Public Research centre

Research activities
Inland Fisheries (including estauries)
Fish ecology and population dynamics; multi species biological models (particularly for amphihaline species: sturgeon, eel lampreysm shads,...)
Fisheries management
Environmental problems and their impacts on fish resources
Sampling techniques and design
Aquaculture
Extensive:
Studies of ecological functions of the exploited fish species, physical impacts on the inshore brackish water domain of the French Atlantic coast
Testing aquaculture development in rural areas
Intensive:
Acipenserdae rearing (reproduction, larvae, juvenile, and breeder production),
Pathology and stress
Restocking Acipenser sturio
Impact of fish farming waste water (environmental problems)
Technical support for fish farmers.

Facilities
Two research vessels (Esturial: catamaran trawler, 12; long and Alpha: craft, 6m long).
Numerous fishing gears, Electronics in board
Field facilities for electro fishing (salt water, fresh water);
Specialised analytical apparatus and Laboratory
Carrying tanks
Computing centre
Information centre
Aquaculture experimental station (breeding and larvae development laboratory with external basins for juveniles and adults, closed circuit), with accommodation for visitors.
Experimental site for extensive aquaculture.

Staff: Scientists: 11 Others: 14

Description
Fisheries and Aquaculture Division investigates engineering and development of fish farming and fisheries in fresh and brackish water.

6. CEMAGREF - Centre National du Machinisme Agricole du Génie Rural des Eaux des Forêts

Name : Centre National du Machinisme Agricole du Génie Rural des Eaux des Forêts
(Institute of Agricultural and Environmental Engineering Research)
Division Biologie des Ecosystèmes Aquatiques
Address : 3 bis, Quai Chauveau
F - 69336 Lyon Cedex 09
Head : Jeanne Garric
Tel. : (33)72.20.87.39 **Fax**: (33)78.45.78.75

Type of institute
Governmental Research centre

Research activities
Fish ecology and ethology, Giomrohology of river streams,
Statistics and modelization.
Ecotoxicology of micro crustaceans and fish, biochemistry.
Fish behaviour, hydraulics and civil engineering
Hydrobiology

Facilities
Research vessels
Field facilities for electrofishing
Rearing and assays tanks
Controlled environmental chambers
On the site: specialized analytical apparatus, computing centre.

Staff: Scientists: 10 Others: 18

Description
The division is involved in applied research for the development of methodologies that can be used for the protection of rivers from an ecological point of view.
The main research subjects are: The modelisation of the capacity of rivers for fish. Ecological management of watersheds. Developing fish by-passes for dams. Impact of pollution on benthic communities. Developing of chronic ecotoxicology. Study of the selfsewage of rivers.

7. Centre Etudes de Projets

Name	:	Centre d'Etudes de Projets
Address	:	14 rue Cardinal de Cabrières F - 34000 Montpellier
Head	:	Hélène Rye/Michel Garrabé
Tel.	:	67604231

Type of institute
Research centre
University Department/Higher Education Institute

Research activities
Multi-disciplinary approach to fisheries management
Aquaculture economics

Facilities
Specialised computer hardware

Staff: Scientists: 4 Others: 1

Research budget (ECU/year): 122.400

8. CNEVA - Lab Central de Recherches Vétérinaires

Name : Centre National d'Etudes Vétérinaires et Alimentaires
Laboratoire Central de Recherches Vétérinaires
Unité d'Ichtyopathologie Continentale
(CNEVA-/LCRV/ICP)
Address : 22 rue Pierre Curie-BP 67
F - 94703 Maisons-Alfort Cedex
Head : A.M. Hattenberger
Tel. : 16(1)49.771300 **Fax**: 16(1)43.689762

Type of institute
Public Research Establishment

Research activities
Ichtyopathologie:
Recherche sur les maladies virales des poissons.
Méthodologie - Mise au point de techniques de mise en Èvidence des agents pathogènes.

Facilities
Equipement pour la virologie, la biologie moléculaire, la bactériologie, la parasitologie et l'histologie.
Installation expérimentale:
50m2 de surface avec 30 aquariums représentant un volume de 2m3 en circuit ouvert (renouvellement 1,3m3/h).

Staff: Scientists: 1 + 1 ingénieur d'études Others: 4

Research budget (ECU/year): 45.000

9. CNEVA - Lab. Pathologie Animaux Aquatiques

Name : Centre National d'Etudes Vétérinaires et Alimentaires
Laboratoire de Pathologie des Animaux Aquatiques
(CNEVA- LPAA)
Address : Centre de Brest-BP 70
F - 29280 Plouzane Brest
Head : F. Baudin Laurencin
Tel. : 98224462 **Fax**: 98055165

Type of institute
Research centre

Research activities

Aquaculture

Disease, immunology, stress
Diseases: viral, bacterial, immunological, parasitic, diseases
Ecopathology
Metabolic and nutritional pathology
Experimental physiology and pathology

Facilities

Laboratories of 300 m2 and equipment which allow the implementation of virology, bacteriology, parasitology, histology, hematology, techniques.
Rearing room of 280 m2 for experimental fish pathology.
Possibility of using fresh water, sea water, or a mixture, with thermoregulation: 20 m3/h in open circuit (200 tanks) and eight 15 m3 tanks in closed circuit.

Staff: Scientists: 6 Others: 9

Research budget (ECU/year): 140.000

Description

1) Viral pathology:
   fish birnavirus and rhabdovirus pancreatic disease
2) Bacterial pathology:
   Vibriosis, Flexibacteriosis, Yersiniosis, Pasteurellosis
3) Immunology and vaccination:
   immune response after immunization
   immunology and endocrinology
   immunology and environment
4) Ecopathology in farmed fish
5) Parasitology: Kudoa (Myxosporidea) in sea-water farmed salmonids

10. Collège de France - Lab. Biologie Marine

Name	:	Collège de France Laboratoire de Biologie Marine de Concarneau
Address	:	Place de la Croix-B.P. 225 F - 29182 Concarneau-Cedex
Head	:	François Gros/Yves Legal
Tel.	:	98970659 **Fax**: 98978124

Type of institute

Research centre
University Department/Higher Education Institute

Research activities
Fisheries management
Stock assessment
Environmental problems and their impact on resources
Aquaculture
Physiology
Growth, nutrition, feed composition, live food production for aquaculture
Upgrading of fishery products
Novel processing methods for by-products, discards, and process wastes.

Facilities
Research vessel
Rearing tank
Field site, electron microscope, specialized analytical apparatus

Staff: Scientists: 15 Others: 8

Research budget (ECU/year): 130.000

Description
Mainly biochemistry, enzymology molecular biology equipments.
Research activities of the Laboratory of Marine Biology are mainly devoted to comparative biochemistry and physiology of marine systems.
- Study of molecular and functional evolution of peptide hormones in crustaceans and molluscs (growth hormone, growth factors, gastrines, calcitonine).
- Ecophysiological adaptations at the enzyme level.

Ongoing research applications deal with:
- fisheries management, environmental assessment by biochemical methods.
- growth improvement for invertebrates aquaculture and biotechnological upgrading of fishery products.

11. COM - Centre d'Océanologie de Marseille URA No. 41 du CNRS

Name : Centre d'Océanologie de Marseille
URA no. 41 du CNRS
OSU - Université d'Aix Marseille 2
Address : Station Marine d'Endoume
Chemin de la Batterie des Lions
F-13007 Marseille

Head : François Blanc
Tel. : 33.91.04.16.00 **Fax**: 33.91.04.16.35

Type of institute
Research centre
University Institute/Higher Education Institute

Research activities
Marine Environment; impact on exosystems, and resources.
Population Dynamics
Most of the Scientific teams are involved in International programs, I.G.B.P., Biodiversity..., and are studying Biogeochemical processes associated with Primary Production, Transport, Migration and Biodegradation of organic matter in open Ocean (J. GOFS) and coastal sea (LOICZ), A special attention is paid on coupling of physical and biological models.
In coastal zones the influence of environmental factors on ecosystems functionning (accumulation of organic matter including hydrocarburem bloom of toxic algae...), Biodiversity, biogeographical distribution and paleoecology of benthic species.

Facilities
Research vessel: access to coastal fleet of 4 ships (15 to 24 meters)
Experimental aquariums
Laboratories: Diving, Chemical and Biochemical analysis; Scanning Microscopy; Remote sensing imagery; Field and Laboratory measurements of hydrobiological and meteorological parameters; Data processing; Physical and biological modelling; Automatized micro organisms cultures; Experimentation on organic matter fluxes in a submarine environment.
Specialised Library.

Staff: Scientists: 62 Others: 50

12. Conseil Supérieur de la Pêche

Name : Conseil Supérieur de la Pêche
Address : 134 avenue de Malakoff
F - 75016 Paris
Head : Jacques Arrignon/D. Viard
Tel. : 33.1.45012020

Type of institute
Government Department / Regional or local authority

Research activities
Fisheries management
  Stock assessment
  Environmental problems and their impact on resources
Fishing techniques
  Selective gears
Aquaculture

Facilities
Rearing tank
Controlled environmental chamber
Data bases, extensive libraries
Specialised computer hardware
Specialised software
Data bank in the process of being set up

Staff: Scientists: 5 Others: 50

Research budget (ECU/year): 143.000

Description
Key axis for the future programmes:
Efficiency of the restocking
Research about the National migratory fish plan in freshwater.
Economics and sociology of the freshwater fisheries.

13. Digital Equipment Europe - European Technical Center

Name : Digital Equipment Europe - European Technical Center
Address : P.O. Box 27
F - 06901 Sophia Antipolis
Head : L.F. Pau/Technology Director
Tel. : 33.9295.5111 **Fax**: 33.9295.5200

Type of institute
Industrial Research Laboratory

Research activities
Fishing techniques, Electronic acoustics, Geographical information systems for fish location.
Upgrading of fishery products, Computer vision and inspection systems for fish products.

Facilities
3 vision computers, special live scan, high speed cameras, networks for trawlers and instrumentation

Description
Research projects and contract/development work for industry (Scandinavia, Iceland, Japan, Canada) in the area of computer vision and expert systems for fish inspection, and also in information systems/networks for ships and plants.

14. Dynecar - Dynamics of Caribbean Marine Ecosystems and Biology of Associated Species

Name : Dynamics of Caribbean Marine Ecosystems and Biology of Associated species
Address : Université des Antilles et de la Guyane
Faculté des Sciences Exactes et Naturelles
B.P. 5592
F - 97159 Pointe-à-Pitre cedex (Guadeloupe)
Head : Prof. J. Portecop
Tel. : (590)93.86.82 **Fax**: (590)93.86.43 **Telex**: 919739 GL

Type of institute
Research Centre
University department

Research activities
Knowledge of the Caribbean coastal marine environment:
Ecology (mangroves, seagrass beds, coral reefs)
Biology of marine organisms (Molluscs, Crustaceans, Fishes), Bacteria/Molluscs symbiosis.
Growth and fish ageing
Growth of crustacean decapods
Biology and population dynamics
Aquaculture: Bivalve Molluscs

Facilities
Research boats
Diving equipment
Fishing nets and traps
Electron microscope scanning
Image processing laboratory
Scientific library

Staff: Scientists: 8 Others: 2

15. Ecole Nationale Supérieure Chimie Paris - Lab. Bioorganique et Biotechnologie

Name : Ecole Nationale Supérieure de Chimie de Paris
Laboratoire de Bioorganique et Biotechnologie
Address : 11, rue Pierre et Marie Curie
F - 75231 Paris Cedex 05
Head : Prof. Le Goffic
Tel. : 33.1.43266409 **Fax**: 33.1.43257975 **Telex**: 270021 F

Type of institute
University Department/Higher Education Institute

Research activities
Upgrading of fishery products
Methods for grading and quality assessment of fish
Novel processing methods for by-products, discards, and process wastes
Quantitative purification of proteins, peptides, amino-acids, lipides, fatty acids from fish, shellfish
Analytical chemistry
Valorisation of wastes (chitin, chitosan)

Facilities
Field site, electron microscope, specialized analytical apparatus
Quantitative separation apparatus
Analytical instruments: IR, UV, NMR
Mass spectrometers
Analytical gas, HPLC
Counter current chromatographs
Preparature liquid chromatography

Staff: Scientists: 2 Others: 2

16. ENSMP - Ecole Nat. Supér. des Mines de Paris - Centre de Géostatistique

Name : Ecole Nationale Supérieure des Mines de Paris (ENSMP)
Centre de Géostatistique
Address : 35 rue St-Honoré
F - 77305 Fontainebleau

Head : P.Petitgas (Chargé de recherche)
Tel. : 16.164224821 **Fax**: 16.164223903

Type of institute
Research centre
University Department/Higher Education Institute

Research activities
Fisheries management
Stock assessment
Environmental problems and their impact on resources

Facilities
Specialised software

Staff: Scientists: 1 Others: 1

17. ID. Mer - Lorient

Name : ID. Mer
Institut Technique pour le Développement
des Produits de la Mer
Address : 6 Bis rue François Toullec
F - 56100 Lorient
Head : P. Allaume
Tel. : 97.838683 **Fax**: 97.371103 **Telex**: 950.089 F

Type of institute
Research centre
Engineering/consulting firm

Research activities
Upgrading of fishery products
Improvement of techniques for handling, storing, processing, packaging fish
Methods for grading and quality assessment of fish
Novel processing methods for by-products, discards, and process wastes

Facilities:
Pilot equipment: for grinding-mixing, cooking-drying; freezing; packing; sterilizing. Pilot reactor for enzymatic extraction.
Complete pilot surimi line
Laboratory equipment: standard chemical analyses viscosity - rheology standard bacteriology tasting room
International computerized documents base.

Staff: Scientists: 6 Others: 2

Research budget (ECU/year): 618.900

18. IFREMER - Environnement et Aménagement Littoral

Name : Direction de l'Environnement et de l'Aménagement du Littoral
Address : 155 rue Jean-Jacques Rousseau
F - 992138 Issy-Les-Moulineaux
Tel. : 46.48.22.25 **Fax**: 46.48.22.24

Type of institute
Marine research and development institute with three main research centers with different research laboratories and several coastal implantations.

Research activities
Centre de Brest - BP 70 - F - 29280 Plouzané
Tel. : 98 22 40 40 Fax: 98 22 45 48
Environmental problems and their impact on marine resources; coastal management (hydrodynamics, sedimentology, bacteriology, marine chemistry ecology, physical and biological modelling).

Centre de Nantes - BP 1049 - F - 44037 Nantes Cedex 01
Tel. : 40 37 40 00 Fax: 40 37 40 73
Environmental problems and their impact on marine resources (pollutants, ecotoxicology, harmful plankton, monitoring).

Centre de Toulon - BP 330 - F - 83507 La Seyne-sur-Mer
Tel. : 94 30 48 00 Fax: 94 30 73 72
Environmental problems and their impact on marine resources (monitoring, water quality modelling).

Facilities
Research laboratories and their equipments including experimental rearing facilities. Research vessels and sampling gear.

Staff: Scientists and Technicians: 200 Others: 20

19. IFREMER - Génie Océanique - Service Navires Pêche Aquaculture

Name : IFREMER - Institut Français de Recherche pour l'Exploitation de la Mer
Direction de l'ingénierie, de la technologie et de l'informatique
Address : Centre de Brest-BP 70
F-29280 Plouzane Cedex
Head : Philippe Marchand
Naval architectural section: Philippe Marchand
Fishing gear technology section: Georges Massart
Aquaculture technology section: Christian Danioux
Tel. : 33.98.22.41.21 **Fax**: 33.98.22.41.35 **Telex**: 940627

Type of institute
Research centre

Research activities
Naval architecture
- Fast vessels design
- New lay-out of fishing vessels studies
- Computer hydrodynamics

Fishing techniques
- Trawl studies (numerical modelization, flume tank studies, at sea trials)
- Static gear
- Purse seine
- Selective gear
- Fish acoustic detection studies
- Electronic/acoustic detection studies
- Towed gear hydrodynamics
- On board fish handling, grading and storing techniques

Aquaculture technology
- Fish cages
- Tools and equipment
- Mussels and oysters culture technology

Facilities
Research vessels
Flume tanks (in Boulogne and Lorient)
Numerical echo-sounder and echo-integration system
Electronic/acoustic instrumentation
Underwater towed vehicle
Specialized computer hardware
Specialized software

Staff: Scientists: 24 Others: 12

20. IFREMER - Service Economie Maritime

Name : IFREMER - Institut Français de Recherche
pour l'Exploitation de la Mer
Direction des ressources vivantes
Service Economie Maritime
Address : 155, rue Jean-Jacques Rousseau
F-92138 Issy les Moulineaux Cedex
Head : Joseph Catanzano
Tel. : 46.48.22.71 **Fax**: 46.48.22.76 **Telex**: 631.914

Type of institute
Research centre

Research activities
Fisheries economics
Aquaculture economics
Markets analysis
Economics and the environment

Facilities
Data bases
Specialized software

Staff: Scientists: 7 Others: 1

21. IFREMER - Ressources Vivantes - Lab. Recherche Aquacole

Name : IFREMER - Institut Français de Recherche
pour l'Exploitation de la Mer
Direction des ressources vivantes
Laboratoire de Recherche Aquacole
Address : BP 70
F - 29280 Plouzané
Head : Jean-François Samain
Tel. : 98.22.43.66 **Fax**: 98.22.45.47 **Telex**: 940.627

Type of institute
Research centre

Research activities
Molluscs (scallops - oysters) Fish (turbot - Salmon) aquaculture
Nutrition aspects, feed composition, live food production
Early stage
Larval rearing
Physiology
Growth
Molluscs Hatchery processings

Facilities
Main hall for nutrition experiments with many tanks
Additional hall for peneid shrimps rearing (240 sq.m)
Experimental feed processing plant (120 sq.m)
Sheltered and thermoregulated, or outside ponds for brood stock rearing for marine fish hatchery
Bivalves research hatchery

Staff: Scientists: 24 Others: 26

22. IFREMER - Ressources Vivantes - Lab. Algoculture

Name : IFREMER - Institut Français de Recherche
pour l'Exploitation de la Mer
Direction des ressources vivantes
Laboratoire Algoculture
Address : BP 1049
F - 44037 Nantes Cedex 01

Head : René Perez
Tel. : 40.37.41.26 **Fax**: 40.37.40.73 **Telex**: 711.196

Type of institute
Research centre

Research activities
Macrophytes Aquaculture, laminariale, red algae
Reproduction
Physiology
Early stage growth
Cryopreservation of gametes

Facilities
Cryopreservation unit
Gametes unicellular culture
Hatchery

Staff: Scientists: 4 Others: 3

23. IFREMER Ressources Vivantes - Lab. Aquaculture Tropicale

Name : IFREMER - Institut Français de Recherche
pour l'Exploitation de la Mer
Direction des ressources vivantes
Laboratoire d'Aquaculture Tropicale
Address : Centre Océanologique du Pacifique
BP 7004 Taravao
Tahiti-French Polynesia
Head : Jean-Louis Martin
Tel. : (689)57.12.74 **Fax**: (689)57.24.77

Type of institute
Aquaculture research centre Fish/Crustacean/Mollusc

Research activities
Reproduction, nursery and growth out (fish and shrimp)
Genetics and conservation of gametes (shrimp)
Pathology, bacterial and virus disease (fish, shrimp and mollusc)
Nutrional requirements and feed composition (fish and shrimp)
Ecophysiological studies (shrimp and mollusc)
Aquaculture/environment interactions (shrimps)
Aquaculture structures, systems, techniques

Facilities
2ha of ponds and concrete tanks
Experimental rooms with tanks from 50.1 to 5m3
shrimp and fish hatcheries, maturation units and spawning room
Laboratories (histology, bacteriology, virology, physiology, chemical analysis, genetics)
Shrimp broodstock (P. vanamei, P. monodon, P. stylirostris, P. indicus)
Fish broodstock (L. calcarifer)

Staff: Scientists: 45 Others: 25

24. IFREMER - Ressources Vivantes - Lab. Biologie et Ecologie Invert. Marins

Name	:	IFREMER - Institut Français de Recherche pour l'Exploitation de la Mer Direction des ressources vivantes Laboratoire Biologie et Ecologie des Invertébrés Marins
Address	:	BP 133-Mus du Loup F - 17390 La Tremblade
Head	:	Maurice Heral
Tel.	:	46.36.18.41 **Fax**: 46.36.18.47 **Telex**: 632.160

Type of institute
Research centre
National level

Research activities
Molluscs aquaculture
Bivalves grounds management
Stock assessment
Physiology
Reproduction
Growth
General pathology
Genetics

Facilities
Research vessel
Flume tank
Simulated tidal tank
Specialised computer hardware
Specialised software
Materials for physiology: multisizer, ATP, HPLC amino acids
Counter for scintilation
Hatchery Nurseries

Staff: Scientists: 17 Others: 19

25. IFREMER - Ressources Vivantes - Lab. Défense et Résistance chez les Invertébrés Marins

Name : IFREMER - Institut Français de Recherche
pour l'Exploitation de la Mer
Direction des ressources vivantes
Laboratoire Défense et Résistance chez les invertébrés
marins (DRIM)-Laboratoire mixte IFREMER/CNRS
Address : Université de Montpellier II
Place Eugène Bataillon
F - 34095 Montpellier Cedex 05
Head : Philippe Roch

Type of institute
Research centre

Research activities
Mollusc/Crustacean
Immunology (oysters and mussels peneids)
Virology structure
Transgeniques methodology

Facilities
Type P2 molecularbiology, ADN sequence, cellular culture, cytology, protein purification, micromanipulation

Staff: Scientists: 6 Others: 1

26. IFREMER - Ressources Vivantes - Lab. L'Houmeau CREMA

Name : IFREMER - Institut Français de Recherche
pour l'Exploitation de la Mer
Direction des ressources vivantes
Laboratoire de Recherche en Ecologie Marine et en
Aquaculture de l'Houmeau CREMA
(Laboratoire mixte IFREMER/CNRS)
Address : BP 5-L'Houmeau
F - 17137 Neuil sur Mer

Head : Alain Herbland
Tel. : 46.50.94.40 **Fax**: 46.50.91.60 **Telex**: 791.931(IMPROCH)

Type of institute
Research centre

Research activities
Fish/Crustacean/Mollusc farming/Phytoplancton/Dynoflagelate
Ecological studies in Coastal Sea and Salt marshes
Physiology
Environmental conditions and ecophysiological mechanisms lading to the bloom of Dinophysis sp
Bioenergy in marsh
Trophic links
Aquaculture/environment interaction

Facilities
Mesocosm
Electron, microscope/scanning
Field site, specialised analytical apparatus
HPLC, gas chromatograph, mass spectrometer, liquid scintillation counter CHN

Staff: Scientists: 9 Others: 6

27. IFREMER - Ressources Vivantes - Lab. Recherches Aquacoles Méditerranéen

Name : IFREMER - Institut Français de Recherche
pour l'Exploitation de la Mer
Direction des ressources vivantes
Laboratoire Méditerranéen de Recherche Aquacole
(unité gérée par un groupement d'intérít économique
Recherche Aquacole GIE-RA)
Address : Chemin de Maguelone
F - 34250 Palavas Les Flots
Head : Alain Febvre
Tel. : 67.68.07.64 **Fax**: 67.68.28.85 **Telex**: 490.419

Type of institute
Research centre
Institute for professional training

Research activities
Fish - Mollusc
Aquaculture
Physiology
Reproduction
Genetics
Early stage of reared species
Growth, nutrition, live food production for aquaculture feed composition
Desease, immunology, stress
Aquaculture-environment interaction
Aquaculture structures, systems, techniques

Facilities
Controlled environmental chamber
Data base, extensive libraries
Specialised software
Field site, electron microscope, specialized analytical apparatus
Experimental and pilot training tanks

Staff: Scientists: 24 Others: 6

28. IFREMER - Ressources Vivantes - Lab. Aquaculture en Marais Noirmoutier

Name : IFREMER - Institut Français de Recherche pour l'Exploitation de la Mer
Direction des ressources vivantes
Laboratoire Aquaculture en Marais de Noirmoutier
Address : BP 59
F - 85530 Noirmoutier en l'Ile
Head : Vincent Buchet
Tel. : 51.39.15.27 **Fax**: 51.39.36.82

Type of institute
Research centre

Research activities
Sea-bass, sea-bream
Aquaculture/environment interaction
Aquaculture structures, systems, techniques
Extensive fish rearing

Facilities
Earth - ponds
Concrete tanks
Plastic tanks
Four wells for underground sea water
Fields analytical laboratory (microscope, binocular, spectrophotometer, oxygenmater, ph meter, salinometer

Staff: Scientists: 2 Others: 6

29. IFREMER - Ressources Vivantes - Lab. Recherche Aquacole Martinique

Name : IFREMER - Institut Français de Recherche
pour l'Exploitation de la Mer
Direction des ressources vivantes
Laboratoire Recherche Aquacole de Martinique
Address : Le Robert-Pointe Fort
F - 97231 Le Robert-Martinique
Head : Jean-Marc Ricard
Tel. : 596.65.11.54 **Fax**: 596.65.11.56 **Telex**: 912.488 MR

Type of institute
Research centre

Research activities
Fish, Sciaenops ocellata
Tilapia in sea water
Reproduction
Early stages of reared species
Growth nutrition, feed composition, live food production for aquaculture

Facilities
Rearing tanks controlled environmental chamber

Staff: Scientists: 3 Others: 5

30. IFREMER - Ressources Vivantes - Lab. Côtier Ressources Aquacoles Nouvelle Calédonie

Name : IFREMER - Institut Français de Recherche pour l'Exploitation de la Mer
Direction des ressources vivantes
Laboratoire Côtier Ressources Aquacoles de Nouvelle Calédonie
Address : BP 2059
Noumea-Nouvelle Calédonie
Head : Christian Galinié
Tel. : (687)28.51.71 **Fax**: (687)28.78.57 **Telex**: 070.6003045

Type of institute
Research centre

Research activities
Scylla serrata stock assessment
Peneides aquaculture
Craaostrea echnata and giant clam rearing methods
Reproduction growth

Facilities
Hatchery
Experimental shrimp ponds

Staff: Scientists: 4 Others: 14

31. IFREMER - Ressources Vivantes - Lab. Côtier Ressources Aquacoles Trinité sur Mer

Name : IFREMER - Institut Français de Recherche pour l'Exploitation de la Mer
Direction des ressources vivantes
Laboratoire Côtier Ressources Aquacoles de la Trinité sur Mer
Address : 12, rue des Résistants
F - 56470 La Trinité sur Mer
Head : Joseph Mazurie
Tel. : 97.30.25.70 **Fax**: 97.30.25.76 **Telex**: 950.571

Type of institute
Research centre

Research activities
Oysters - mussells, clams, scallops
Bivalves culture grounds management
Stock assessment - carrying capacity
Growth
Genetics tests
Pathology
Zoosanitary survey

Facilities
Rearing tanks
Research vessel
Field site, diving equipments
Specialised analytical apparatus

Staff: Scientists: 4 Others: 8

32. IFREMER - Ressources Vivantes - Lab. Côtier Ressources Aquacoles Port en Bessin

Name : IFREMER - Institut Français de Recherche pour l'Exploitation de la Mer
Direction des ressources vivantes
Laboratoire Côtier Ressources Aquacoles de Port en Bessin
Address : Avenue du Général de Gaulle-BP 32
F - 14520 Port en Bessin
Head : Jean-Pierre Joly
Tel. : 31.51.13.00 **Fax**: 31.51.13.01

Type of institute
Research centre

Research activities
Oysters - mussels
Bivalves culture grounds management
Stock assessment
Growth
Genetics tests
Optimisation of sapt collecting techniques and rearing cycle

Facilities
Rearing tanks
Thermoregulated tanks for remote setting experiments
Laboratory (29 sq.m) equipped for biochemical analysis
Specialised computer hardware
Specialised software

Staff: Scientists: 2 Others: 2

33. IFREMER - Ressources Vivantes - Lab. Dynamique Syst. Production

Name : IFREMER - Institut Français de Recherche pour l'Exploitation de la Mer
Direction des ressources vivantes
Laboratoire Dynamique des systèmes de production
Address : rue de l'Ile d'Yeu-BP 1049
F - 44037 Nantes Cedex 01
Head : Jean-Louis Durand
Tel. : 40.37.40.00 **Fax**: 40.37.40.75 **Telex**: 711.196

Type of institute
Research centre

Research activities
Assessment models
Assessment and management under uncertainties
Exploitation systems analysis

Facilities
Data bases
Sun station

Staff: Scientists: 5 Others: 5

* New laboratory since 1/01/1994. In 1993, this unit is included in the MAERHA laboratory.

34. IFREMER - Ressources Vivantes - Lab. Ecologie Halieutique

Name : IFREMER - Institut Français de Recherche
pour l'Exploitation de la Mer
Direction des ressources vivantes
Laboratoire Ecologie Halieutique (ECOHAL)
Address : rue de l'Ile d'Yeu-BP 1049
F - 44037 Nantes Cedex 01
Head : Yves Desaunay
Tel. : 40.37.40.00 **Fax**: 40.37.40.75 **Telex**: 711.196

Type of institute
Research centre

Research activities
Fish - shellfish
Resource biology and ecology
Environmental problems and their interactions with resources and recruitment
Multi species biological models
Stock variability assessment

Facilities
Research vessels
Data bases
Specialised software
Specialised analytical apparatus

Staff: Scientists: 26 Others: 13

35. IFREMER - Ressources Halieutiques - Lab. Mathématiques Appl. à l'Exploitation Ress. Halieutiques et Aquacoles

Name : IFREMER - Institut Français de Recherche
pour l'Exploitation de la Mer
Direction des ressources vivantes
Laboratoire de Mathématiques appliquées à l'Exploitation
des Ressources Halieutiques et Aquacoles (MAERHA)
Address : rue de l'Ile d'Yeu-BP 1049
F - 44037 Nantes Cedex 01
Head : Jean-Louis Durand
Tel. : 40.37.40.00 **Fax**: 40.37.40.75 **Telex**: 711.196

Type of institute
Research centre

Research activities
Methodological research in numerical analysis, statistics, biological models
Population dynamics

Facilities
Research vessels
Data bases
Computers
Statistical software

Staff: Scientists: 4 Others: 3

36. IFREMER - Ressources Vivantes - Lab. Sclérochronologie

Name : IFREMER - Institut Français de Recherche
pour l'Exploitation de la Mer
Dir des ressources vivantes
Laboratoire de sclérochronologie des animaux aquatiques
(LASAA) (IFREMER-ORSTOM)
Address : Sainte Anne du Portzic-BP 70
F - 29280 Plouzane
Head : LoÔc Antoine
Tel. : 98.22.43.70 **Fax**: 98.22.45.47 **Telex**: 940.627

Type of institute
Research centre

Research activities
Studies on calcified pieces of marines species, determination of biological modifications (age, growth...)

Facilities
Image analysis, software
Artificial video standing software

Staff: Scientists: 5 Others: 1

37. IFREMER - Ressources Vivantes - Lab. Ressources Halieutiques Boulogne sur Mer

Name : IFREMER - Institut Français de Recherche
pour l'Exploitation de la Mer
Direction des ressources vivantes
Laboratoire Ressources Halieutiques de Boulogne sur Mer
Address : BP 699
F - 62321 Boulogne sur Mer

Head : Jean-Paul Delpech
Tel. : 21.99.56.00 **Fax**: 21.99.56.01 **Telex**: 131.565

Type of institute
Research centre

Research activities
Fisheries studies
  Assessment of North Sea and English channel stocks
  Fleet activities studies
Management advice
  Deep sea species study
National fisheries statistic system support

Facilities
Research vessel
Data bases sampling results, fleet activities
Specialised software

Staff: Scientists: 5 Others: 8

38. IFREMER - Ressources Vivantes - Lab. Côtier Ressources Halieutiques Brest

Name : IFREMER - Institut Français de Recherche
pour l'Exploitation de la Mer
Direction des ressources vivantes
Laboratoire Côtier Ressources Halieutiques de Brest
Address : BP 70
F - 29280 Plouzane
Head : Patrick Berthou
Tel. : 98.22.43.70 **Fax**: 98.22.45.47 **Telex**: 940.627

Type of institute
Research centre

Research activities
Shellfishes stocks of
  North Brittany
  West channel
Fisheries studies
  Stock assessment
  Fleet activities study
National fisheries statistics system support
Management advice

Facilities
Research vessel
Data bases (sampling results, molluscs, crustaceans and fishes fleet activities)
Specialised software

Staff: Scientists: 4 Others: 7

39. IFREMER - Ressources Vivantes - Lab. Côtier Ressources Halieutiques Cayenne

Name : IFREMER - Institut Français de Recherche
pour l'Exploitation de la Mer
Direction des ressources vivantes
Laboratoire Côtier Ressources Halieutiques de Cayenne
Address : BP 477
F - 97302 Cayenne
Guyane Française
Head : Philippe Moguedet
Tel. : (594)30.22.00 **Fax**: (594)30.80.31 **Telex**: 910.358

Type of institute
Research centre

Research activities
Fishes (lutjanidae) - shrimps
Fisheries studies
  Stock assessment
  Fleet activities
Management advice

Facilities
Research vessel
Data bases

Staff: Scientists: 4 Others: 2

40. IFREMER - Ressources Vivantes - Lab. Côtier Ressources Halieutiques La Réunion

Name : IFREMER - Institut Français de Recherche
pour l'Exploitation de la Mer
Direction des ressources vivantes
Laboratoire Côtier Ressources Halieutiques de la Reunion
Address : BP 60
F - 97822 Le Port Cedex-La Reunion
Head : Jean-Pierre Minet
Tel. : (262)42.03.40 **Fax**: (262)43.36.84 **Telex**: 916.452

Type of institute
Research centre

Research activities
Fisheries studies
- Biology of main species
- Monitoring of fisheries, specially in the viscounty of AFD
- Development of new activities (tuna fisheries)

Management advice to local administration
Cooperation on fishery research with neighbouring island (Mayotte)

Facilities
Research vessel
Data bases

Staff: Scientists: 1 Others: 3

41. IFREMER - Ressources Vivantes - Lab. Côtier Ressources Halieutiques La Rochelle

Name : IFREMER - Institut Français de Recherche
pour l'Exploitation de la Mer
Direction des ressources vivantes
Laboratoire Côtier Ressources Halieutiques de La Rochelle

Address : Place du Séminaire-BP 7-L'Houmeau
F - 1737 La Rochelle
Head : André Forest
Tel. : 46.50.93.50 **Fax**: 46.50.93.79 **Telex**: 791.931

Type of institute
Research centre

Research activities
Fishes, pectinids fisheries management
Fisheries studies
  Assessment of Bay of Biscay stocks
  Fleet activities
  Age structured dynamic models
  Technical interactions
Management advice

Facilities
Research vessel
Data bases
Specialised software

Staff: Scientists: 7 Others: 2

42. IFREMER - Ressources Vivantes - Lab. Côtier Ressources Halieutiques Lorient

Name : IFREMER - Institut Français de Recherche
pour l'Exploitation de la Mer
Direction des ressources vivantes
Laboratoire Côtier Ressources Halieutiques de Lorient
Address : 8 rue François Toullec
F - 56100 Lorient
Head : Catherine Talidec
Tel. : 97.87.73.10 **Fax**: 97.83.41.06

Type of institute
Research centre

Research activities
Fish and shellfish studies
Assessment of stocks from western approaches, Celtic Sea, Bay of Biscay
Fleet activities
Fisheries statistics system support
Management

Facilities
Research vessel
Data bases (Celtic Sea etc...)
Specialised sofware

Staff: Scientists: 7 Others: 14

43. IFREMER - Ressources Vivantes - Lab. Côtier Ressources Halieutiques Port en Bessin

Name	:	IFREMER - Institut Français de Recherche pour l'Exploitation de la Mer Direction des ressources vivantes Laboratoire Côtier Ressources Halieutiques de Port en Bessin
Address	:	Avenue du Général de Gaulle-BP 32 F - 14520 Port en Bessin
Head	:	Michel Lemoine
Tel.	:	31.51.13.00 **Fax**: 31.51.13.01 **Telex**: 171.150

Type of institute
Research centre

Research activities
Fisheries studies
Assessment of English channel stocks
Multidisciplinary approach to fisheries modelling
Management advice
Resource biology and ecology
Stock identity, migrations, diversity, interaction with marine environment
National fisheries statistics system support

Facilities
Research vessel
Data bases (sampling results, fleets activities)
Specialised software

Staff: Scientists: 5 Others: 3

44. IFREMER - Ressources Vivantes - Lab. Côtier Ressources Halieutiques Sète

Name : IFREMER - Institut Français de Recherche
pour l'Exploitation de la Mer
Direction des ressources vivantes
Laboratoire Côtier Ressources Halieutiques de Sète
Address : 1, rue Jean Vilar-BP 4240
F - 34200 Sète

Head : Jacques Bertrand
Tel. : 67.74.77.67 **Fax**: 67.74.7090 **Telex**: 490.503

Type of institute
Research centre

Research activities
Fisheries studies
  West Mediterranean stocks assessment
  Multi-disciplinary approach to fisheries management
  Ecology of resources
Management advice
Fishing techniques
  Fishing gear and materials
  Fishing methods and operations
  Technical support to fishermen and organisations

Facilities
Research vessel (sea and lagoon)
Specialised computer hardware
Specialised software
Data bases, extensive libraries

Staff: Scientists: 3 Others: 4

45. IFREMER - Ressources Vivantes - Lab. Côtier Ressources Halieutiques St. Pierre et Miquelon

Name : IFREMER - Institut Français de Recherche
pour l'Exploitation de la Mer
Direction des ressources vivantes
Maboratoire Côtier Ressources Halieutiques de St Pierre et Miquelon
Address : Quai de l'Abysse-BP 4240
F - 97500 St. Pierre et Miquelon
Head : André Battaglia
Tel. : (508)41.30 83 **Fax**: (508)41.49.36 **Telex**: 914.130

Type of institute
Research centre

Research activities
Fishes, (Gadus mohrua Hyppoglosoldes platessoides)
Molluscs (Clamys islandicus, Placopecten magellanicus)
Fisheries studies
Stock assessment
Fleet activities
Management advice

Facilities
Research vessel
Data bases

Staff: Scientists: 1 Others: 4

46. IFREMER - Ressources Vivantes - Lab. Côtier Ressources Halieutiques Tahiti

Name : IFREMER - Institut Français de Recherche
pour l'Exploitation de la Mer
Direction des ressources vivantes
Laboratoire Côtier Ressources Halieutiques de Tahiti
Address : Orstom-BP 520
Papéete-Polynésie Française
Head : René Abbes
Tel. : (689)43.98.87 **Fax**: (689)42.95.55

Type of institute
Research centre

Research activities
Fisheries studies
  Biological and behavioral studies of tuna resources
  Advice and monitoring of the tuna fisheries development
Management advice to local authorities

Facilities
Data bases

Staff: Scientists: 1

47. IFREMER - Ressources Vivantes - Lab. Biochimie et Molécules Marines

Name	:	IFREMER - Institut Français de Recherche pour l'Exploitation de la Mer Direction des ressources vivantes Laboratoire Biochimie et Molécules Marines
Address	:	rue de l'Ile d'Yeu-BP 1049 F - 44037 Nantes Cedex 01
Head	:	Patrick Durand
Tel.	:	40.37.40.67 **Fax**: 40.37.40.71 **Telex**: 711.196

Type of institute
Research centre

Research activities
Molecule valorisation methods
Novel processing methods for by products discards and process wastes

Facilities
Amino acid analyses, chromatograph (HPLC, FPLC, GC)
Micro-ultra filtration
dry distillation polyvalante unit

Staff: Scientists: 6 Others: 4

48. IFREMER - Ressources Vivantes - Lab. Etudes Technico-Réglementaires

Name : IFREMER - Institut Français de Recherche
pour l'Exploitation de la Mer
Direction des ressources vivantes
Laboratoire Etudes Technico-réglementaires
Address : rue de l'Ile d'Yeu-BP 1049
F - 44037 Nantes Cedex 01
Head : Henri Loreal
Tel. : 40.37.41.52 **Fax**: 40.37.40.71 **Telex**: 711.196

Type of institute
Research centre

Research activities
Technical support to administration and manufacturers
Studies to determine product quality standards
Data bases
Bibliography on quality, nutrition, reglementation

Facilities
Bibliomer bulletin

Staff: Scientists: 2 Others: 1

49. IFREMER - Ressources Vivantes - Lab. Génie Alimentaire

Name : IFREMER - Institut Français de Recherche
pour l'Exploitation de la Mer
Direction des ressources vivantes
Laboratoire Génie Alimentaire
Address : rue de l'Ile d'Yeu-BP 1049
F - 44037 Nantes Cedex 01
Head : Jean-Luc Vallet
Tel. : 40.37.40.64 **Fax**: 40.37.40.71 **Telex**: 711.196

Type of institute
Research centre

Research activities
Upgrading of fishery products
Methods for grading and quality assessment of fish and aquaculture products
Improvement of techniques for storing and transformation

Facilities
Pilot equipment extrusion, smoking, coking, freezing
Bacteriological analysis

Staff: Scientists: 5 Others: 7

50. IFREMER - Ressources Vivantes - Lab. Qualité et Physico-Chimie

Name : IFREMER - Institut Français de Recherche
pour l'Exploitation de la Mer
Direction des ressources vivantes
Laboratoire qualité et physico chimie
Methods standardization and validation
Address : rue de l'Ile d'Yeu-BP 1049
F - 44037 Nantes Cedex 01
Head : Monique Etienne
Tel. : 40.37.40.00 **Fax**: 40.37.40.71 **Telex**: 711.196

Type of institute
Research centre

Research activities
Product quality control
Process standardisation and validation

Facilities
DSC (Differential Scanning Colorimeter)
Texturormeter
Cryotome

Staff: Scientists: 4 Others: 5

51. INRA - Dept. Hydrobiologie et Faune Sauvage

Name : Institut National de Recherche Agronomique (INRA)
Département d'Hydrobiologie et Faune Sauvage

Address : Campus de Beaulieu Av. du Général Leclerc
F - 35042 Rennes Cedex
Head : Bernard Jalabert
Tel. : 99.285021 **Fax**: 99.285034 **Telex**: INRARBE 741.659 F

Type of institute
Research centre

Research activities
Fisheries management
Stock assessment
Multi-species biological models
Multi-disciplinary approach to fisheries management
Environmental problems and their impact on resources
Aquaculture
Upgrading of fishery products

Facilities
Research vessel
Flume tank
Rearing tank
Controlled environmental chamber
Specialised software
Field site, electron microscope, specialized analytical apparatus

Staff: Scientists: 76 Others: 90

Research budget (ECU/year): 1.000.000

Description
Programmes: hydrobiology defined as the science dealing with investigations to understand and to control the different life processes in the aquatic environment.
Two mains issues are at stake: Food resources and leasure activities and quality of life.
Main topics: Environmental biology: salmonid rivers, Alpine great lakes, ponds in temperate and typical zones, wet lands. Population biology: stock estimation.
Determinism of recruitment. Management strategies.
Biology of organisms: Nutrition and feeding. Genetic improvement.
Physiology and endocrinology - Pathology and ecotoxicology.
Ten out of the twenty five INRA'S Departments devote also part of their activities to these fields of research. INRA has at its disposal several experimental hatcheries, rearing systems and fish farms, and collaborates with other organisations (IFREMER, CEMAGR F,CSP) to set up common experimental facilities.

52. INRA - Laboratoire de Génétique des Poissons

Name : Institut National de Recherche Agronomique (INRA)
Laboratoire de Génétique des Poissons
Address : Domaine de Vilvert
F - 78352 Jouy en Josas Cedex
Head : Daniel Chourrout
Tel. : 33.1.34.65.23.91 **Fax**: 33.1.34.65.23.90

Type of institute
Research centre

Research activities
Fishery
  Population genetics
Aquaculture
  Genetics

Facilities
Library
Computer facilities
Laboratories
Rearing tanks
Controlled environmental chamber
Specialised software
Field site
Electron microscope
Specialized analytical material

Staff: Scientists: 6 Others: 12

Research budget (ECU/year): 200.000

53. INRA - Laboratoire de Physiologie des Poissons

Name : Institut National de Recherche Agronomique (INRA)
Laboratoire de physiologie des poissons
Address : Campus de Beaulieu
F - 35042 Rennes Cedex
Head : Bernard Breton
Tel. : 33.99.28.50.02 **Fax**: 33.99.28.50.20

Type of institute
Research centre

Research activities
Fisheries
Stock assessment
Fisheries biology
Population dynamics
Environmental aspects
Socioeconomic aspects
Fishing techniques
Aquaculture
Early stages
Larval rearing
Physiology, reproduction, growth, adaptation
Environmental aspects
Upgrading of fisheries products
Hygiene/quality

Facilities
- Library
- Laboratories
  the laboratories are equipped for:
  biochemistry of protein and lipide, low pressure chromatography units, HPLC, gas chromatography, electrophoresis and transfer systems, ultracentrifuges and centrifuges, spectrophotometers, microbalances.
- Immuno assays procedures:
  . gamma counter
  . fl counters
  . Top count counters
- Cell biology: thermoregulated culture room unit
- Molecular biology: 2 special laboratories
- Rearing tanks
- Others: Thermoregulated recirculating water systems, 6 units with several tanks, 12 individual tanks, 3 aquarium rooms.
- Fish farm for fish production and field experimentation

Staff: Scientists: 14 Others: 22

Research budget (ECU/year): 350.000

54. INRA - Molecular Virology and Immunology Unit Fish Diseases Teams

Name : Institut National de Recherche Agronomique (INRA)
Molecular Virology and Immunology Unit
Fish Diseases Teams
Address : Centre de Recherches de Jouy en Josas
Domaine de Vilvert
F - 78352 Jouy en Josas Cedex
Head : Hubert Laude
Team leaders : P. de Kinkelin, C. Michel
Tel. : 33.1 34.65.21.21 **Fax**: 33.1.34.65.26.21

Type of institute
Research centre

Research activities
Aquaculture
Pathology

Facilities
- Library
- Computer facilities
- Laboratories
  Virology, Molecular Biology, Biochemistry of proteins and nucleic acids
- Electron microscope
- Others: Thermo regulated experimental facilities for fish

Staff: Scientists: 30 Others: 36
(10 for fish disease teams)

Research budget (ECU/year): 404.300

Description
Automatic DNA sequensor (applied biosystem 373 A) plus robot (applied biosystem catalyst 800)
HPLC devices
Ultracentrifuges
DNA synthetisers (Biosearch 86000 and applied system cyclone 392)

55. INRA - Station d'Hydrobiologie de St. Pee sur Nivelle

Name : Institut National de Recherche Agronomique (INRA)
Station d'Hydrobiologie de St. Pee sur Nivelle
Address : BP 3
F - 64310 St. Pee sur Nivelle
Head : Olivier Clement
Tel. : 59.51.59.51 **Fax**: 59.54.511.52

Type of institute
Research centre

Research activities
Fisheries
- Stock assessment
- Fisheries biology
- Population dynamics
- Environmental aspects
- Fishing techniques
- Electric fishing, migratory fish counters and electrical barriers

Aquaculture
- Early stages
- Larval rearing
- Nutritional aspects
- Genetics
- Environmental aspects

Upgrading of fisheries products
- Hygiene/quality

Facilities
Library
Computer facilities
Laboratories
Flume tanks
Rearing tanks
Controlled environmental chamber
Specialised software
Field site
Specialised analytical material

Staff: Scientists: 25 Others: 29

Research budget (ECU/year): 350.000

56. INRA - Station d'Hydrobiologie Lacustre

Name : Institut National de Recherche Agronomique (INRA)
Station d'Hydrobiologie Lacustre
Address : 75 Avenue de Corzent-BP 511
F - 74203 Thonon Cedex
Head : Daniel Gerdeaux
Tel. : 33.50.26.78.00 **Fax**: 33.50.26.07.60

Type of institute
Research centre

Research activities
Stock assessment
- Fisheries Biology
- Population dynamics
- Environmental aspects

Aquaculture
- Larval rearing
- Nutritional aspects
- Physiology
- Environmental aspects

Facilities
- Library
- Computer facilities
- Laboratories
- Research vessels 3 boats (8m, 6,50m, 6,50m)
- Rearing tanks: 25 in cold water (7-10∞C)
- Field site: Lake Geneva, Annecy, Bourget
- Others: Echosounding acoustics

Staff: Scientists: 10 Others: 24

Research budget (ECU/year): 100.000

57. INTECHMER

Name : Institut National
des Techniques de la Mer (INTECHMER)
Address : Digue de Collignon
F - 50100 Cherbourg
Head : J.Cl. Guary
Tel. : 33203765 **Telex**: 306022 F

Type of institute
Research centre
University Department/Higher Education Institute
Higher Education Institute

Research activities
Aquaculture
Growth, nutrition, feed composition, live food production for aquaculture
Upgrading of fishery products
Methods for grading and quality assessment of fish
Novel processing methods for by-products, discards, and process wastes

Facilities
Rearing tank
Field site, electron microscope, specialized analytical apparatus
Analytical apparatus, IR Spectrometry CPG, HPLC etc.

Staff: Scientists: 6

Research budget (ECU/year): 29.000

Description
Major items of the research:
Rearing trials of rainbow trout fry in salt water.
Artificial feeding for bivalves (Ruditopes philippinarum)
Optimization of the rotifere rearing (B. plicatilis)
Optimization of the production of economic interest molecules using micro marine algae.

58. IUT - La Rochelle - Lab. Biologie Biochemie Marine

Name : Institut Universitaire de Technologie de la Rochelle (IUT)
Laboratoire de Biologie Biochimie Marine
Address : Rue de Roux
F - 17026 La Rochelle
Head : J. Tardy
Fax : 46443142

Type of institute
University Department/Higher Education Institute
Processing Industry/ agro-food / diets producer
fish/crustac/mollusc farming

Research activities
Fisheries management
Environmental problems and their impact on resources
Aquaculture
Physiology
Early stages of reared species
Growth, nutrition, feed composition, live food production for aquaculture

Facilities
Controlled environmental chamber
Field site, electron microscope, specialized analytical apparatus
Microscope à interférence et à fluorescence

Staff: Scientists: 2 Others: 5

Research budget (ECU/year): 5.000

Description
The centre has joint research programmes with the laboratory IFREMER of Ronces-les Bains on diploïd and polyploïd obtained through genetic manipulation.
It is looking to establish collaboration with Spanish and British laboratories working on malacology (embryology, determinism of the metamorphoris, nutrition, growth and reproduction ecophysiology).

59. IUT - St. Nazaire - Lab. Génie des Procédés

Name : Institut Universitaire de Technologie de Saint-Nazaire
Laboratoire de Génie des Procédés
Address : B.P. 420
F - 44606 St. Nazaire Cedex
Head : M. F. Quemeneur
Tel. : 40703750 **Fax**: 33.40531646

Type of institute
University Department/Higher Education Institute

Research activities
Fisheries management
Environmental problems and their impact on resources
Aquaculture
Growth, nutrition, feed composition, live food production for aquaculture
Aquaculture economics

Facilities
Equipment for separation processes

Staff: Scientists: 3 Others: 3

Research budget (ECU/year): 11.500

Description
Major research activities:
Depollution of SURIMI process-plant effluents.
Fractionation and analysis of the sarcoplasmic proteins contained in the SURIMI wastewaters. Concentration of marine microalgal biomass (aquafarming, bivalve nursery). concentration of marine microalgae through different harvesting techniques: centrifugation and cross-flow microfiltration. Technico-economical comparison of methods. Recovery and separation of fish soluble proteins by means of membrane processes. Equipments for separation processes: 6 laboratory scale units (Filtration, microfiltration, ultra-filtration, electrodialysis). 1 Pilot plant unit (microfiltration) 1 laboratory scale unit (centrifugation). Analysis Equipment: High performance liquid chromatography. Pollution level measurements (susp. matters, COD, Kjeldahl Nitrogen, turbidity...)

60. Lab. Vétérinaire Départemental - Arras

Name : Laboratoire Vétérinaire Départemental
Address : SP 18
F - 62022 Arras Cedex
Head : J. Catel
Tel. : 33.21514654

Type of institute
Fish/Crustacean/mollusc farming
Government Department/Regional or local authority
public admin/regional

Research activities
Fisheries management
Environmental problems and their impact on resources
Aquaculture
Disease, immunology, stress

Facilities
Equipment for Viriology-Bacteriology-Parasitology-Seriology concerning salmonides

Staff: Scientists: 1 Others: 1

61. Lab. Vétérinaire Départemental - Lons-le-Saunier

Name : Laboratoire Vétérinaire Départemental
Address : BP 376-Bd Th. Vernier
F - 39016 Lons-le-Saunier Cedex
Head : M. Morand Marc
Tel. : 33.84242441 **Fax**: 33.84246865

Type of institute
Government Department / Regional or local authority

Research activities
Fisheries management
Stock assessment

Facilities
Equipment for parasitology, virology, bacteriology, and soon, for histology - Eight tanks are available

Staff: Scientists: 2 Others: 4

Research budget (ECU/year): 15.000

Description

Major Research activities

1. relationship between fish, bacteria and organic manure (slurry, dung,...)
2. the biological cycle of some parasites: Sanguinicola, Sphaerospora renicola, Gryporhynchus, etc...
3. pathogenicity of iridovirus of the catfish, Ictalurus nebulosus
4. dosage of radionucleides in fish and their environment.

62. Musée Histoire Naturelle - Lab. Ichtyologie Générale et Appl.

Name : Musée d'Histoire Naturelle
Laboratoire d'Ichtyologie Générale et Appliquée
Address : 43 rue Cuvier
F - 75231 Paris Cedex 05
Head : Prof. Billard Roland
Tel. : (1)4079 3749 **Fax**: (1)40 79 3771 **Telex**: MUSNAHN 202641 F

Type of institute

University Department/Higher Education Institute

Research activities

Fisheries management
Stock assessment
Aquaculture
Reproduction and conservation of gametes
Genetics (Cytogenetics)
Aquaculture/environment interaction
Aquaculture structures, systems, techniques

Facilities

Rearing tank (field site)
Data bases, extensive libraries
Specialised computer hardware
Field site,, electron microscope, specialized analytical apparatus
National reference collection of fish (world coverage)

Staff: Scientists: 24 Others: 14

Research budget (ECU/year): 100.000

Description

Main research activities:
- Phylogeny of fishes inferred from molecular data (ribosomic genes sequences) and comparison with morphological and anatomical data.
- Systematic and knowledge base systems.
- Data base system to identify fish species and stock assessment.
- Biosystematic of fishes and stock assessment in the southern Ocean fisheries (statistical area 58).
- Systematic, Biology, Ecology and population structure of freshwater tropical species.
- Fish biology and structure of fish assemblage in continental waters (the fish as a biotic integrity water indicator).
- Impact of fish predation on the planktonic populations in fertilized farmed ponds, application to pond fish farming.
- Methodologies for wastes recycling in pond and production of plankton and fish.
- The control of reproduction and keeping of freshwater fish gametes.

63. Musée Histoire Naturelle - Lab. Physiologie Générale et Comparée

Name : Musée d'Histoire Naturelle
Laboratoire de Physiologie Générale et Comparée
Address : 7, rue Cuvier
F - 75231 Paris cedex 05
Head : Prof. Y.A. Fontaine
Tel : 40.79.36.15 **Fax**: 40.79.36.18

Type of institute

Research centre
University Department/Higher Education Institute
Government Department

Research activities

Fish Physiology (endocrinology)
Reproduction
Disease, stress
Aquaculture/environment interaction

Facilities

Research aquarium, electron microscope, specialized analytical apparatus, cell culture and molecular biology equipment.

Staff: Scientists: 11 Others: 15

Research budget (ECU/year): 105.000

Description
Research:
Basic research on fish physiology and biochemistry specially in the European Eel, Anguilla anguilla. Neuroendocrine control of reproduction in fish. Thyroid function and adaptation to salinity.
Regulation of calcium metabolism. Physiopathological effects of waterpollutions. Parasitosis of the eel (Anguilla anguilla) by the Nematodes Anguillicola.
Facilities:
Physicochemistry apparatus (radioactivity counters, spectrophotometers, spectrofluorometer). Equipment for molecular biology and cell culture (freezer - 80° C, autoclave, laminar flow hood, inverted microscope, oligonucleotide synthetizer). Histology (image analyzer, microphotometer, microscope, electron microscope, microtomes and ultramicrotome).

64. SANOFI - France Aquaculture

Name	:	France Aquaculture
Address	:	10 rue de Chaillot F - 75116 Paris
Head	:	A. Michel (Directeur Général)
Tel.	:	47202511 **Fax**: 1.47230204 **Telex**: 610249 F

Type of institute
Research centre
Engineering/consulting firm
International Distribution
Fish/crust/mol-Proc. indust./diets producer

Research activities
Aquaculture
- Physiology
- Reproduction
- Early stages of reared species
- Growth, nutrition, feed composition, live food production for aquaculture
- Aquaculture structures, systems, techniques
- Aquacuture economics
- Aquaculture products

Facilities
Rearing tank
Data bases, extensive libraries
Feed production line
Field site, specialized analytical apparatus

Staff: Scientists: 10 Others: 40

Research budget (ECU/year): 430.000

Description
France aquaculture is a subsidiary of SANOFI (70%) and IFREMER (30%) France aquaculture gathers a set of means in the world of aquaculture to design and implement, consult and assist, train and transfer its techniques and know-how.
Main activities :
Consultancies
Studies of regional planning
Project studies
Engineering and work control
Supply of equipment
Supply of products and specialized feed
Technical assistance
Management of exploitations
International financing

65. SEPIA Internationale

Name	:	SEPIA International
Address	:	14 Avenue Gustave Eiffel F - 78182 St Quentin Yvelines Cedex
Head	:	Gilles Brunel
Tel.	:	(1)30.96.49.00 **Fax**: (1)30.96.49.99 **Telex**: 600.734F

Type of institute
Public company

Research activities
Aquaculture
Fisheries
Environment
Commercialisation
Research

Facilities
Marine Fish and Fresh Water hatcheries
Intensive seabass, seabream and turbot farms
Integrated shrimp farms
Research units

Staff: Scientists: 10 Others: 80

66. Univ. Blaise Pascal Clermont II - Equipe d'Hydrobiologie

Name : Equipe d'Hydrobiologie - Université Blaise Clermont II
Laboratoire Zoologie-Protistologiem URA CNRS 138
Address : F - 63177 Aubière Cedex
Head : Christian Amblard
Tel. : 73.40.74.83 **Fax**: 73.40.76.70

Type of institute
Research centre
University Department/Higher Education Institute

Research activities
Fisheries
- Stock assessment in lakes and reservoirs
- Fisheries biology
- Population dynamic (growth fecondity, mortality)

Aquaculture
- Early stages
- Larval rearing
- Nutritional aspects

Facilities
Library
Computer facilities
Physicochemitry apparatus
Inverted, epifluorescence and electron microscopes

Staff: Scientists: 7 Others: 4

Description
1. Impact of fish predation on planktonic population in lakes and reservoirs: feeding activity and food consumption (Salvelinus alpinus, Rutilus rutilus, Perca fluviatilis, Gymnocephalus cernua, Esox lucius...)
2. Intensive production of fingerlings and nutritional aspects (lipid metabolism) during larval development of Esox lucius.

67. Univ. Bordeaux I - Laboratoire d'Océanographie Biologique d'Arcachon

Name : Université Bordeaux I
Centre d'Océanographie et de Biologie marine
Address : 2 rue du Professeur Jolyet
F - 33120 Arcachon
Head : Prof. P. Caumette
Tel. : 56.83.10.22 **Fax**: 56.83.51.04

Type of institute
Research centre
University Department/Higher Education Institute

Research activities
Marine microbiology
Biogeochemical oceanography
Productivity of benthic ecosystem on the continental shelf of the Southern Bay of Biscay
Spatio-temporal structure of benthic communities
valuation of the production
Trophic organisation
Structure and dynamics of the communities in the eutrophic litoral ecosystems
Larval development and recruitment processes
Meiofauna
Invertebrate macrofauna communities
Physiology of fish reproduction

Facilities
Small research vessel
Controlled environmental chambers
Rearing tanks
Respiromectic measure line
Ultra centrifuges
Scintillation counter
Specialized analytical apparatus
Field site

Description
The "Laboratoire d'Oceanographie Biologique", part of the "Centre d'Océanographie et de Biologie Marine d'Arcachon", is linked to the University Bordeaux I and is integrated in several national programs supported by the CNRS and in several international projects, mainly from the Commission of the European Communities. Its main research activity is devoted to:
The study of organic matter in the sea and, more specifically, to the process of mineralization of the detritic matter through biochemical activities of aerobic and anaerobic bacteria in littoral waters and sediment-water interface.
The other part of the "Centre d'océanographie'is constituted by the "Laboratoire de Neutrobiologie", the activity of which is focused on the study of paucineuronal systems of invertebrates.

68. Univ. Caen - Centre Régional Etudes Cotières - Lab. Luc-sur-Mer

Name : Université de Caen
Centre Régional d'Etudes Cotières
Station Marine de Luc-sur-Mer
Address : Rue Charcot-BP 02
F - 14530 Luc-sur-Mer
Head : C. Larsonneur
Tel. : 33.31.973154 - 31.973673 **Fax**: 31.975588

Type of institute
Research centre
University Department/Higher Education Institute
fish/crust/moll.far/sea-ur/

Research activities
Fisheries management
- Stock assessment
- Environmental problems and their impact on resources
- Fishing techniques
- Testing techniques/instruments

Aquaculture

Facilities
Research vessel
Rearing tank
Controlled environmental chamber
Data bases, extensive libraries

Staff: Scientists: 15

Description
Research centre specialized in sea-urchin and seaweeds besides the research on fish biology and marine geology.
Research vessel: the 'Cotes de Normandie', 16 m in length armed since 1985 by the CNRS;
Laboratories with sea-water circuit.
Rooms for seaweeds culture.
Room which can be sterilized, for organ culture.
A laboratory for photographic work.

69. Univ. Claude Bernard Lyon I - Institut Michel Pacha - Station Maritime de Physiologie

Name : Université Claude Bernard Lyon I
Station Maritime de Physiologie "Institut Michel Pacha"
Address : 1337 Corniche Michel Pacha
F - 83500 La Seyne sur Mer

Head : Gérard Brichon (Maître de conférences)
Tel. : 94.94.82.02 **Fax**: 94.06.78.57

Type of institute
Research centre
University Department/Higher Education Institute
Associated to Laboratoire de Physiologie Générale et Comparée Lyon I

Research activities
Fisheries management
 Environmental problems and their impact on resources
Aquaculture
 Physiology and metabolism (Lipids)
 Growth, nutrition, feed composition
 Multispecies biological problems for aquatic animals (Crustaceans, molluscs, fishes)
Upgrading of fisheries products: quality of the products (lipids state)

Facilities
Library (technical books, specialized journals, current contents on diskette for Apple McIntosh, Agriculture, biology and environmental sciences, life sciences J1200)
Computer facilities McIntosh (4) and IBM (1)
Research vessel, rearing tank, tanks for metabolic studies, controlled environmental chamber, specialized analytical apparatus for the study of lipids metabolism (TLC) LC, GC, HPLC, Scintillation spectrometer, proportional counter, spectrophotometers, etc.
Concessions with different marine farms producing red-sea bream and sea-bass and with IFREMER (Mediterranean Center)
Accomodations: lodging is possible in the station

Staff: Scientists: 6 Others: 4

Description
Studies of major metabolisms in fishes, crustaceans and molluscs, in different ecophysiologic conditions (temperature, salinity, quality of the diet) and at various stages of development. The laboratory is particularly involved in the study of complex lipids metabolism (phospholipids and sphingolipids), metabolic and enzymologic responses of aquatic animals to different pollutants and thermal shock (whole animals and isoloated specialized cells).
Role of lipids in the thermoregulation process in aquatic birds (in collaboration with Laboratoire de Physiologie Générale et Comparée.

70. Univ. de Nice - Sophia Antipolis - Lab. Physiologie Cellulaire et Comparée

Name : Université de Nice - Sophia Antipolis
Laboratoire de Physiologie Cellulaire et Comparée
Address : FR Sciences Parc Valrose-B.P. 71
F - 06108 Nice Cedex 2
Head : Prof. M. Bornancin
Tel. : 93.52.99.36 **Fax**: 93.52.99.48

Type of institute
Research centre
University Department/Higher Education Institute
Government Department

Research activities
Aquaculture:
- Fish physiology
- Osmoregulation
- Hormonal control
- Quality of gametes
- Fertilization
- Cell biology

Facilities
Aquaculture installations for marine fishes
Research laboratory, Electron Microscopy Center
Cell culture, hormonal dosage, radio-isotopic techniques

Staff: Scientists: 15 Others: 5

Research budget (ECU/year): 103.000

Description
Study of major physiological functions such as reproduction, development, growth and osmoregulation of fishes (trout, sea bass, turbot...) in relation to the environment. Cell biology of marine invertebrates.

71. Univ. Languedoc - Sciences et Techn. - Station Biol. Marine et Lagunaire

Name : Université des Sciences et Techniques du Languedoc
Station de Biologie Marine et Lagunaire
Address : Quai de la Daurade
F - 34200 Sète
Head : Prof. A. Raibaut
Tel. : 1667743670 **Fax**: 1667543079 **Telex**: USTMONT 490944 F

Type of institute
Research centre
University Department/Higher Education Institute

Research activities
Aquaculture
- Reproduction
- Early stages of reared species
- Growth, nutrition, feed composition, live food production for aquaculture
- Aquaculture/environment interaction

Facilities
Rearing tank
Mesocosm
Field site, electron microscope, specialized analytical apparatus

Staff: Scientists: 3 Others: 8

Description
The Station de Biologie Marine et Lagunaire in Sète is currently undergoing major restructuration and redevelopment.
The installation of a consequent marine molecular genetics department is planned for the coming months. This laboratory will study both natural and cultured populations; especially fish and molluscs.
The second major research theme concerns lagoon and coastal ecosystems and environmental problems linked to their exploitation notably by aquaculture.
Other projects envisaged include the nutrition of larval fish and the ecophysiology of larval crustaceans.

72. Univ. Limoges - Lab. Immunophysiologie

Name : Université de Limoges
Laboratoire de Immunophysiologie Générale et Comparée
Address : UER des Sciences - 123 av. Albert-Thomas
F - 87606 Limoges cedex
Head : Prof. P. Deschaux
Tel. : 55794622 **Fax**: 55457201

Type of institute
Research centre
University Department/Higher Education Institute

Research activities
Fisheries management
- Multi-species biological models
- Environmental problems and their impact on resources

Aquaculture
- Physiology
- Disease, immunology, stress
- Molecular biology

Facilities
Laboratory of cell culture, of immunology cellular and humoral, of molecular biology

Staff: Scientists: 4 Others: 4

Research budget (ECU/year): 30.000

Description
Actually research involved in MHC in fish (molecular level).

73. Univ. Lyon I - Claude Bernard - Lab. Physiologie Générale et Comparée

Name	:	Université Cl. Bernard-Lyon I Lab. de Physiologie Générale et Comparée
Address	:	11, bld du 11 Novembre 1918 F - 69622 Villeurbanne Cedex
Head	:	G. Zwingelstein/ Pierre Joly
Tel.	:	72448000 **Fax**: 72448466 **Telex**:: 330722 F

Type of institute
University Department/Higher Education Institute

Research activities
Fisheries management
Environmental problems and their impact on resources
Aquaculture
Physiology
Growth, nutrition, feed composition, live food production for aquaculture

Facilities
Research vessel
Flume tank
Controlled environmental chamber

Staff: Scientists: 6

Description
Major topics:
1. studies of major metabolism in fishes, crustacea or mollusca in different ecophysiologic conditions.
2. Influence of temperature, salinity and diet supplementation.
3. Influence of different pollutants.

74. Univ. Lyon I - Lab. Biologie Animale et Ecologie U.R.A. CNRS 1451 - Ecologie des Eaux Douces et des Grands Fleuves

Name : Université Cl. Bernard-Lyon I
Lab. de Biologie Animale et Ecologie (U.R.A. CNRS 1451)
Ecologie des Eaux Douces et des Grands Fleuves
Equipe d'Ichtyologie
Address : 43, bld du 11 Novembre
F - 69622 Villeurbanne Cedex
Head : Prof. M. Bournaud
Tel. : 7244.82.18 **Fax**: 7243.11.41

Type of institute
Research Centre, University Department/Higher Education Institute

Research activities
Fish behaviour, habitat use by young and adult stages, population dynamics, genetic diversity, influence of perturbations under natural conditions and conditions modified by man.

Facilities
Both in Lyon (University) and Arles (Biological Station).
Department and University Libraires
A Macintosh computer network with over 40 machines, connected with the Computer Centre of the University and with E-mail Network.
Experimental grounds in the University Campus, with artificial ponds and flowing channels, Infrared video, organic carbon and nitrogen analysers, regulated-temperature chambers. Laboratory for conventional enzyme electrophoresis and University Centre for Molecular Taxonomy involving DNA Investigations, Several boats, lorries and cars. Electrofishing gear.

Staff: Scientists: 6 Others: 3

Research budget (ECU/year): 76.000

Description
Habitat strategies in fluvial fish at the organism and population levels within space and time. Investigation of presence, abundance, behaviour, migrations, dynamics and genetic structure in the field. Consequences of natural and man-made habitat changes.
Experiments in artificial ponds and rivers. Coordination of three French and European cooperative programmes.

75. Univ. Lyon - Catholique - Inst. Reg. Rech. Appl. Aquaculture

Name : Université Catholique de Lyon
Institut Régional de Recherches
Appliquées en Aquaculture (IRRA)
Address : 25 rue du Plat
F - 69288 Lyon Cedex 02
Head : J. Goubier
Tel. : 72.32.50.38

Type of institute
Research centre
University Department/Higher Education Institute

Research activities
Fisheries management
Multi-disciplinary approach to fisheries management
Fishing techniques
Testing techniques/instruments
Aquaculture
Early stages of reared species
Aquaculture/environment interaction
Aquaculture structures, systems, techniques

Facilities
Rearing tank
Hatchery, Experimental ponds, Monitor for field datas

Staff: Scientists: 4 Others: 6

Research budget (ECU/year): 260.000

Description
IRRA is an academic organization admitting graduate students, visiting scientists and post-doctoral fellows, and dealing with international research and opportunities for technological transfers.
IRRA is involved in research programs and technical assistance.
Research programmes concerned with breeding, growth and nutrition of promising aquacultural organisms as well as water management and socio-economical aspects of aquacultural practices.
Technical assistance in France is devoted to prospection and adaption of new technologies to improve and develop aquaculture, in relation with farmers and manufacturers.

In Eastern Europe and developing countries, technical assistance is focused on technological transfers and foreign students' training.
IRRA is also involved in aquacultural projects, achieving technical and economical feasability studies, as well as specific equipment for water quality monitoring, aeration,...

76. Univ. Marseille III - Fac. Sciences et Techn. St. Jérome - CERAM

Name	:	Université de Marseille III Fac. des Sciences et Techniques de St. Jérome Centre Etudes des Ressources Animals Marines (CERAM)
Address	:	Av. Escadrille Normandie Niemen F - 13397 Marseille
Head	:	N. Vincente, (Directeur CERAM.)/P Rancurel
Tel.	:	91.288441 **Fax**: 91.288030 **Telex**: FAC ST JE 402876 F

Type of institute
University Department/Higher Education Institute
Government Department / Regional or Local Authority Fishing

Research activities
Fisheries management
 Stock assessment
 Multi-disciplinary approach to fisheries management
 Environmental problems and their impact on resources
Upgrading of fishery products
 Novel processing methods for by-products, discards, and process wastes

Facilities
Research vessel
Field site, electron microscope, specialized analytical apparatus

Staff: Scientists: 11

Research budget (ECU/year): 180.000

Description
- The research programme is focused on studies on deep-sea squales (centrophorus) for biological as well as commercial purposes.
- Ecophysiology of molluscs (Bivalves and Gastropodes).
- Pollution by micropollutants.

77. Univ. Montpellier II - Dept. Aquaculture et d'Halieutique

Name : Université de Montpellier II
Sciences et Techniques du Languedoc
Département d'Aquaculture et d'Halieutique
Address : F - 34095 Montpellier Cedex 5
Head : J. P. Quignard
Tel. : 67.143030 **Fax**: 67.543079 **Telex**: USTMONT 490944

Type of institute
University Department/Higher Education Institute
Engineering/consulting firm
Research centre

Research activities
Fisheries management
  Stock assessment
  Environmental problems and their impact on resources
  Fishing techniques
  Testing techniques/instruments
Aquaculture

Facilities
Research vessel
Rearing tank
Mesocosm
Controlled environmental chamber
Data bases, extensive libraries
Field site, electron microscope, specialized analytical apparatus
Calculation centre

Staff: Scientists: 19 Others: 15

78. Univ. Montpellier II - Lab. Biologie Animale

Name : Université de Montpellier II
Sciences et Techniques du Languedoc
Laboratoire de Biologie Animale
Address : Place E. Bataillon
F - 34095 Montpellier Cedex 5
Head : R. Connes
Tel. : 67.143675 **Fax**: 67.543079 **Telex**: USTMONT 490944

Type of institute
Research centre
University Department/Higher Education Institute

Research activities
Aquaculture

Facilities
Controlled environmental chamber
Field site, electron microscope, specialized analytical apparatus

Staff: Scientists: 5 Others: 2

Description
Topic: Ontogenetic, nutritional and pathological studies on economic interest fish.
Main objective: understanding of physiological processes occuring during ontogeny of the sensory, digestive and pancreatic hormonal control systems to improve fish farming.
Major research projects:
(1) - Transition to the exotrophic phase: yolk absorption, early intestinal and hepathic functions, initial control of pancreatic hormones.
(2) - Ontogeny of sensory (eyes, taste buds, olfactory organ, lateral line) and digestive (gut, liver, pancreas) systems of fish larvae reared on natural-like (mesocosme) or artificial conditions.

79. Univ. Montpellier II - Lab. Ichthyologie

Name : Université de Montpellier II
Sciences et Techniques du Languedoc
Laboratoire d'Ichthyologie
Address : Place E. Bataillon
F - 34095 Montpellier Cedex 5
Head : J.P. Quignard
Tel. : 67.631160 **Fax**: 67.543079 **Telex**: USTMONT 490944

Type of institute
Research centre
University Department/Higher Education Institute

Research activities

Fisheries management
- Stock assessment
- Environmental problems and their impact on resources
- Fishing techniques
- Testing techniques/instruments
- Ichthyology-Biology and dynamic of populations

Facilities

Research vessel
Data bases, extensive libraries
Calculation Center(CNUSC) Tel.: 67141414 Telex: 490439

Staff: Scientists: 7 Others: 2

80. Univ. Montpellier II - Station Méditerranéenne de l'Environnement Littoral

Name	:	Université de Montpellier II Sciences et Techniques du Languedoc Station Méditerranéenne de l'Environnement Littoral
Address	:	1, Quai de la Daurade F - 34200 Sète
Head	:	Prof A. Raibaut
Tel.	:	16.67.46.33.70 **Fax**: 16.67.46.02.56

Type of institute

University Centre of Research and Higher Education Institute

Research activities

Reproduction and early stages of reared species (fish and shellfish)
Growth, nutrition, feed composition
Genetic markers of fish and shellfish populations
Aquaculture/environment interaction
Parasitological ecology and modelisation

Facilities

Photonic microscope, specialized analytical apparatus (PCR, HPLC, Multiparameter probe...)
Computer and image analysis
Large scale rearing tanks
Coastal experimental conchyliculture site
Research vessel

Staff: Scientists: 9 Others: 9

Description
The restructuring of the Station Méditerranéenne de l'Environnement Littoral in Sète has now been ongoing for two years. Research is now being developed arround two major topics:
1) molecular genetics of marine organisms
2) the coastal environment
The first theme has seen the launch in 1992 of a major programme to study genetic markers (RAPDs) in the sea bass and penaeid schrimps. This programme is set to flourish in 1993. The second axis has seen the installation of several themes, including the development of solar-powered water circulators to improve shellfish culture, the mathematical modelling of marine population dynamics and epidemiology, the experimental ecology of (Salmonella) and the molecular biology of metaalothionein in bivalves.
Other research include the nutrition of larval fish (sea bream) and the ecophysiology of larval crustaceans (lobster).

81. Univ. Nantes - Lab. Biologie Marine

Name : Université de Nantes
Laboratoire de Biologie Marine
Address : 2 rue de la Houssinière
F - 44072 Nantes Cedex
Head : J. Marchand
Tel. : 40373295

Type of institute
University Department/Higher Education Institute

Facilities
Field site, electron microscope, specialized analytical apparatus

Staff: Scientists: 3 Others: 0.5

Research budget (ECU/year): 22.000

Description
Bioecology of the Demersal Fish in Estuarine Ecosystems
1. Study of the nursery grounds of the Loire estuary: distribution, growth and feeding habits of the juveniles of amphihaline (flounder) and euryhaline (sole, whiting, bib, bass, etc) species.

2. Study of the food webs of the Loire estuary: interrelationships opposite to the dynamics of the benthic communities.
3. Specific study on the sole Solea solea: hydrological conditions of settlement of post-larvae (interannual comparisons related to meteorological conditions) their behaviour (distribution and vertical migration) and their feeding activity. Overall project leader: IFREMER and CNRS(France): Program on Recruitment Determinism 5 (PNDR) (from 1985). This program is linked to the CLIMAPECHE program which concerns recruitment studies.
4. Study of the impact on fishes of human activities: mortality of juvenile populations of smelt and flounder, caused by power plant activity and by small-mesh fisheries (glass-eel and shrimp fisheries).
5. Study of the mullet population(Liza sp.) in the Loire estuary: problem of their mortality linked to the water quality (turbidity maximum) during their reproductive migration.

82. Univ. Paris VI - P. et M. Curie - I.N.S.U - Observatoire Océanologique Villefranche-sur-Mer

Name	:	Université de Paris VI (P. et M. Curie) Observatoire Océanologique
Address	:	BP 28 F - 06230 Villefranche-sur-Mer
Head	:	Prof. Soyer
Tel.	:	93.763890 **Fax**: 93.763893

Type of institute
Research centre
University Department/Higher Education Institute

Research activities
Fisheries management
Environmental problems and their impact on resources, recruitment
Fishing techniques
Testing techniques/instruments
Electronics/acoustics
Aquaculture
Reproduction
Genetics and conservation of gametes
Early stages of reared species

Facilities
Research vessel
Controlled environmental chamber
Data bases, extensive libraries
Specialised computer hardware
Specialised software
Field site, electron microscope, specialized analytical apparatus

Staff: Scientists: 52 Others: 78

Description
The 'Observatoire Océanologique de Villefranche is linked to the University Pierre et Marie Curie and to the CNRS (particularly to the Institut National des Sciences de l'Univers: INSU).
Its research fields are:
Biology of marine cells (M. Christian Sardet)
Ecology of the marine plankton (M. Paul Nival)
Biochemical oceanography (M. Patrick Mayzaud)
Marin Chemistry and Physics (M. Andre Morel)
Underwater geodynamics (M. Jean Mascle)
These different teams cooperate with the big French and foreign research centres, as well as with international organisations (UNESCO, EEC,...)
The 'Observatoire Océanologique' provides also a teaching for the students (from the second cycle level to the thesis).
VESSELS : Two oceanological vessels : The Korotneff (19,7 m in length), the Catherine Laurence (21,22 m in length).
Two additional small boats to collect plankton and animals by diving.

83. Univ. Paris-Sud XI - Lab. Physiologie Cellulaire et Metabolique Poissons

Name : Université de Paris-Sud XI et URA 1134 CNRS
Laboratoire de Physiologie Cellulaire
et Métabolique des Poissons
Address : Campus d'Orsay-Bâtiment 447
F - 91405 Orsay Cedex
Head : J.M. Vernier
Tel. : 1.69417933 **Fax**: 64451992 **Telex**:: 602.166F

Type of institute
Research centre
University Department/Higher Education Institute

Research activities
Aquaculture
Physiology
Reproduction
Growth, nutrition, feed composition, live food production for aquaculture
Disease, immunology, stress

Facilities
Field site, electron microscope, specialized analytical apparatus

Staff: Scientists: 7

Research budget (ECU/year): 32.700

Description
Main contracts:
"Utilisation tissulaire des lipides plasmatiques chez la truite".(1987)
"Transfert transépithélial de la GH bovine (bGH) au niveau de l'intestin postérieur chez la truite arc-en-ciel" (1988).
Improvement of yield and food conversion in salmonids and in maricultured fish.
The scientific field of this laboratory concerns:
"Fish nutrition and metabolism".
- Macromolecular absorption: protein absorption at the posterior intestine level, and their nutritional and immunological implications.
- Plasma lipoproteins: synthesis and secretion at anterior intestines, liver and kidney level; metabolic implications; role in hormonal transport and yolk deposition.

84. Univ. Perpignan - Lab. Biologie Marine

Name : Université de Perpignan
Faculté des Sciences - Laboratoire de Biologie Marine
Address : Avenue de Villeneuve
F - 66025 Perpignan Cedex
Head : J. Truslé
Tel. : 68.51.00.51. **Fax**: 68671221 **Telex**: 505005

Type of institute
Research centre
University Department/Higher Education Institute

Research activities
Fisheries management
Stock assessment
Environmental problems and their impact on resources
Aquaculture
Reproduction
Disease, immunology, stress

Facilities
Field site, electron microscope, specialized analytical apparatus

Staff: Scientists: 4 Others: 10

Research budget (ECU/year): 7.500

Description
The main topic of the laboratory is "Biology of marine and brackish water fishes":
a) recruitment and age of glass eels (using otolithometry): comparison of European population.
b) sex-differentiation and germinal potentialities (primordial germ cells) in gonochoristic fishes (grey mullets, eels) and in hermaphroditic fishes (sex-inversion in groupers, sea-brams and clownfishes).
c) liver pathology induced by parasites (Helminths metacercariae) pollutants (herbicides and heavy metals) and artificial feeding (pellets).
d) ichtyosarcotoxism: ciguatera: effect of toxin from Dinoflagellates on visceral organs of tropical fishes (groupers, snappers, parrotfishes).
e) ichtyoplankton: egg and larval development of Soleidae.

85. Univ. Rennes I - Equipes Biologie Moléculaire et Physiologie des Régulations de l'URA CNRS 256

Name : Université de Rennes I - Equipes Biologie Moléculaire et Physiologie des Régulations de l'URA CNRS 256
Biologie Moléculaire et Physiologie des Régulations
Address : Complexe Scientifique de Rennes-Beaulieu
Université de Rennes I
F - 35042 Rennes Cedex
Head : Y. Valotair/ P. Jégo
Tel. : 99.28.61.35/99.28.61.33

Type of institute
Research centre
University Department/Higher Education Institute

Research activities
Aquaculture
  Molecular biology and Physiology in Salmonids reproduction
Ecotoxicology

Facilities
Cells culture chamber; analytical apparatus for molecular biology, biochemistry and physiology

Staff: Scientists: 9 Others: 12

Research budget (ECU/year): 40.000

Description
The main research activities are: hormonal regulation of genes involved in the reproduction: vitellogenin, estradiol receptor, gonadotropin releasing hormone, proopiomelanocortin, glucocorticoid receptor, tyrosine hydroxylase. Genes or cDNAs are cloned and sequenced and used as probes to study; RNA levels upon hormonal treatments. Hepatocytes culture is also used to study the interaction between xenobiotics and reproduction.

86. Univ. Rennes I - Station de Biologie Marine - Ile de Bailleron

Name	:	Université de Rennes I - UER des Sciences et de l'Environnement Station de Biologie Marine de l'Ile de Bailleron
Address	:	F - 56860 Sène
Head	:	Prof. Jacques Daguzan
Tel.	:	97.26.41.04 **Fax**: 99.28.67.00

Type of institute
Research centre
University Department/Higher Education Institute

Research activities
Fisheries management
- Stock assessment
- Multi-species biological models
- Environmental problems and their impact on resources

Aquaculture

Facilities
Research vessel
Simulated tidal tank
Rearing tank
Field site, electron microscope, specialized analytical apparatus

Staff: Scientists: 9 Others: 4

Description
Biology and dynamics of SEPIA OFFICINALIS population in the Morbihan Gulf. - Crustacean biology and dynamics in the Morbihan Gulf. - Surveillance of water quality and study of the phytoplanktonic populations of the Morbihan Gulf. - Functioning of the Atlantic coastal lagoons : bird-environment relations. The centre has a guest house for visiting students, for whom it organizes stages.

87. Univ. Strasbourg Louis Pasteur - Lab. Psychophysiologie

Name	:	Université Louis Pasteur Laboratoire de Psychophysiologie, section comport. des poissons
Address	:	7, rue de l'Université F - 67000 Strasbourg
Head	:	M. Anthouard
Tel.	:	88358213

Type of institute
Research centre
University Department/Higher Education Institute

Research activities
Aquaculture
- Growth, nutrition, feed composition, live food production for aquaculture
- Aquaculture structures, systems, techniques

Facilities
Rearing tank
Data bases, extensive libraries
Specialised software

Staff: Scientists: 1 Others: 1

Research budget (ECU/year): 2.200

Description
The main research topics are:
- Studies of the factors which could modify the level of trophic activity of a population of fish. - Use of self-feeding method for determining the food intake rhythm and needs of different aged populations, in fresh water and marine species.
- Analyse of behavioural data which could be used for increasing the rentability of rearing methods.
- Use of an automatic device for data acquisition, storage and treatment of food consumption by population fish, in rearing situation.

In cooperation with:
Université des Sciences et Techniques du Languedoc Roussillon:
Laboratoire de Biologie Marine, and Station de Biologie Marine et Lagunaire de Sète:
Professeur PARIS
Instrumentation: 18 experimental tanks, controlled 24h/24h.
36 specialised apparatus(self feeders) in relation with a computer
A specialised software, for scrutinisation of 48 independant input lines,
at a rate of 2000 cycles per second.

88. Univ. Toulouse - ENSAT Lab. Ichtyologie

Name : Université de Toulouse
ENSAT/Laboratoire d'Ichtyologie
Address : 145 avenue du Muret
F - 31076 Toulouse
Head : Prof. M. Beleau/Prof. M. Moreau
Tel. : 61428398 **Fax**: 33.61423029

Type of institute
Research centre
University Department/Higher Education Institute
Engineering/consulting firm
National/international level

Research activities

Fisheries management

- Stock assessment
- Multi-species biological models
- Environmental problems and their impact on resources

Aquaculture

- Growth, nutrition, feed composition, live food production for aquaculture

Aquaculture/environment interaction

- Field studies on fish population and human impact on environment.

Facilities

Data bases, extensive libraries
Specialised computer hardware
Specialised software
Field sampling equipment

Staff: Scientists: 22 Others: 4

Research budget (ECU/year): 144.000

State of Research

GERMANY

In the Federal Republic of Germany, research and teaching is allocated to the individual states of the federation. The Federal Government finances its own research only for the need of the legislation and for internal measures, e.g. quota allocation for fishing.

Generally, the regulation of marine fisheries, including management of stocks, is the responsibility of the Federal Ministry of Food, Agriculture and Forestry(BML) which coordinates the relevant activities of all public institutions of research and administration concerned.

The German Scientific Commission for Exploration of the Sea(DWK) advises the BML, so it is involved in the coordination of research at the national level. It is also the liaison to international organizations, such as ICES or NAFO. The members of the DWK represent a wide range of fisheries research activities at the level of Bund and Länder, including relevant university institutes. The European Inland Fisheries Advisory Commission of FAO-EIFAC-consultants BML about three subcommissions:

* Fish population and management
* Aquaculture and fish diseases
* Fish and water pollution

Research at national level is conducted at the Federal Research Centre for Fisheries (BFAFi) in Hamburg, which concentrates research in fisheries. The centre is attached to BML, it has three research vessels and modern technology equipment. The BFAFi is divided into five institutes:

* Institute for Sea Fisheries
* Institute for Baltic Fisheries
* Institute for Fishing Technology
* Institute for Biochemistry and Technology
* Institute for Fisheries Ecology

Coordinated fisheries research activities of an immediately practical nature (e.g. stock monitoring for TAC ...) rely initially on data collected by the ICES and NAFO organizations as well as from the northern part of the EC marine zone and the waters around Greenland. Surveys based on these data are conducted by scientists and research vessels of the BML and BFAFi fisheries institutes.

Research activities extend beyond the commercial fisheries of these particular geographic sectors into more general themes. Broad research beyond the capacity of BML - Fisheries Research can be achieved in cooperation with specialized marine institutes of different universities or by contracts.

The responsibility for the sector of inland fisheries lies with the Bundesländer. These have their own research centres. However, most institutes involved in fisheries research are part of agriculture or veterinary faculties.

Links with the industry depend on the sector, and its extend of industrialization, e.g. aquaculture being at an artisanal state, cooperation with industry is rare. The general trend is of little industry/research cooperation, even though a variety of industry/science meetings permits a direct exchange of information between the two sectors.

Industry finances its own research centres, but there is generally no information available on their topics and results.

University degrees can only be obtained in fishery biology. Specialized advanced level training for professional fisheries biologists, working at government fisheries research institutes, is only provided in exceptional circumstances for selected staff.

List of Institutes from GERMANY

1. Alfred Wegener Institut für Polar und Meeresforschung
2. Bayerische Landesanstalt für Fischerei
3. Bayerische Landesanstalt für Wasserforschung - Munchen
4. Bayerische Landesanstalt für Wasserforschung - Wielenbach
5. Brandenburgische Fischereiforschungsstation
6. Bundesamt für Seeschiffahrt und Hydrographie
7. Bundesansalt für Gewässerkunde
8. Bundesforschungsanstalt für Fisherei
9. Bundesforschungsanstalt für Viruskrankheiten der Tiere
10. Bundesgesundheitsamt - Institut für Veterinärmedizin
11. Deutsches hydrographisches Institut
12. Federal Research Centre for Nature Conservation and Landscape Ecology
13. Fischereiamt Berlin
14. Fischereiforschungsstelle des Landes Baden-Württenberg
15. Forschungsinstitut Senckenberg
16. Forschungszentrum Jülich - Arbeitsgruppe Modellierung für Umweltforschung und Lebenswissenschaften (ATÖ)
17. Institut für Binnenfischerei e.v.
18. Institut für Fischqualitat GMBH, Bremerhaven
19. Institut für Gewässerökologie und Binnenfischerei Abteilung Biologie und Ökologie der Fische
20. Institut für Gewässerökologie und Binnenfischerei Abteilung Fischzucht und Fischpathologie
21. Institut für Polarkölogie Landesforschungsanstalt für Landwirtschaft und Fischerei Mecklenburg/Vorpommern
22. Landesansalt für Fischerei Nordrhein-Westfalen
23. Niedersächsisches Ministerium für Ernährung Landwirtschaft und Forsten/Staatlicher Fischseuchenbekämpfungsdienst Niedersachsen und Fischgesundheitsdienst
24. Polarmar
25. Sächsische Landesanstalt für Landwirtschaft
26. STAGAtech GbR
27. Technische Universität Berlin
28. Tierärztliche Hochschule Hannover
29. Univ. Berlin - FUB-FB Biologie WE5
30. Univ. Bochum - Inst. für spezielle Zoologie und Parasitologie
31. Univ. Bonn - Inst. für Lebensmittelchemie
32. Univ. Bonn - Inst. für Anatomie, Physiologie und Hygiene der Haustiere
33. Univ. Göttingen - Inst. für Tierphysiologie und Tierernährung
34. Univ. Göttingen - Inst. für Tierzucht und Haustiergenetik
35. Univ. Hamburg - Inst. für Hydrobiologie und Fischereiwissenschaft
36. Univ. Hamburg - Inst. für Meereskunde
37. Univ. Hamburg - Zoologisches Institut und Zoologisches Museum

38. Univ. Heidelberg - Zoologisches Institut
39. Univ. Karlsruhe - Inst. für Zoologie II
40. Univ. Kiel - Inst. für Meereskunde - Abt. Fischereibiologie
41. Univ. Kiel - Institut für Polarkölogie
42. Univ. Konstanz - Fak. Biologie - Inst. Limnologie
43. Univ. München - Inst. für Zoologie und Hydrobiologie
44. Univ. München - Techn. - Institut für Ernährungsphysiologie
45. Univ. München - Techn. - Lehrstuhl für Tierhygiene - Fachgebiet Fischbiologie
46. Univ. Oldenburg - Inst. Chemie und Biol. des Meeres (ICBM)
47. Univ. Rostock - Inst. für Ostseeforschung
48. Univ. Stuttgart - Inst. für Siedlungswasserbau
49. Univ. Tübingen - Zoologisches Institut
50. Versuchsanstalt für Wasserbau und Schiffbau

1. Alfred Wegener Institut für Polar und Meeresforschung

Name : A. Wegener Institut für Polar und Meeresforshung Sektion Biologie I
Address : Columbusstrasse
D - 27568 Bremerhaven
Head : Prof. Dr. W. Arntz
Tel. : 49.471.4831.300 **Fax**: 49.471.4831.149

Type of institute
Research Centre

Research activities
Fisheries ecology

Facilities
2 research vessels (Antarctic, North Sea)

Staff: Scientists: 2 Others: 1 (fisheries only)

Description
Research projects: Processes controlling recruitment of clupeiform fish in different regions, (cooperation with Lisboa University, Instituto Español de Oceanografía, Plymouth Marine Laboratory, Hamburg University, Institut für Ostseeforschung Warnemünde).
Ecosystem Research Wadden Sea, part: Developing of monitoring strategies for fish and decapods (cooperation with University of Bremen).

2. Bayerische Landesanstalt für Fischerei

Name : Bayerische Landesanstalt für Fischerei
Address : Weilheimer Str. 8
D - 82319 Starnberg/See (Obb)
Head : Dr. M. von Lukowicz
Tel. : 08151.2692.0 **Fax**: 8151.269270

Type of institute
Research centre
Institute for professional training/Secondary school

Research activities
Fisheries management
Stock assessment
Multi-species biological models

Multi-disciplinary approach to fisheries management
Environmental problems and their impact on resources
Fishing techniques
Aquaculture

Facilities
Data bases, extensive libraries
Specialised computer hardware
Specialised software
Diverse laboratories with specialized analytical apparatus
Rearing Tanks
2 hatcheries with breeding equipment and rearing facilities for warm and cold freshwater fish.
Cooling chamber for constantly adjusted temperature of water
3 pond farms (experimental): carp and trout and many other species of fish.

Staff: Scientists: 9 Others: 25

3. Bayerische Landesanstalt für Wasserforschung - München

Name	:	Bayerische Landesanstalt für Wasserforschung (Bavarian State Agency for Water Research)
Address	:	Kaulbachstrasse, 37 D-8000 München 22
Head	:	Dr. W. Mühlhölzl
Tel.	:	089.2180.2291 **Fax**: 089.542.0211

Type of institute
Research centre

Research activities
Applied research work in:
- Hydrobiology including microbiology
- Chemistry of water and sewage, especially the analysis of dangerous inorganic and organic substances
- Sewage treatment technology
- Fish biology incl. fish pathology
- Limnology
- Radiology (radio aquatic ecology).

Staff: Scientists: 20 Others: ca. 100

Research budget (ECU/year): 1.500.000

4. Bayerische Landesanstalt für Wasserforschung - Wielenbach

Name	:	Versuchsanlage Wielenbach Abteilung fishereibiologie
Address	:	Demollstrasse, 31 D-82407 Wielenbach
Head	:	M. Bohl
Tel.	:	0881.1850 **Fax**: 881.41.318

Type of institute
Research centre - Specific research activities in fish biology

Research activities
Specific research activities in fish biology
Fisheries management
 Stock assessment
 Environmental problems and their impact on resources
Aquaculture
 Fish farm effluents
 Pond ecology, fish diseases and therapy

Facilities
Rearing tank
Controlled environmental chamber
Data bases, extensive libraries
153 ponds for salmonids and cyprinids. Fish test toxicology plant.
Warm and cold water hatcheries.
Aquaria room, toxicological, histological and water chemistry laboratories.

Staff: Scientists: 7 Others: 22

Research budget (ECU/year): 280.000

Description
Research projects: Reproduction of endangered species. Research on the stock situation of native freswater crayfish species. Rearing fingerlings for stocking. Ecological investigations of brooks. Sublethal effects of toxic substances on fish after long-term exposition (NOEC). Effect of fish production on water quality. Investigation on the occurrence of sediments-dependent fish species. Fish farm effluents and treatment. Impact of calcium hypochlorite on biocenosis and water chemistry in carp ponds.

5. Brandenburgische Fischereiforschungsstation

Name : Brandenburgische Fishereiforschungsstation
Address : Jägerhof
D 0-1501 Potsdam-Sacrow, P.O. Großglienicke
Head : Dr. R. Knösche

Type of institute
Research centre

Research activities
De-eutrophication of lakes
Fisheries management
  Fishery yield assessment
Fishing techniques
  Fishing methods and gears
Aquaculture
  Biotechnique, engineering, effluents treatment

Facilities
Experimental closed recirculating fish farm
Experimental lake of 100 ha

Staff: Scientists: 3 Others: 9

Description
Research projects: Intensive aquaculture with eel, sturgeon and catfish especially in closed recirculating systems.
Aquacultural effluents and their treatment.
De-eutrophication of stratified lakes by hypolimnic aeration.
Fishery yield assessment in North German lakes.

6. Bundesamt für Seeschiffahrt und Hydrographie

Name : Bundesamt für Seeschiffahrt und Hydrographie
Address : Bernhard-Nocht-Straße 78
PO Box 30 12 20
D - 2000 Hamburg 36
Head : Präs. u. Prof. Dr. Ehlers

Type of institute
Marine Reserach and Service Centre
Monitoring

Research activities
Fisheries management
Environmental problems and their impact on resources

Facilities
Data bases, extensive libraries
Research vessels; hydrographic equipment
Large computing facilities
German Oceanographic Data Center

Staff: Scientists: 50 Others: 70

Research budget (ECU/year): 3.400.000

Description
General hydrography and description oceanography, fishery research support. Applied oceanographic research North Sea, Baltic, NE Atlantic.
Marine services (among others: ice service, tidal prediction, storm surge forecast).
Pollution monitoring (including coordination) North Sea, Baltic.
German Oceanographic Data Bank/Center.
Operational hydrodynamic model(s) North Sea and Baltic.
Marine climate observation (ship of opportunity programme, fixed stations).

7. Bundesansalt für Gewässerkunde

Name : Abteilung Ökologie and Abteilung Qualitative Gewässerkunde
Address : Kaiserin-Augusta-anlagen 15-17
D - 56068 Koblenz
Head : Prof. V. Wetzel
Tel. : 0261.1306.300 **Fax**: 0251.1306.302

Type of institute
Research centre

Research activities
Environmental impact assesements
Fish-biological studies
Ecotoxicology
Investigations of micropollutants
Application-orientated consulting

Facilities
Electric fishing equipment
Controlled environmental chamber

Ecotoxicological laboratory
Fish test equipments
Photometry
Infrared spectroscopy
HPLC-Liquid chromatography
UV-VIS spectroscopy
Atomic absorption spectrometry

Staff: Scientists: 5 Others: 3

Description
Research projects
Developing methods for the evaluation of environmental impact assessments. Ecotoxicological studies. Application of fish tests for substance screening. Effect monitoring of rivers with biomonitors. Mapping of fish habitats. Controlling the effectiveness of fish ladders for migrating fishes. Determination of contaminants (micropollutants) in fish flesh.

8. Bundesforschungsanstalt für Fischerei

Name : Bundesforschungsansalt für Fischerei
(Federal Research Center for Fisheries)
Address : Palmaille 9
D - 22767 Hamburg
Head : Dr. Wolfgang Schreiber (on a 2 years basis)
Tel. : 040.38.90.50 **Fax**: 040.38.90.51.29 **Telex**: 21.57.16 bfafid

Type of institute
Research centre, 5 departments (Sea Fisheries, Baltic Fisheries, Fishing Technology, Biochemistry and Technology, Fisheries Ecology)

Research activities
Fisheries management
 Stock assessment, recruitment, mortality, multi-species models
 Environmental problems
Fishing techniques
 Fishing gear technology and echo-sounding
Aquaculture
 Aqua- and mariculture
Characteristics and upgrading of fishery products
 Quality and health aspects, standardization and legislation

Facilities

3 research vessels (2 sterntrawlers, 77,5 and 35,4 m l.o.a., 1 sidetrawler, 17,5 m l.o.a.)
Recycling systems for aquaculture
Analytic laboratories
Material testing laboratory
Documentation service
Database, extensive libraries
UW-video-system, processing of video-materials
Echo-integration system

Staff: Scientists: 81 Others: 206

Research budget (ECU/year): 7.030.600

Description

1. Institute for Sea Fisheries

Stock assessment of commercially exploited fish stocks in North Atlantic, Greenland and North Sea; population dynamics; ecological investigations of marine fish assemblages; multi-species interactions; fisheries oceanography; biological monitoring of North Sea and Atlantic fish stocks.

2. Institute for Baltic Fisheries

Stock assessment of commercially exploited fish stocks in the Baltic; ecological investigations of baltic marine and brackish fish assemblages; multi-species interactions; mathematical modelling; aquaculture, sea ranching, population dynamics.

3. Institute for Fishing Technology

Development and improvement of fishing gear and catching methods, size and species selection, echo-integration, impact of towed gear on benthos and bottom structure, evaluation of technical measures for protection of living resources.

4. Institute for Biochemistry and Technology

Composition of fishery products (main ingredients, essential minor components, contaminants), impact of processing on quality and composition, quality assessment and development of standards (national/EC/FAO), evaluation of legal projects.

5. Institute for Fisheries Ecology

Research on radionucleids, heavy metals and chlorinated hydrocarbons in selected compartments of aquatic ecosystems (spatial distribution, temporal development); effects of marine pollution on fish diseases and malformations.

9. Bundesforschungsanstalt für Viruskrankheiten der Tiere

Name : Bundesforschungsanstalt für Viruskrankheiten der Tiere
Address : Paul-Ehrlich-Str. 28-P.O.Box 1149
D - 72001 Tübingen
Head : P.J. Enzmann
Tel. : 7071.603350 **Fax**: 7071.603201 **Telex**: 17707131 D

Type of institute
Research centre
Government Department

Research activities
Aquaculture
Disease, virology, immunology, pathology, diagnostics.

Facilities
Research vessel
Controlled environmental chamber
4 aquaria with 300 l. each
2 aquaria with 200 l. each
4 aquaria with 80 l. each
8 research vessels with 3.000 l. each
Each tank is built as a recirculating unit with controlled environmental conditions

Staff: Scientists: 6 Others: 5

Research budget (ECU/year): 50.000

Description
Research projects:
Pathogenetic and immunological studies on viral diseases of freshwater fish.
Epidemiological studies on fish diseases.
Biomedical and serological studies on fishpathogenic Rhabdoviruses.
Molecular Biology of fishpathogenic Rhabdoviruses.

10. Bundesgesundheitsamt - Institut für Veterinärmedizin

Name	:	BGA - Institut für Veterinärmedizin Abteilung: Zoonosen und Tierseuchenforschung, Fachgebiet Fischkrankheiten
Address	:	Diedersforfer Weg 1 D - 12277 Berlin
Head	:	Prof. Dr. K. Gerigk

Type of institute
Governmental Office with research activities

Research activities
Bacterial infections in fish (diagnostic techniques)
Fish pathology and Histology
Chemical, physical and Bacteriological water analyses
Fish immunology regarding bacterial diseases

Facilities
2 microbiological laboratories, incl. 1 flow box
3 fish laboratories with experimental rearing units (bath closed recirculating systems and flow through system)
1 large scale recirculating unit

Staff: Scientists: 2 Others: 1

Description
Development and standardization of serological techniques (ELISA, IFAT, BLOT) for detection of bacterial pathogenes. Analysis of bacterial lipopolysaccharides and cell surface proteins by electrophoresis. Standardization of trout rearing in closed recirculating systems for infectivity trials. Virulence studies (bacterial pathogens).
Immunization studies (bacterial pathogens).

11. Deutsches hydrographisches Institut

Name : Deutsches hydrographisches Institut
Address : Bernhardt Nocht Strasse 78-PO Box 301220
D - 2000 Hamburg 36
Head : P. Ehlers/Zickwolff
Tel. : 040.31.9001 **Fax**: 40.3190.5150 **Telex**: 211138 bmvhhd

Type of institute
Research centre

Staff: Scientists: 32 Others: 42

12. Federal Research Centre for Nature Conservation and Landscape Ecology

Name : Federal Research Centre for Nature Conservation and Landscape Ecology
Address : Konstantinstr. 110
D - 5300 Bonn 2

Head : Prof. Dr. U. Bohn
Tel. : 0228.8491.0

Type of institute
Institution (under the auspices of the Federal Ministry for Environmental Protection, Nature Conservation and Nuclear Safety).

Research activities
Nature conservation
Animal ecology
Landscape management

Facilities
Rearing tanks
Flow channel
Ponds

Staff: Scientists: 2 Others: 1

Description
Projects: ecology of endangered freshwater fish species; development of breeding methods for endangered fish species, Red Data Book of fishes in Germany; reintroduction of extinct species; field research on the influence of stream channelization upon fishes.

13. Fischereiamt Berlin

Name : Fischereiamt Berlin
Address : Havelchaussee, 149/151
D - 14055 Berlin
Head : Dr. U.A. Grosch
Tel. : 49.30.3006990 **Fax**: 49.30.3041805

Type of institute
Applied fisheries research
Consulting in fisheries
Fishery authority

Research activities
Analysis of environmental residues in aquatic organisms
Analysis of water
Distribution and status of native fish
Lake fishery management
Stock assessment
Environmental problems and their impact on resources

Influencing water quality by fishery
Fishing techniques
Gear technology

Facilities
4 patrol and research launches (10.5 to 15 m.; 100 to 490 kW) all equipped for sterntrawling and with several echosounding systems
Sonar and echo-integration system
Bottom and midwatertrawls, seine nets, electric fishing equipment
Specialised seagoing computer hard and software
Shipborne permanent water quality detection system
Rearing and experimental tanks (flow through system, 35m| holds)
Fish test toxicology plant
Analytic laboratory (e.g. titration, flow injection analysis, gas chromatography)
Databases
25 tons slipway

Staff: Scientists: 3 Others: 12

Description
Research projects:
Environmental control of residues in aquatic organisms, waters and soils, control of biotic and abiotic factors for fish, influences of fishing on water quality, fishery management in multi-used lowland lakelike stream-systems.

14. Fischereiforschungsstelle des Landes Baden-Württenberg

Name : Fischereiforschungsstelle des Landes Baden-Württemberg
Address : Mühlesch 13
D - 88085 Langenargen
Head : Dr. R. Berg
Tel. : 07543.93080 **Fax**: 07543.930820

Type of institute
Research centre

Research activities
Fisheries management
Stock assessment
Environmental problems and their impact on resources
Fishing techniques
Aquaculture

Facilities
Research vessel
Rearing tank
Data bases, extensive libraries
Specialised software

Staff: Scientists: 5 Others: 3

Description
- Ultrasonic telemetry of fish migration in Lake Constance and small lakes.
- Radiotelemetry of fish in rivers.
- Fish stock monitoring in Lake Constance.
- Changes in the fish populations of rivers due to man made changes of river structure and trophic status.
- Crayfish stocks in Baden-Württemberg: actual distribution and reasons for the decline.
- Investigations on fish-zooplankton interactions in shallow eutrophicated lakes.
- Rearing of fish:
  a) optimization of dry diet rearing of coregonid (Coregonus lavaretus L.) larvae including physiological and genetical aspects of rearing.
  b) development of rearing methods of endangered species.

15. Forschungsinstitut Senckenberg

Name : Forschungsinstitut Senckenberg
Section Ichtyologie II und Fischölogie & Section of Aquatic Ecology
Address : Senckenberganlage 25
D - 6000 Frankfurt M
Head : Prof. Dr. A. Lelek/Prof. Dr. W. Tobias
Tel. : 069.7542368/7542244 **Fax**: 69.746238

Type of institute
Research centre in close cooperation with
University Departments in Frankfurt/M and Göttingen.

Research activities
Fisheries management
Stock assessment
Multi-species biological models
Multi-disciplinary approach to fisheries management
Environmental problems and their impact on living resources and water quality
Inventory and protection of species communities

Fishing techniques
Productivity of aquatic systems
Predator-prey relationship in running waters

Facilities
Data bases, extensive libraries
Field site, electron microscope, specialized analytical apparatus
Subarctic field station in the region of NE-Norway.
Research boat equipped for electrofishing and routine limnological work.
Data collection of macroinvertebrate and fish from large rivers Main and Rhine, partly Danube and several sensible streams.
Libraries for limnologie, ichthyologie and fisheries.

Staff: Scientists: 7 Others: 7

Research budget (ECU/year): 200.000

Description
Research projects:
Prediction and maintenance of water quality including prognoses for the sustainable fish yield (in the Tropics).
Analysis and prognoses of fish fauna on the Rhine and Main rivers as well as on their adjacent waters involving cooperation with France, Netherlands and Switzerland. Rehabilitation of waterbodies and reintroduction of vanished species-salmon, sturgeon, etc.

16. Forschungszentrum Jülich - Arbeitsgruppe Modellierung für Umweltforschung und Lebenswissenschaften (ATÖ)

Name : Forschungszentrum Jülich - Arbeitsgruppe Modellierung für Umweltforschung und Lebenswissenschaften (ATÖ)
Address : Postfach 1913
D - 52425 Jülich
Head : Dr. F. Drepper
Tel. : 49.2461.612305 **Fax**: 49.2461612470

Type of institute
Research centre

Research activities
Fisheries management
Stock assessment
Multi-species biological models

Multi-disciplinary approach to fisheries management
Aquaculture
Aquaculture economics

Facilities
Data bases, extensive libraries
Specialised software

Staff: Scientists: 3 Others: 1

Research budget (ECU/year): 4.100

17. Institut für Binnenfischerei e.v.

Name : Institut für Binnenfischerei e.v.
Address : Jägerhof
D - 14476 Potsdam-Sacrow, P.O. Großglienicke
Head : Dr. R. Knösche

Type of institute
Research centre

Research activities
De-eutrophication of lakes
Fisheries management
Fishery yield assessment
Fishing techniques
Fishing methods and gears
Aquaculture
Biotechnique, engineering, effluents treatment

Facilities
Experimental closed recirculation fish farm
Experimental lake of 100 ha

Staff: Scientist: 4 Others: 11

Description
Research projects:
Intensive aquaculture with eel, sturgeon and catfish especially in closed recirculating systems. Aquacultural effluents and their treatment. De-eutrophication of stratified lakes by hypolimic aeration. Fishery yield assessment in North German lakes.

18. Institut für Fischqualitat GmbH, Bremerhaven

Name : Institut für Fischqualität GmbH
Address : Freiladestr.1
D - 2850 Bremerhaven
Head : Erich Hollmann (Geschäftsführer)
Prof. Harz (Wissenschaftlicher Leiter)

Research activities
Upgrading of fishery products
Improvement of quality standards of fishery products
Production research

Facilities
Special equipment for food law investigations, quality assessment and product development.
Capacity assessment

Staff: Scientists: 4 Others: 8

Research budget (ECU/year): 100.000

Description
Consulting of fishery economy with respect to high quality products.
Development and introduction of a quality seal.
Carrying out of chemical, micro-biological and biological examinations as well as researches.
Development of technological procedures.

19. Institut für Gewässerökologie und Binnenfischerei Abteilung Biologie und Ökologie der Fische

Name : Institut für Gewässerökologie und Binnenfischerei
Abteilung Biologie und Ökologie der Fische
Address : Müggelseedamm 310
D - 12587 Berlin-Fridrichshagen
Head : Dr. F. Kirschbaum

Type of institute
Research centre
Higher Education Institute

Research activities
Population dynamics and ecology of fish communities
Genetics of population
Environmental control of reproductive cycles
Reproductive biology and breeding of endangered species
Life cycle of fishes and early life history stages
Embryotoxicology
Hydroacoustic methods in fisheries research

Facilities
Experimental tanks and ponds, fishing gears and boats for lakes and rivers, microscopical equipment, electrofishing equipment, split-beam echosounder, computer-based image analysis system, field station 80 km north of Berlin, electrophoretic equipment.

Staff: Scientists: 8 Others: 8

Description
Research projects:
Environmental control of reproductive cycles of both temperate and tropical species - Permeability of chorion of different fish species. - Influence of abiotic and biotic factors on population dynamics. - Reproduction biology and breeding of various endangered species. - Food and habitat preference of various stages of the life cycle of fishes. - Lake fishery management, - Population genetics of cyprinid fishes. - Fish migration in lowland rivers. - Spatial and temporal distribution patterns in lakes and rivers. - Diel activity cycles and diurnal migration patterns.

20. Institut für Gewässerökologie und Binnenfischerei Abteilung Fischzucht und Fischpathologie

Name : Institut für Gewässerökologie und Binnenfischerei
Abteilung Fischzucht und Fischpathologie
Address : Müggelseedamm 310
D - 12587 Berlin-Friedrichshagen
Head : Prof. Dr. W. Steffens

Type of institute
Research centre
Higher education institute

Research activities
Fish culture
Fish reproduction and rearing
fish breeding and genetics

Nutrition and feeding of fish
Aquaculture and environment
Fish pathology
Fish parasitology in lakes, rivers and aquaculture
Ecobacteriology of freshwater fish
Etiology of fish diseases
Fish toxicology
Embryotoxicology of fish
Bioaccumulation of xenobiotics in fish
Hazard assessment of xenobiotics
Reproductive toxicology in fish

Facilities
Experimental basis with recycling systems
Experimental ponds
Laboratory equipment

Staff: Scientists: 8 Others: 8

Description
Basic research on reproduction, rearing and nutrition of freshwater fish species (e.g. sturgeon). - Culture of endangered fish species for restocking of natural waters. - Influence of aquaculture on water metabolism. - Basic genetic research in fresh water fish species. - Selective breeding of salmonids. - Parasitological investigations in freshwater fish. - Environmental impact on certain tissues and reproduction of teleost fish.

21. Institut für Polarkölogie Landesforschungsanstalt für Landwirtschaft und Fischerei Mecklenburg/Vorpommern

Name : Institut für Polarkölogie Landesforschungsanstalt für Landwirtschaft und Fischerei Mecklenburg/Vorpommern
Address : Südstrasse, 10a
D - O-2382 Born
and : An der Jögerbök, 2
D - O-2510 Rostock
Head : H.-J. Jennerich/Dr. E. Anders
Tel. : 038234.297 **Fax**: 0381.810202/0381.82055

Type of institute
Research centre
Fishery authority
Consulting in fisheries and in fish fauna

Research activities
Fisheries management in fresh- and brackish waters
Fishery yield assessment
Environmental problems and their impact on resources
Aquaculture in fresh- and brackish waters
Fish reproduction, fish culture
Fish genetics
Fish pathology

Facilities
Fresh- and brackish water hatcheries for warm- and cold water fishes
Controlled environmental chamber
Experimental basis with recirculating systems and rearing tanks
Bacteriological laboratory
Research vessel

Staff: Scientists: 5 Others: 8

Description
Research projects:
Investigations on fish populations, environment and fishery management in rivers, lakes and bays. Selection and breeding of trouts. Rearing fingerlings for stocking of different species.

22. Landesansalt für Fischerei Nordrhein-Westfalen

Name : Institut für Fischerei Nordrhein-Westfalen
Address : Heinsberger Str. 53
D - W-5942 Kirchhundem-Albaum
Head : Dr. Harald Ungemach

Type of institute
Research and consulting in fisheries and in fish fauna
Training of fish farmers and sport fishermen
Fish health service

Research activities
Chemical analysis of environmental residues in aquatic organisms, analysis of water
Investigation of fish losses in natural waters
Ecological studies of fish populations
Consulting on construction of fishways
Consulting on fisheries in preserves
Evaluation of reserves for fish
Actual distribution and status of native species of fish

Assessment and evaluation of fish stocks
Protection and re-implantation of endangered native fishes, mussels and crayfish
Trout culture
Fish physiology and immunology
Fish pathology

Facilities
Gas chromatography
Atom-absorption spectrometry
Photometry
Ion chromatography
Fishing gear and boats for rivers and lakes
Electric fishing equipment
Echo sounder
30 experimental ponds (0.7 ha)
Cold-water hatchery
Microbiological laboratory, incl. virology haematological and physiological laboratory with cytofluorimetric analyser
Iso-electric focussing electrophoretic equipment
Electronic cellcounter and analyser
Experimental basins for controlled rearing of infected freshwater fishes

Staff: Scientists: 7 Others: 13

23. Niedersächsisches Ministerium für Ernährung Landwirtschaft und Forsten/Staatlicher Fischseuchenbekämpfungsdienst Niedersachsen und Fischgesundheitsdienst

Name : Niedersächsisches Ministerium für Ernährung Landwirtschaft und Forsten/Staatlicher Fischseuchenbekämpfungsdienst Niedersachsen und Fischgesundheitsdienst
Address : Eintrachtweg 17
D - 3000 Hannover 1
Head : Dr. Dr. H.J. Schlotfeldt

Tel. : 49.0511.281112 **Fax**: 49.0511.1063309

Type of institute
Government Department/regional authority
Diagnostic and research centre
Training of fish farmers foreign guests and sport fishermen

Research activities
Diseases of fish (viral, bacterial, parasitic, environment dependent), diagnosis, prophylaxis, immunology, stress.
Field trials and field applicability of therapeutics and vaccines for licensing purposes
Aquaculture/environment interaction
Environmental problems and their impact on resources

Facilities
Laboratories and equipment for fish necropsy and zhole frame of complementary diagnostic examinations (virology, bacteriology, parasitology, serology, water aulity and residues)
experimental aquaria facilities
specialised computer software development
Gas chromatograhy, atom-absorption spectrometry, photometry, ion chromatography
Data bases, extensive library

Staff: Scientists: 3 Others: 7

Description
Diagnostic support of fish farmers
Development of diagnostics for fish pathogens
Pathogenicity of fishpathogenic viruses
Field trials and efficiency tests of therapeutants and fish vaccines, also for registration purposes
Water quality and disease
Residues - drugs, heavy metals and chlorinated hydrocarbons
Relation between pollution and fish diseases
Pracical training of fish farmers, foreign guests (mainly young fish pathologists) and sport fishermen
National and international ex-/import certification

24. Polarmar

Name : OPC OEKOMAR-Polarmar Consulting GmbH Fisheries Dept.
Address : Columbus Center
Obere Bürger 20
D - 27568 Bremerhaven
Head : Dr. Wolfgang Welsch
Tel. : 49.0.471.47053 **Fax**: 49.0.471.45711

Type of institute
Engineering/consulting firm

Staff: Scientists: 1 Others: 1

25. Sächsische Landesanstalt für Landwirtschaft

Name : Sächsische Landesanstalt für Landwirtschaft
Referat Fisherei
Address : Hauptstr. 12a
D-O - 8613 Königswartha
Head : Dr. G. Füllner
Tel. : 035931.20206 **Fax**: 035931.20209

Type of institute
Research centre
Institute for professional training/Secondary school
Training of fish farmers and sport fishermen
Fishery authority

Research activities
Fisheries management
Pond management and environmental problems
Biotechniques for production of pond fishes
Chemical analysis of environmental residues in fishes and sediments
Renaturation of running waters

Facilities
Experimental pond farm (experimental) for carp and many other species
Data bases, extensive library
Special software
Special analysis equipment

Staff: Scientists: 3 Others: 17

Description
Research projects:
Production of economically important species in complement to carps in ponds. (tench, sheat, whitefish, eel-pout, sturgeon).
Management of high eutrophic ponds with rigorous considerations of environment protection.
Protection and rearing of endangered species.
Stock taking of species and elaboration of a "Fish map of Saxony"
Chemical analysis of heavy metals and organic residues in fish and pond/river sediments.
Favourable: 30 experimental ponds with a total surface of 8,00 ha.

26. STAGAtech GbR

Name : STAGAtech Gbr
Address : Richard-Wagner-Str. 31, (Technologiezentrum), 18119 Warnemünde
Head : Dr. O. Gabriel

Type of institute
Technology Center

Research activities
Fishing methods and gears, especially mechanisation of longlining, environmental problems;

Facilities
Universal longlining system for all kinds of investigations (influence of baits, line parameters, hooks a.o.): special winch for maritime measurement systems;

Staff: Scientists: 2

27. Technische Universität Berlin

Name : Technische Universität Berlin
Faculty of Food-Science and Biotechnology
Chair for meat and fish technology
Address : Philippstraße, 13
D - O-1040 Berlin
Head : Prof. Dr. Heinz Sielaff
Tel. : 030.2895245/2895238 **Fax**: 030.2895278

Type of institute
2 departments (meat-technology and fish-technology)

Research activities
Function properties of fish-proteins
Basic research of fish technology and preservation/conservation
Impact of processing on quality

Facilities
Analytic laboratories
Technological laboratory
Material testing laboratory

Staff: Scientists: 5 Others: 5

Description
Investigation of functional properties of sea fishes, baltic fishes and other fishes, basic research for fish-technology and fish-preservation/conservation, impact of raw material and processing on quality and durability of fish products.

28. Tierärztliche Hochschule Hannover

Name : Tierärztliche Hochschule Hannover
Fachgebiet Fischkrankheiten und Fischhaltung
Address : Bischofsholer Damm 15
D - 3000 Hannover 1
Head : Prof. W. Körting
Tel. : 0511.856.8889

Type of institute
Research centre
University Department/Higher Education Institute

Research activities
Fisheries management
Environmental problems and their impact on resources, pond management
Aquaculture
Fish diseases, immunology, stress, aquaculture in ponds and intensive systems

Facilities
Specialised computer hardware
Specialised software
Rearing tanks

Staff: Scientists: 7 Others: 3

Description
Research projects: Myxosporea in fresh water fishes.
Pancreas disease in atlantic salmon.
Pyruvatekinase in the metabolism of the carp tapeworm.
Pathogenicity and developmental cycles of protozoan pathogens of commercial fishes.
Glutathione-S-transferases in fish parisitic helminths.
Coccidia in the common carp.
Investigation on VHSV persistence and excretion by experimentally infected rainbow trout.

29. Univ. Berlin - FUB-FB Biologie WE5

Name : FUB-Freie Universität Berlin
FB Biologie WE5
Address : Tietzenweg 85-87
D - 1000 Berlin 45
Head : Dr. P. Doering
Tel. : 030.8334029

Type of institute
University Department/Higher Education Institute

Research activities
Fisheries management
Stock assessment
Environmental problems and their impact on resources

Staff: Scientists: 2

Description
Research projects:
Fish stocks of lake Tegel-stock assessment to document the hypertrophic conditions as a basis for a later assessment of the impact of a phosphate stripping plant (from 1985 to 1987).
Ongoing research about Fourier shape analysis of otoliths as an objective ageing method.

30. Univ. Bochum - Inst. für spezielle Zoologie und Parasitologie

Name : Ruhr Universität Bochum - Fakultät für Biologie
Institut für spezielle Zoologie und Parasitologie
Address : Universitätstrasse 150
D - 44801 Bochum 1
Head : Dr. Dr. H. Mehlhorn

Tel. : 0234.7004563 **Fax**: 234.7094114

Type of institute
Research centre
University Department/Higher Education Institute

Research activities

Aquaculture
  Growth, nutrition, feed composition, live food production for aquaculture
  Disease, immunology, stress
Upgrading of fishery products
  Consumer response and demand studies
Fisheries management
  Environmental problems and their impact on resources

Facilities

Rearing tanks
Specialised computer hardware
Field site, electron microscope, specialized analytical apparatus
Relations to Pharma Companies

Staff: Scientists: 14 Others: 8

Description

Research projects: Lifecycles of fish parasites.
Development and test of new drugs against parasites and fungi.

31. Univ. Bonn - Inst. für Lebensmittelchemie

Name : Rheinische Friedrich Wilhelms Universität Bonn
Professur für Lebensmittelchemie
Address : Endenicher Allee 11-13
D - 53115 Bonn 1
Head : Prof H. Büning-Pfaue
Tel. : 0228.732361 **Fax.** : 0228.733757

Type of institute

University Department/Higher Education Institute

Research activities

Aquaculture
  Disease, immunology, stress
  Upgrading of fishery products
  Consumer response and demand studies
  Improvement of techniques for handling, storing, processing fish
Fisheries management
  Environmental problems and their impact on resources

Facilities
Research vessel
Specialised software
HPLC,GC,GC/MS u.a.
Specialised analytical apparatus

Staff: Scientists: 2 Others: 6

Description
Studies of different fish medication
Development of methods to determine drug residues in fish (skin, meat, liver kidney).
Assessment of withdrawal times; statistical evaluation of the residue behaviour (decay curves; models) of drugs and their metabolites (decay curves with confidence and tolerance belts). Determination of "bound residues".

32. Univ. Bonn - Inst. für Anatomie, Physiologie und Hygiene der Haustiere

Name : Rheinische Friedrich Wilhelms Universität Bonn
Institut für Anatomie, Physiologie und Hygiene der Haustiere
Address : Katzenburgweg 7-9
D - 53115 Bonn 1
Head : Edmund Greuel/Otmar Schmitz
Tel. : 0228.735112

Type of institute
Research centre
University Department/Higher Education Institute

Research activities
Aquaculture
- Reproduction
- Growth, nutrition, feed composition, live food production for aquaculture
- Aquaculture/environment interaction
- Aquaculture structures, systems, techniques
- Upgrading of fishery products
- Methods for grading and quality assessment of fish

Facilities
Research vessel
Rearing tank
Research vessel for pilot plant recirculating systems using up-flow-filters on rotating biological contactors.

Staff: Scientists: 2

Research budget (ECU/year): 44.200

Description
Research projects: Breeding of rainbow trout in recirculating fish system.
Studies on denutrification in totally closed recirculating systems.
Methods of quality assessment of fish. Studies on durability of smoked vacuum packed fish.

33. Univ. Göttingen - Inst. für Tierphysiologie und Tierernährung

Name	:	Institut für Tierphysiologie und Tierernährung der Universität Göttingen
Address	:	Oskar-Kellner-Weg 6 D - 3400 Göttingen-Weende
Head	:	Günther/E. Mathies/H. Meyer-Burgdorff
Tel.	:	0551.393330 **Fax**: 551.399612 **Telex**: 96703 unigoe

Type of institute
University Department/Higher Education Institute

Research activities
Aquaculture
Aquaculture/environment interaction
Growth, nutrition, feed composition, live food production for aquaculture
Physiology (protein and energy metabolism)

Staff: Scientists: 5 Others: 3

Research budget (ECU/year): 3.000

Description
The institute is working with three species: Carp (C.carpio), rainbow trout (O.mykiss) and tilapia (O.niloticus).
Main research activities are listed as follows:
- Protein and energy metabolism in tilapio (O. niloticus).
- Protein and energy metabolism in carp, utilization of crystalline amino acids, protein turnover.
- Digestibility of certain agricultural by-products, incorporation in diets for carp and tilapia.
- Growth of intensively cultured Piaractus brachypomus.

- Development of feeds for rainbow trout in order to reduce water pollution by minimized nitrogen - and phosphorus excretion.
- Fishmeal replacement by hydrolysed feathermeal or lupine seeds in carp grower diets.
- Magnesium supply to carp and tilapia.
- Biotin requirement in carp and rainbow trout.
- Immune response of rainbow trout fed different diets.

34. Univ. Göttingen - Inst. für Tierzucht und Haustiergenetik

Name : Universität Göttingen
Institut für Tierzucht und Haustiergenetik
Address : Albrecht Thaer Weg 3
D - 3400 Göttingen
Head : Prof. H. J. Langholz/PD Dr. G. Hörstgen-Schwark/Prof. W. Holtz
Tel. : 0551.395604 **Fax**: 0551.395587 **Telex**: Unigoe 96703

Type of institute
University Department/Higher Education Institute

Research activities
Aquaculture
Genetics
Physiology
Reproduction
Production technolgy
Toxicology

Facilities
Warm-water recirculation systems
Laboratory aquaria units
Research station for rainbow trout

Staff: Scientists: 5 Others: 3

Description
At the Institute following aquaculture candidates are included in research activities: Rainbow trout (ONCORHYNCHUS MYKISS), African catfish (CLARIAS GARIEPINUS), Tilapia (OREOCHOMIS NILOTICUS), Zebrafish (BRACHYDANIO SERIO). Main research activities: Selection on late maturity and high growth rate in tilapia; - polyploidy and gynogenesis in tilapia, rainbow trout and Zebra fish; - selection on higher growth performance in rainbow trout; - evaluation of product quality in African

catfish, tilapia and rainbow trout; - controlled reproduction of new aquaculture species (Heterobranchus ssp.); - conservation of gametes; - optimization of production techniques; - evaluation of testfish populations (Zebra fish) for ecotoxicological studies.

35. Univ. Hamburg - Inst. für Hydrobiologie und Fischereiwissenschaft

Name	:	Universität Hamburg Inst. für Hydrobiologie und Fischereiwissenschaft
Address	:	Olbersweg 24 D - 22767 Hamburg 50
Head	:	W. Nellen/H. Kausch
Tel.	:	040.41236600 **Fax**: 040.41236618

Type of institute
University Institute as part of the "Zentrum für Meeres- und Klimaforschung" (Centre of Marine and Climate Science; ZMK) of the University of Hamburg.

Research activities
Fisheries management
Fish recruitment processes / plankton research
Multi-species biological investigations and models
Environmental problems and their impact on resources
Aquaculture
Basic ecophysiology
Energy budgets of fish species

Facilities
Research vessel and boats
Research aquaria for fresh and saline water conditions
Controlled environmental chambers
Specialised computer hardware
Specialised software
Electron microscope, specialized analytical apparatus

Staff: Scientists: 12 Others: 11

Research budget (ECU/year): 143.000

36. Univ. Hamburg - Inst. für Meereskunde

Name	:	Universität Hamburg Inst. für Meereskunde
Address	:	Troplowitzstrasse 7 D - 22529 Hamburg 54
Head	:	Prof. Dr. J. Sündermann
Tel.	:	040.41.23.26.05 **Fax**: 040.41234044 **Telex**: 212586 d

Type of institute
University Department/Higher Education Institute

Research activities
Fisheries management
Environmental problems and their impact on resources

Facilities
Research vessel

Staff: Scientists: 2 Others: 1

Research budget (ECU/year): 98.300

37. Univ. Hamburg - Zoologisches Institut und Zoologisches Museum

Name	:	Universität Hamburg Zoologisches Institut und Zoologisches Museum
Address	:	Martin Luther-King-Platz 3 D - 20146 Hamburg
Head	:	Prof. Dr. L. Renwrantz/Prof. Dr. Chr. Schemmel
Tel.	:	4123.3858 **Fax**: 49.40.4123.3937

Type of institute
Higher Education Institute/University Department
Research centre

Research activities
Aquaculture
Species discrimination by means of electrophoretic techniques (e.g. immunbiological studies)
Environmental impact by uncontrolled release or escape of aquacultured species, completion of a tilapine database and reference collection of tilapias

Facilities
Cold- and warmwater aquaria
Specialised electrophoretic equipment
Specialised computer hard- and software
Ichthyological reference collection

Staff (over all): Scientists: 34 Others: 70

Research budget (for aquaculture only) (ECU/year): 120.000.00 (GTZ funded; 1990-1994)

38. Univ. Heidelberg - Zoologisches Institut

Name	:	Universität Heidelberg Zoologisches Inst. 1
Address	:	Im Neuenheimer Feld 230 D - 69120 Heidelberg 1
Head	:	Prof. Dr. Volker Storch / Dr. Günter Vogt
Tel.	:	06221.562656

Type of institute
Research centre
University Department/Higher Education Institute

Research activities
Aquaculture
- Physiology
- Reproduction
- Early stages of reared species
- Aquaculture/environment interaction
- Disease, immunology, stress
- Nutrition

Facilities
Field site, electron microscope, specialized analytical apparatus

Staff: Scientists: 4 Others: 5

Research budget (ECU/year): 35,000

Description
The main research projects deal with:
Aquaculture of Penaeus monodon.
Pathological effects of feed ingredients in Penaeus monodon.
Transformation of digestive tract cells by monodon baculovirus.

Influence of contaminants in the digestive tract of Penaeus monodon.
Larval growth of turbot.
Toxicology in freshwater fish.
Vitamine c influence on fish.
Influence of acid water on fish.
Digestive enzyme synthesis in Astacus astacus.

39. Univ. Karlsruhe - Inst. für Zoologie II

Name	:	Universität Karlsruhe Inst. Zoologie II
Address	:	Kaiserstr. 12 D - 76131 Karlsruhe
Head	:	Dr. W. Hanke
Tel.	:	0721.6083990 **Fax**: 0721.6083992

Type of institute
University Department/Higher Education Institute

Research activities
Aquaculture
Physiology
Early stages of reared species
Endocrinology

Facilities
Specialized analytical apparatus, microscopy equipment for cell culture

Staff: Scientists: 3 Others: 2

Research budget (ECU/year): 24.600

Description
Main research projects concern:
Endocrine control of stress response of fish.
Endocrine control of gluconeogenesis of fish.
Lipid nutrition in adult and larval fish.

40. Univ. Kiel - Inst. für Meereskunde - Abt. Fischereibiologie

Name : Universität Kiel - Institut für Meereskunde
Abteilung Fischereibiologie
Address : Düsternbrooker Weg 20
D - 2300 Kiel 1
Head : Prof. Dr. D. Schnack
Tel. : 0431.5973910 **Fax**: 431.565876 **Telex**: 17431793

Type of institute
National Research Centre
University Department/Higher Education Institute

Research activities
Fisheries management
Stock assessment
Multi-species biological models
Environmental problems and their impact on resources
Ecology of early life history stages, recruitment problems
Aquaculture
Physiology
Reproduction
Early stages of reared species
Growth, nutrition, feed composition, live food production for aquaculture
Aquaculture/environment interaction
Aquaculture structures, systems, techniques

Facilities
Research vessel
Rearing tank
Controlled environmental chamber
Specialised computer hardware
Specialised software
Image analysis

Staff: Scientists: 22 Others: 10

Description
Research projects: - Biological characterisation of fish and invertebrate stocks, including squids and bivalves in different parts of the world ocean. Community studies in different habitats, assessment and modeling of trophic interactions.
- Ichthyoplancton survey programmes for stock assessment and recruitment studies. Seasonal production and ecology of egg and larval stages; development

and application of acoustical, optical and sampling methods for small scale studies on ichthyoplancton distribution and ecology.
- Fish disease and parasites in coastal and estuarine areas.
- Behaviour, metabolic demands and growth of species in relation to aquaculture system design; effects of group size and reduced oxygen levels on behaviour and metabolism.
- Impacts on the environment by fish farming in net cages; optimization of feeding systems.

41. Univ. Kiel - Institut für Polarökologie

Name : Universität Kiel - Institut für Polarökologie
Address : Wischhofstr., 1-3, Gebäude 12
D - 24148 Kiel
Head : Prof. Hempel/prof. Kappen/prof. Spindler
Tel. : 0431.720.87.0 **Fax**: 0431.720.87.20

Type of institute
Research institute at the university

Research activities
Basic ecological research in polar regions on plankton, benthos, fish, sea ice biota (marine group) and lichens (terrestrial group).
Marine group: taxonomic studies, community analyses, image analyses, population dynamics, respiration measurements, lipid biochemistry, blood physiology, enzyme activity measurements.
Terrestrial group: vegetation studies, microclimate measurements, CO_2 gas exchange measurements, chlorophyll fluorescence measurements, microcalorimetry, fluorescence microscopy.

Facilities
Analytical laboratories
Aquarium coldrooms for the cultivation of polar organisms

Staff: Scientists: 20 Others: 4

Description
The terrestrial research concentrates on ecophysiological investigations of polar cryoptogams and on microbiological studies of Antarctic soils, while the marine working group is involved in interdisciplinary research on structures and processes of polar ecosystems, which includes cryopelagic, pelagic and benthic organisms. Major geographical working areas are the Weddell Sea in the Antarctic and the Fram Strait in the Arctic for the marine studies as well as the maritime Antarctic, Greenland and Svalbard (Spitsbergen) for the terrestrial studies.

A large fraction of the scientific investigations is embedded in national and international, usually interdisciplinary research programs. Besides research, the IPO scientists are involved in teaching at the university. They are also active members of national and international planning committees dealing with Arctic and Antarctic research.

42. Univ. Konstanz - Fak. Biologie - Inst. Limnologie

Name : Fakultät für Biologie / Limnologisches Institut der Universität Konstanz-Fischereiabteilung

Address : Postfach 5560
D - 7750 Konstanz 1

Head : Dr. A. Appenzeller

Tel. : 07531.883531 **Fax**: 07531.883533

Type of institute
University Department/Higher Education Institute

Research activities
Distribution and abundance of fish using hydroacoustic methods
Spatial statistics
Application of acoustic methods to fisheries management
Fisheries techniques
Behavioural studies of fish
Multi-disciplinary analysis of aquatic ecosystems

Facilities
EK 500 Simrad hydroacoustic equipment
Research vessels
Fishing gear
Aquaculture facilities
Specialized analytical lab facilities

Staff: Scientists: 1 Others: 2

Description
Research projects:
Seasonal distribution of pelagic and littoral fish species in lake of Konstanz using acoustic methods;
Covariation of fish abundance with abiotic and biotic factors;
Population dynamics of Coregonid fish;
Commercial fishing patterns in lake of Konstanz;
Patterns of fish distribution in the littoral zone;
Predator-Prey relationships of littoral species.

43. Univ. München - Inst. für Zoologie und Hydrobiologie

Name : Institut für Zoologie und Hydrobiologie
Address : Kaulbachstrasse 37,
D - 80539 München FRG
Head : Prof. Dr. W. Hoffmann
Tel. : 089.21802282 **Fax**: 089.2805175

Type of institute
Research centre
University Department/Higher Education Institute

Research activities
Fisheries management
Environmental pollution
Biology of endangered species
Aquaculture
Early stages of reared species
Disease, pathology, prophylaxis, treatment
Aquaculture/environment interaction

Facilities
Data bases, extensive libraries
Field site, electron microscope, specialized analytical apparatus
Parasitological, bacteriological and virological centre

Staff: Scientists: 7 Others: 9

Description
Research projects:
Diseases caused by bacteria and virus, parasite-host interaction, development of myxosporeans, histology of different species.

44. Univ. München - Techn. - Institut für Ernährungsphysiologie

Name : Technische Universität München
Institut für Ernährungsphysiologie
Address : D - 85354 Freising-Weihenstephan
Head : M. Kirchgessner
Tel. : 8161.713696

Type of institute
University Department/Higher Education Institute

Research activities
Aquaculture
  Growth, nutrition, feed composition, live food production for aquaculture
  Quality of the product fish

Facilities
Research vessel
Controlled environmental chamber

Staff: Scientists: 2 Others: 2

Description
Influence of fat and Vitamin E or of carbohydrates on growth, and carcass quality of fish, especially of carp (Cyprinus carpio L.) - optimizing feed rations - main future work: digestibility studies, measurement of energy content of feedstuffs, energy metabolism of fishes - requirement of amino acids of carp - minimizing of N excretion.

45. Univ. München - Techn. - Lehrstuhl für Tierhygiene - Fachgebiet Fischbiologie

Name	:	Technische Universität München Lehrstuhl für Tierhygiene, Fachgebiet Fischbiologie
Address	:	D-8050 Freising-Weihenstephan
Head	:	Prof.H. Stein
Tel.	:	08161.713767

Type of institute
University Department/Higher Education Institute

Research activities
Fisheries management
  Environmental problems and their impact on resources
Aquaculture
  Genetics and conservation of gametes

Facilities
Rearing tank
Field site, electron microscope, specialized analytical apparatus
Hatchery including: Breeding facilities, rearing tanks, fishery equipment including electric fishing equipment, boats
Laboratory, about 10 km of small rivers

Electron microscopes are present in the central laboratory of electron microscopy of the university, and equipment are available for research.

Staff: Scientists: 2 Others: 1

Description

Research projects:

Research on the effect of hydroelectric power and pollution on fish populations, including endangered species. Genetics: production of gynogenetic salmonids, polyploidization and production of triploid hydrids. Conservation ofgametes: cryopreservation of semen (including endangered species), long term preservation of eggs and embryos.

46. Univ. Oldenburg - Inst. Chemie und Biol. des Meeres (ICBM)

Name : University of Oldenburg ICBM,
Chemical Ecology Research Group

Address : Carl v. Ossietzkystr. 9
D - 2900 Oldenburg

Head : Prof. Dr. E. Zeeck

Tel. : 49.441.798.3781 **Fax**: 49.441.798.3384

Type of institute

Research centre

University Department/Higher Education Institute

Research activities

Research of the impact of fish farming (e.g. sewage sludge reduction)

Environmental problems and their impact on resources (e.g. aquatic (marine and freshwater)biotoxins)

Fish and crustacean farming

Demand studies and consulting

Staff: Scientists: 6 Others: 1

47. Univ. Rostock - Inst. für Ostseeforschung

Name : Institut für Ostseeforschung an der Universität Rostock
Sektion Biologische Meereskunde

Address : Seestraße. 15
D - 2530 Warnemünde

Type of institute
Research Institute

Research activities
Ecosystem analysis
Pollution research
Monitoring

Description
The institute started in 1992

48. Univ. Stuttgart - Inst. für Siedlungswasserbau

Name	:	Universität Stuttgart Institut für Siedlungswasserbau, Abt. Biologie
Address	:	Bandtäle 1 D - 70569 Stuttgart
Head	:	G. Haider
Tel.	:	0711.6853733 **Fax**: 0711.6853729

Type of institute
University Department/Higher Education Institute

Research activities
Fisheries management
  Stock assessment
  Environmental problems and their impact on resources
Aquaculture
  Upgrading of fishery products
  Methods for grading and quality assessment of fish

Facilities
Rearing tank
Field site, electron microscope, specialized analytical apparatus

Staff: Scientists: 2 Others: 1

Research budget (ECU/year): 60.000

Description
Propagation and stock improvement in rainbow trout, mainly problems of hybridisation and gynogenesis;
Intensive culture, mainly problems of recirculation systems;

Environmental problems and their impact on fishfauna in the rivers Neckar and Rems.
Problems in hatching and rearing of sterlet.

49. Univ. Tübingen - Zoologisches Institut

Name	:	Universität Tübingen - Fakultät für Biologie Lehrstuhl Tierphysiologie - Zoologisches Institut
Address	:	Auf der Morgenstelle 28 D - 7400 Tübingen 1
Head	:	Prof. Dr. W. Pfeiffer
Tel.	:	07071.292619

Type of institute
University Department/Higher Education Institute

Research activities
1) The fright reaction in ostariophysean fish, the effects of the alarm substance on fish, schooling in fishes.
2) Sound production in catfish. Fish physiology and behaviour.

Facilities
Aquaria, tanks, special recording equipment, specialised software, the libraries of the department and of the University of Tübingen.

Staff: Scientists: 5 Others: 1

Research budget (ECU/year): 21.000

Description
Main project:
The helminths which parasitize the fish of Lake Constance: Their seasonal and location dependent propagation and their effect on fish industry.
For 24 months fishes from various places on Lake Constance are regularly investigated for parasites. On the basis of these data it will be possible to report on how the decreasing eutrophication in the lake influences the health of the fish and the quality and the quantity of the fish harvest and to suggest possible preventative measures against parasitic diseases. The focus of the project is on taking fishes from various places on the lake and subjecting them immediately to ectoparasites. A more detailed investigation for endoparasites is conducted at the University of Tübingen, Institut für Biologie III.
The results are stored and analysed by computer. The goal of this procedure is to produce a map indicating parasitic infestation according to location. In addition frequency and distribution of parasites are studied in relation to the season.

50. Versuchsanstalt für Wasserbau und Schiffbau

Name	:	Versuchsanstalt für Wasserbau und Schiffbau - Berlin Model Basin -	
Address	:	Müller-Breslau-Str. (Schleuseninsel) D - 1000 Berlin 12	
Head	:	Prof. Dr. -Ing. H. Schwanecke	
Tel.	:	030.31841 **Fax**: 030.3123074	**Telex**: 184010 VWS D

Type of institute
Research centre
Governmental institute
Model basin/Water quality laboratory
Consulting firm

Research activities
1. Dept. of Naval Architecture:
   - propulsion and seakeeping tests for fishery and oceanographic research vessels
   - manoeverability tests
   - seakeeping-under-operation-tests
   - operation and optimization of fishing and research gears
   - hydroacoustic measurements of towing noises
2. Dept. of Environment and Ocean Engineering:
   - development of oceanographic and environmental monitoring instrumentation
   - modelling of particle motion
   - mariculture/environment interaction
   - environmental problems and their impact on resources
   - soil mechanisms of recent coastal and deepsea depositions

Facilities
Special test basins and flumes (longest LxBxH = 250x8x5,5 m3)
Wave tanks
Chemical and soil mechanical laboratories
Workshops
Specialized computer hardware
Specialized sofware

Staff: Scientists: 22 Others: 65

Research budget (ECU/year): 4.000.000

Description
Research projects:
1. Dept. of Naval Architecture
   - fastgoing, surface effect and multi hull ships

- impact of propeller noises on environment
- vibration and noise of towed oceanographic equipment

2. Dept. of Environment and Ocean Engineering
   - particle motion and transport in coastal water
   - development of maritime shear stress meter for the investigation of bed load transport
   - development of soft coastal and deepsea mining equipment
   - environmental impact from placer and deepsea mining
   - development of coastal and deepsea soil mechanical instrumentation

State of Research

GREECE

Development and management of the Greek fishery industry is of primary importance to the National Development Plan.

Fishery research in Greece depends on the Ministry of Industry, Energy and Technology and is done through several institutions. The ministry funds two research institutions: the National Centre for Marine Research (NCMR) of Athens, and the Marine Biology Institute of Crete (IBMC). The Fishery Laboratory of Pirea- under the Ministry of Agriculture - also takes part in fishery research, as well as specialised institutes of different Greek universities.

The results of the research programmes are sent to NCMR and IBMC, however, it is the Ministry of Agriculture which takes all the decisions concerning fishery management (e.g. TAC, improvement of the economic situation...).

A semi-independent public society (ETANAL) is responsible for the management of all the fishery quays of the Greek coast. ETANAL is partially controlled by the Ministry of Agriculture and the Agricultural Bank of Greece.

All import, export and transformation activities are controlled by the Ministry of Commerce.

Apart from funds provided by the Ministry of Agriculture, financial support comes from the EC and occasionally from private sources.

Lack of coordination of the many bodies involved in research and development and the lack of integration of support service decrease effectiveness of research.

The Ministry of Industry, Energy and Technology finances all the activities of the two main marine research institutions (NCMR, IBMC). On the other hand, the Ministry of Agriculture finances independently research projects which are implemented by various research teams within these national research institutions. Furthermore, the ministries sometimes finance other research projects carried out in universities or in private sectors. But the two government bodies operate without direct coordination.

One of the main problems arising from the organisational structure is the lack of a system for collecting reliable statistics. Various authorities are involved in collecting fisheries data in Greece, and each have somewhat different priorities when collecting the data, and thus give different pictures of the fishery sector.

The University of Athens has a programme leading to a Master degree either in Fishery or in Oceanography. The University of Crete has a Technological Institute of intermediate level, to educate technicians.

However, staffing of relevant institutions with trained personnel is still far from being sufficient, especially as the needs for fishery research is progressively increasing.

A great effort is being made to advance aquaculture in Greece, as much of the demand has to be satisfied through imports. However, the existing laws for aquaculture installation and operation, though sufficient, should be changed in order to fit the new existing methods.

There exists a nationwide Development Program for Aquaculture, but it undergoes changes every year. It plans to establish four research centres at government level by 1992, and organise and coordinate cooperation between the different bodies involved in aquaculture research. The main problem in developing aquaculture is the competition from tourism and other coastal activities.

List of Institutes from GREECE

1. Cephalonian Fisheries SA
2. Development Agency for Amvrakikos
3. Ecloserie Edessa
4. Epiros SA - Agricultural Development Company
5. Epsilon International SA
6. GIST - Graduate Industrial School Thessaloniki
7. Greek Centre of Aquaculture Fisheries and Hydrobioresearch
8. Hatchery Station of Preveza
9. Hellenic Aquacultures SA
10. Ichtyka - Centre Aquaculture Acheloos
11. IMBC - Inst. Marine Biology of Crete and University of Crete - Dept. Marine Genetics
12. IMBC - Inst. Marine Biology of Crete - Dept. Aquaculture
13. IMBC - Inst. Marine Biology of Crete - Dept. Fisheries
14. Ministry of Agriculture - Dept. Hygiene and Food Technology
15. Ministry of Agriculture - Fisheries Dept. of Preveza
16. Ministry of Agriculture - Fisheries Laboratory - Piraeus
17. Muncipal Company of the Ioannina Lake - Fish Farming Station
18. National Centre for Marine Research - Inst. Marine Biol. Resources
19. National Centre for Marine Research - Inst. of Inland Water
20. National Centre for Marine Research - Inst. of Oceanography
21. Organotecnica Group
22. Station Ichtyogenetique Louros
23. Tei Messologi
24. Univ. Athens - Agricultural Fac. Animal Production - Lab. App. Hydrobiology
25. Univ. Athens - Biochemistry Cell and Molecular Biology and Genetics
26. Univ. Athens - Techn. Dept. - Naval Architech and Marine Engineering
27. Univ. Patras - Biological Dept.
28. Univ. Patras - Section of Animal Biology
29. Univ. Patras - Zoological Laboratory
30. Univ. Thessaloniki - Aristotle - Dept. Genetics
31. Univ. Thessaloniki - Aristotle - Dept. of Veterinary - Lab. Ichthyology
32. Univ. Thessaloniki - Aristotle - Dept. Zoology
33. Univ. Thessaloniki - Aristotle - Lab. of Fresh Water Fish and Wildlife

1. Cephalonian Fisheries SA

Name : Cephalonian Fisheries SA
Address : Livadi
GR - 28200 Lixouri-Cephalonia
Head : Mr. A. Romanos
Tel. : 671.92600 **Fax**: 671.92339 **Telex**: 312399 CEFI GR

Type of institute
Fish Farming

Research activities
Aquaculture
Reproduction
Larval rearing
Growth, nutrition, feed composition, live food production for aquaculture
Disease, immunology, stress
Aquaculture structures, systems, techniques

Facilities
Different rearing installations for fish larvae and their food
Experimental room

Staff: Scientists: 11 Others: 74

Description
Research projects:
A combined hatchery and ongrowing farm involved in the production of sea bass and bream. Research projects place emphasis on the nutritional aspects of live feed for the hatchery and feeds for the ongrowing and their effects on fish growth. Studies of the microbiological environment of the hatchery and ongrowing are also being carried out.

2. Development Agency for Amvrakikos

Name : Development Agency for Amvrakikos
Address : Karyotaki 1
GR - 48 100 Preveza
Head : D. Xylogiannis
Tel. : 0689.26104 **Fax:** 0682.22511

Type of institute
Fish/Crustacean/mollusc farming
Engineering/consulting firm
Research centre

Research activities
Fisheries management
  Multi-disciplinary approach to fisheries management
  Environmental problems and their impact on resources
Aquaculture
Upgrading of fishery products
Consumer response and demand studies

Facilities
Data bases, extensive libraries
Specialised computer hardware
Field site, specialized analytical apparatus

Staff: Scientists: 2 Others: 3

Description
The aim is to develop and protect the area around Amvrakiks promoting aquaculture and aquatourism, supporting local development programmes. It provides consultation concerning the modernization of lagoon management and intensive aquaculture as well as coordination of investments for aquaculture in the private sector.

3. Ecloserie Edessa

Name : Ecloserie Edessa
Préfecture de PELA
Address : GR - 58200 Edessa
Head : Dr. A. Paleokostas
Tel. : 381.23595 **Fax:** 0030 46321878

Type of institute
Fish/Crustacean/mollusc farming
Government Department/Regional or local authority

Research activities
Aquaculture
  Reproduction
  Genetics and conservation of gametes
  Early stages of reared species
  Aquaculture structures, systems, techniques

Facilities
Rearing tank

Staff: Scientists: 4 Others: 4

Research budget (ECU/year): 10.900

Description
We test fish food. We try new kinds of fish in the Mediterranean climate. We advise fish farmers. We test new methods of fish development. We conduct seminars for new fish farmers.

4. Epiros SA - Agricultural Development Company

Name	:	Epiros - Agricultural Development Company Ioannina Lake
Address	:	9, Vlachlidi str. GR - 45332 Ioannina
Head	:	S. Koufos
Tel.	:	(0651)36686 **Fax**: (0651)33419 **Telex**: 322/255 EPIR

Type of institute
Engineering/consulting firm
Fish/Crustacean/mollusc farming

Research activities
Fisheries management
Environmental problems and their impact on resources
Aquaculture
Growth, nutrition, feed composition, live food production for aquaculture
Aquaculture structures, systems, techniques

Facilities
Rearing tank
Controlled environmental chamber
Specialised computer hardware
Specialised software
Field site, specialized analytical apparatus

Staff: Scientists: 3 Others: 5

Research budget (ECU/year): 16.400

Description

Research projects concerning:

Artificial feeding of varias species of sea-fish.

Ecological investigation.

Improvement of the structure of floating cages and evaluation of various by-products for fish food.

Facilities:

Free use full-automatic biochemical analyses.

Net cages for sea-fish.

5. Epsilon International SA

Name	:	Epsilon International SA
Address	:	Kifisias 16
		GR - 15125 Marousi
Head	:	Prof. M. Bonazountas
Tel.	:	30.1.6822140 **Fax**: 30.1.6842240 **Telex**: 221926 GR

Type of institute

Research centre

Engineering/consulting firm

Research activities

Fisheries management

- Stock assessment
- Multi-disciplinary approach to fisheries management
- Environmental problems and their impact on resources

Aquaculture

Upgrading of fishery products

Facilities

Controlled environmental chamber

Sea lagoon research facility

Staff: Scientists: 10 Others: 5

Research budget (ECU/year): 300.000

6. GIST - Graduate Industrial School Thessaloniki

Name : Graduate Industrial School of Thessaloniki
GIST-Business Adm. Dept./Research Committee
Address : 45, Tsimiski street
GR - Thessaloniki
Head : Prof. SP. Zevgaridis
Tel. : 003031.279375 **Fax**: 003031.279339 **Telex**: 410497 GIS GR

Type of institute
University Department/Higher Education Institute

Research activities
Fisheries management
Multi-disciplinary approach to fisheries management
Aquaculture
Aquaculture economics
Upgrading of fishery products
Consumer response and demand studies

Staff: Scientists: 3 Others: 4

7. Greek Centre of Aquaculture Fisheries and Hydrobioresearch

Name : Greek Centre of Aquaculture Fisheries and Hydrobioresearch
Address : 15 Oitis
GR - 121 33 Peristeri Athens
Head : Dr Constantine Rigopoulos
Tel. : 5752233 **Fax**: 30.1.5728303

Type of institute
Research centre, Fish, Crustacean, Molluscs
Consulting firm, private group, professional training

Research activities
Fisheries management
Environmental problems and their impact on resources
Aquaculture
Aquaculture/environment interaction
Aquaculture structures, systems, techniques, economics

Facilities
Data base, library, instruments

Staff: Scientists: 2 Others: 1

Description
Studies, methods and research on: fish, mollusc, and crustacean culture, environment, problems, areas, techniques.

8. Hatchery Station of Preveza

Name : Hatchery Station of Preveza
Address : Pogonitsa
GR - 48100 Preveza

Type of institute
Government Department

Research activities
Aquaculture
- Reproduction
- Early stages of reared species
- Growth, nutrition
- Aquaculture structure, systems, techniques

Facilities
Rearing tank
Hatchery facilities for sea water fishes and Penaeids species.

Staff: Scientists: 3 Others: 15

Description
It is a new Government Hatchery Station (2.500 sq.m. buildings) in which all the techniques for the cultivation of the Mediterranean eurayaline fin fishes annd penaeids are applied.

9. Hellenic Aquacultures SA

Name : Hellenic Aquacultures SA
Address : Filonos street 98 B
GR - 185 36 Pireus
Head : Mr John Pikrammenos
Tel. : 4528118 **Fax**: 1.4528377 **Telex**: 241101 GR

Type of institute
Engineering/consulting firm
Fish/Crustacean/mollusc farming

Research activities
Aquaculture
- Aquaculture/environment interaction
- Aquaculture structures, systems, techniques

Staff: Scientists: 2 Others: 2

10. Ichtyka - Centre Aquaculture Acheloos

Name : Centre d'Aquaculture d'Acheloos
Ichtyka A.E.
Address : Neochori - Messolonghi
GR - 300.01 Neoxori
Head : Dr. Sp. Klaoudatos
Tel. : 0632.91959 **Fax**: 0632.91998 **Telex**: 312477 ACEA GR

Type of institute
Institute for professional training/Secondary school
Fish/Crustacean/mollusc farming
Fishing

Research activities
Fishing techniques
Aquaculture

Facilities
Rearing tank

Staff: Scientists: 7 Others: 15

Research budget (ECU/year): 100.000

Description

The Aquaculture Centre of Acheloos is a self-financed Research centre in which all the well known techniques for the cultivation of the Mediterranean euryaline fin fishes are applied, such as sea bass, sea bream, mullets, eels in floating cages, earthern ponds. The major part of the income from the above activities are spent to support research programmes in order to improve the field of aquaculture in Greece.
The main research programmes carried out by the staff of the centre are:

1. Artificial reproduction, larval rearing and on growing for the Mediterranean species of sea bass (Dicentrarhus labrax) and sea bream (Sparus aurata) as well as for Penaeids species (Penaeus Kerathurus, Penaeus Japonicus).
2. Lagoon management
3. Collection, acclimatisation and rearing of the wild fries of the euryhaline species sea bass, sea bream, mullets, eels...
4. Impact of fish farming activities on the marine environment.

11. IMBC - Inst. Marine Biology of Crete and University of Crete - Dept. Marine Genetics

Name : Institute of Marine Biology of Crete
Sector Marine Genetics
Address : PO Box 2214
GR - 71003 Iraklion - Crete
Head : Dr. E. Zouros
Tel. : 081.242022 **Fax**: 081.241882 **Telex**: 262268 IMBC GR

Type of institute
Research centre/Higher education

Research activities
Genetic improvement of aquacultured species
Genetic differentiation of populations of fished species

Facilities
Molecular (DNA) and biochemical (allozyme) work
population genetics theory/statistics
(see also other facilities under IMBC)

Staff: Scientists: 5 Others: 2

Research budget (ECU/year): 110.000

Description
Genetic-molecular tools, such as single-locus hypervariable matters of nuclear and mitochondrial DNA, are used in marker-assisted detection of genetic differences in aquaculture-important traits, such as enhanced growth rates, reduced mortality, tolerance to environmental stress, etc. The same tools can be used as "tags" of marketed fish or shellfish strains for registry and protection of proprietary right. Another area of applications is the study of natural or semi-natural populations of economically important species of fish and shellfish, such as dispersal, population mixing, colonization, hybridization, adaptation potential and biodiversity. Current projects inculde: establishing of genetic tools for selection and population studies in the gilthead sea bream, sea bass and African catfish - population differentiation of anchovy populations in the Mediterranean - genetic differentiation between Atlantic and Mediterranean populations of sword-fish - the physiological and genetic determinants of fast growth rate in triploid oysters.

12. IMBC - Inst. Marine Biology of Crete - Dept. Aquaculture

Name : Institute of Marine Biology of Crete
Sector of Aquaculture
Address : PO Box 2214
GR - 71003 Iraklion - Crete
Head : Prof. M. Kentouri
Tel. : 081.242022 **Fax**: 081.241882 **Telex**: 262268 IMBC GR

Type of institute
Institute for professional training
Research centre

Research activities
Aquaculture
- Reproduction
- Early stages of reared species
- Growth,nutrition, feed composition, live food production for aquaculture
- Aquaculture structures, systems, techniques

Facilities
Rearing tank
Mesocosm
Data bases, extensive libraries
Specialised software

Staff: Scientists: 10 Others: 6

Description

1. Comparison of cost effectiveness of different feeding techniques for sea bass. Prof. M. Kentouri -3 years -6 scientists.
2. Studies on fish feeding behaviour and applications to aquaculture, namely self-feeders and feeding technology. Prof. M. Kentouri - long term project - 4 scientists.
3. Organisation and dynamics of marine pelagic food chains and applications to marine larval rearing. Dr. P. Divanah - long term project - 6 scientists.
4. Biological bases of aquaculture of new species (Pagrus pagrus, Dentex dentex, Epinephelus aeneus). Prof. M. Kentouri and Dr. P. Divanah - long term project - 6 scientists.
5. Growth performances and production costs of sea brass, sea bream, and stripped sea bream in Crete -Greece. Dr. P. Divanah - 3 years - 3 scientists.

13. IMBC - Inst. Marine Biology of Crete - Dept. Fisheries

Name	:	Institute of Marine Biology of Crete Sector of Fisheries		
Address	:	PO Box 2214 GR - 71003 Iraklion - Crete		
Head	:	N. Tsimenidis		
Tel.	:	081.242022	**Fax**: 081.241882	**Telex**: 262268 IMBC GR

Type of institute

Research centre

Research activities

Fisheries management

Stock assessment

Multi-species biological models

Multi-disciplinary approach to fisheries management

Facilities

Research vessel

Data bases, extensive libraries

Specialised software

Staff: Scientists: 14 Others: 7

Research budget (ECU/year): 254.100

Description
Acquisition of a general view of the state, abundance and distribution of the small pelagic fishes by means of acoustic surveys.
Study of the ecology and populations dynamics of the demersal and tuna resources on the Greek sea.
Development of new fishing techniques for a tuna fishery under the conditions of the Greek sea.
Facilities:
- 26 m length research vessel equiped with scientific echosounders and echoprocessing system.

14. Ministry of Agriculture - Dept. Hygiene and Food Technology

Name	:	Ministry of Agriculture Centre of Institute of Athens Veterinary
Address	:	Iera Odos 75 GR - 11855 Athens
Head	:	E. Georgiou
Tel.	:	3468931

Type of institute
Research centre
Government Department/Regional or local authority

Research activities
Upgrading of fishery products
Improvement of techniques for handling, storing, processing, packaging fish
Methods for grading and quality assessment of fish
Consumer response and demand studies

Facilities
Controlled environmental chamber
Data bases, extensive libraries

Staff: Scientists: 3 Others: 3

15. Ministry of Agriculture - Fisheries Dept. of Preveza

Name : Ministry of Agriculture
Preveza Dept. of Fisheries
Address : Perdikari 1
GR - 48100 Preveza
Head : A. Kentrou
Tel. : 0682 22012

Type of institute
Government Department/Regional or local authority

Research activities
Aquaculture
Fisheries management
Protection of fishstock
Waters resolving

Facilities
Field site-specialized analytical apparatus

Staff: Scientists: 3 Others: 1

Description
Establishment of reproduction periods on mullet and stripped bream (collecting ovaries-scales- otoliths).
Experimental breeding of seabass seabream and mussels.
Transfer of salmon from freshwater to floating cages.
Continualy resolvings fo the quality of Ambrakikos Gulf, lagoons and River of Louros in corporation with DAFA (Development Agency for Ambrakikos).
Studies and propositions about Lagoon management, attendance of the results of this lagoon management, and attendance and registration of the fish production of lagoons and Ambrakikos Gulf.

16. Ministry of Agriculture - Fisheries Laboratory - Piraeus

Name : Ministry of Agriculture
Fisheries Laboratory
Address : Karaoli - Dimitriou 15
GR - 185 31 Piraeus
Head : Mrs. E. Diapouli
Tel. : 01.4119352 **Fax** 01.4120178

Type of institute
Research centre

Research activities
Fisheries management
   Stock assessment
Fishing techniques
   Fishing gear designs

Staff: Scientists: 4 Others: 2

Research budget (ECU/year): 50.000

Description
Research projects:
Assessment of the main demersal fish stocks of Saronikos gulf with particular emphasis on hake (merluccius merluccius) and red mullet (Mullus barbatus).
Cataloguing of fishing gear designs.

17. Municipal Company of the Ioannina Lake - Fish Farming Station

Name : Municipal Company of the Ioannina Lake
Fish Farming Station
Address : Anatoli - Lorentzu Mavili
GR - 45 332 Ioannina
Head : HDr. Pashos Yannis
Tel. : 0651.39770 **Fax**: 651 33419/651 39770

Type of institute
Institute for professional training
Consulting firm
Fish farming

Research activities
Aquaculture
   Reproduction
   Early stages of reared species
   Aquaculture structures, systems, techniques
   Genetics

Facilities
Hatchery facilities for fresh water fishes
Rearing tank

Staff: Scientists: 3 Others: 2

Research budget (ECU/year): 15 to 30.000

Description
Research projects:
Reproduction of cyprinidae, catfish and ornamental fish.
Genetics of Cyprinidae.

18. National Centre for Marine Research - Inst. Marine Biol. Resources

Name	:	National Centre for Marine Research Institute of Marine Biological Resources
Address	:	Aghios Kosmas-Hellinikon GR - 166 04 Athens
Head	:	N. Friligos/E.Papathanassiou
Tel.	:	01.9820214 **Fax**: 01.9833095 **Telex**: 224135 GR

Type of institute
Research centre

Research activities
Fisheries management
 Stock assessment
 Multi-disciplinary approach to fisheries management
 Environmental problems and their impact on resources
Fishing techniques
 Electronics/acoustics
Aquaculture

Facilities
Research vessel
Rearing tank
Controlled environmental chamber
Data bases, extensive libraries
Specialised software
The research vessel "AEGAIO", constructed in Dec.85, is 51m long, has a tonnage of 597 gross tonnes, a crew of 22 and can accomodate 17 scientists. She is well equipped to carry out research in physical, chemical, biological and geological oceanography, as well as in fisheries. The Centre has constructed a 3.5 x 3.5 x 2.2 m controlled environmental chamber.

Staff: Scientists: 8 Others: 10

Description

Programmes:

National research centre covering several areas of fisheries and aquaculture research. The fisheries management programme includes the investigation of the distribution and abundance of pelagic and demersal fisheries resources in waters adjacent to Greece. The main research areas in fishing techniques concern the development of electronic and acoustic equipment. The research programmes of relevance to aquaculture include projects on nutritional and infectious disease of intensively cultured bass and bream, diet formulation for rainbow trout and marine fish, the intensive culture of penaeid prawns and the development of aquaculture in the Amvrakikos Gulf. The latter two projects are especially concerned with aquaculture/environment interactions.

19. National Centre for Marine Research - Inst. of Inland Water

Name	:	National Centre for Marine Research Institute of Inland Water, Fisheries Section
Address	:	Aghios Kosmas GR - 166 04 Hellinikon
Head	:	Dr Ch. Daoulas
Tel.	:	01.9821354 **Fax**: 9833095 **Telex**: 224135

Type of institute

Research centre

Research activities

Biology and ecology of freshwaters fishes
Management of freshwater fisheries

Staff: Scientists: 3 Others: 1

Description

The fisheries section of the Institute of Inland Water covers all aspects of fish biology and ecology and their applications on fisheries management. Special emphasis is given to studies of the reproduction and the early life-history of fishes. Research is particularly focused on morphological descriptions of early life-stages of endemic species and the study of their ontogenetic development, based on field work and rearing experiments. Future plans involve the study of threatened fish populations and the prospects of their conservation.

20. National Centre for Marine Research - Inst. of Oceanography

Name : National Centre for Marine Research
Institute of Oceanography
Address : Agios Kosmas
GR - 166 04 Hellinikon
Head : N. Friligos/B. Katsiki
Tel. : 01.9820214 **Fax**: 9833095 **Telex**: 224135 NCMR

Type of institute
Research centre

Research activities
Fisheries management
Environmental problems and their impact on biological and physical resources
Chemical Oceanography
Physical Oceanography
Biological Oceanography and Marine Ecology
Goelogical Oceanography and Geophysics

Facilities
Controlled environmental chamber
Data bases, extensive libraries
Research vessel "Aegean" (52 m long)
SCUBA diving
Equipped laboratories (surface 2100 m2)
Computer equipment and specialized software

Staff: Scientists: 86 Others: 50

Research budget (ECU/year): 3.000.000

Description
Research programmes:
1. Open Seas oceanography Programme (Aegean, Ionian, Cretan, Libyan and Levantine). Its purpose is to obtain an overall knowledge of the marine environment and the continental shelf of the Eastern Meditereranean.
2. MEDPOL/UNEP programme. It concentrates on the monitoring of chemical, biological and physical parameters of the open sea waters around Greece.
3. Oceanographic studies of Greek embayments. They concern areas that have high priority because of their proximity to industrial zones

and large urban centers, describing the state of the environment and its capacity to receive industrial and domestic effluents.

4. Monitoring of coastal water pollution. The monitoring concerns chemical polluants and biological indices. The choice of the monitoring parameters is made by taking into account local conditions and international agreements and protocols (EEC, UNEP etc.)

21. Organotecnica Group

Name : Organotecnica Group
Address : 12 Sofouli str.
GR - 154 51 Neo Psychico
Head : K. Zevgaridis, MBA
Tel. : 30.1.6721778 **Fax**: 30.1.6473234 **Telex**: 224004 GR

Type of institute
Consulting firm

Research activities
Aquaculture
Aquaculture economics
Upgrading of fishery products
Improvement of techniques for handling, storing, processing, packaging fish
Consumer response and demand studies

Facilities
Data bases, extensive libraries
Extensive libraries of publications and chipping of the last 25 years on institutional, marketing, statistical and economic matters concerning fisheries

Staff: Scientists: 4 Others: 6

Research budget (ECU/year): 48.800

Description
Research projects:
Multidisciplinary approach of the management of the fisheries "Ktemaston lake" and "Karpissi fish Reproduction Centre".
Feasibility and Technical study for aquaculture for sea fishes.

Prefeasibility study for regional aquaculture development of the Katerini-Aliakmonas Coast.
Standardization of a package of feasibility studies for eel aquaculture in brackish water.

22. Station Ichtyogénétique Louros

Name : Station Ichtyogénétique Louros
Address : Hani Terovou
GR - 455 00 Ioannina
Head : Mme Theocharis Vassiliki/Mme B. Theoxari
Tel. : 0654.91297 **Telex**: 322130 GR

Type of institute
Government Department/Regional or local authority

Research activities
Aquaculture
- Physiology
- Reproduction
- Genetics and conservation of gametes
- Early stages of reared species
- Growth, nutrition, feed composition, live food production for aquaculture

Facilities
Simulated tidal tank
Rearing tank
Data bases, extensive libraries
Specialised computer hardware
Specialised software
Field site - specialized analytical apparatus
micro-stereoscope
Ground equipment

Staff: Scientists: 4 Others: 5

Research budget (ECU/year): 24.800

23. Tei Messologi

Name : Tei Messologi (Technological and Educational Institution/School of Aquaculture and Fisheries)
Address : Nea Ktiria
GR - 30 200 Messologi
Head : Prof. Fanis Vorinakis
Tel. : 0631.22161.4 **Fax**: 25183.23204 **Telex**: 312457

Type of institute
University Department/Higher Education Institute

Research activities
Fisheries management
- Multi-species biological models
- Environmental problems and their impact on resources
- Water chemistry and invertebrate monitoring of inland waters

Aquaculture
- Growth model developments for marine species
- Phyto and zooplankton rearing
- Wild fish fry exploration as seed for aquaculture production
- Technological innovations and development in pilot scale

Quality and Hygenic control of fisheries-Fish Pathology
- Upgrading of fishery products, quality assesments and chemical analysis of fish feed and fisheries. Microbiological, parasitological and histopathological examinations of sick fish, invertebrates and water

Facilities
Specialised computer hard- and software
Image system analysis
Atomic absorption spectrophotometry
Water and plankton analysis lab.
Specialised field apparatus for fish water and plankton sampling
Rearing tanks and aquaria
Fisheries processing unit
Histopathological section
Microbiology lab.
Lab. for chemical analysis and quality assessments of fisheries

Staff: Scientists: 7 Technicians: 3

24. Univ. Athens - Agricultural Fac. Animal Production - Lab. App. Hydrobiology

Name	:	Agricultural University of Athens Faculty of Animal Production Laboratrory of Applied Hydrobiology
Address	:	Iera Odos 75 Votanikos GR - 118 55 Athens
Head	:	Prof. Sofronios Papoutsoglou
Tel.	:	01.3470656

Type of institute
University Department/Higher Education Institute
Processing Industry/agro-food/diets producer
Research centre
Fish/crustacean/mollusc farm

Research activities
Aquaculture
Physiology
Reproduction
Early stages of reared species
Growth, nutrition, feed composition, live food production for aquaculture
Aquaculture/environment interaction
Aquaculture structures, systems, techniques

Facilities
Rearing tank
Specialised software
Closed water systems (fresh and sea water)

Staff: Scientists: 8 Others: 8

Research budget (ECU/year): 110.000

Description
Research projects:
Toxicology-Aquaculture and protection of Mediterranean environment.
Artificial diets for sea water fish.
Factor influencing the growth rate of fresh water fish and sea water molluscs.
Controlled reproduction and early rearing of sea water fish.
Biological control of aquatic plants in irrigation channels.
Experience in specialized RIA Techniques for thyroid and steroid hormone analyses.

Trace elements analysis.
Electrophonetic studies for proteins.

25. Univ. Athens - Biochemistry Cell and Molecular Biology and Genetics

Name : University of Athens
Biology Department - Biochemistry Section
Address : Panestimioupolis Kouponia
GR - 15701 Athens
Head : M.A. Dessypris
Tel. : 30.1.7227874

Type of institute
University Department/Higher Education Institute

Research activities
Aquaculture
  Physiology
  Reproduction
Upgrading of fishery products
  Consumer response and demand studies
Biochemistry

Facilities
Field site, electron microscope, specialized analytical apparatus

Staff: Scientists: 3 Others: 1

Research budget (ECU/year): 10.900

Description
Programmes/studies:
Combined studies on biochemistry and physiology of reproduction of salmonids, esp. Rainbow trout. Studies on the influence of photoperiodic regimes on the maturation process. Biochemistry based on hormonal determinations as well as on proteins (including electrophoresis), calcium, magnesium, iodine. Parallel studies on water, food of fishes. Classical and microscopic approaches for comparative studies.

26. Univ. Athens - Techn. Dept. - Naval Architech. and Marine Engineering

Name : National Technical University of Athens
Department of Naval Architecture and Marine Engineering
Address : 42, Patission street
GR - 106 82 Athens
Head : Apostolos Papanikolaou
Tel. : 30.1.3613991 **Fax**: 30.1.3616918 **Telex**: 221682 GR

Type of institute
Research centre
University Department/Higher Education Institute

Research activities
Fishing techniques
Technological and hydrodynamic analysis and design of fishing vessels
Improvement of fishing vessels

Facilities
Data bases, extensive libraries
Specialised computer hardware
Specialised software
Ship Towing Tank (available on request)

Staff: Scientists: 15 Others: 20

Research budget (ECU/year): 40.000

Description
Research projects:
Development of a methodology for the determination of the technical and economic characteristics of optimal fishing vessels, directed towards the know-how for the design of improved fishing vessels in Greece. The current-research leads to the development of the required PC software for the preliminary design and cost evaluation of medium fishing vessels in Greece.
Hull form optimization of fishing vessels for improved hydrodynamic performance at sea (theoretical and experimental work).
Technical and economic specifications for fishing vessels, including research vessels.

27. Univ. Patras - Biological Dept.

Name : University of Patras
Biological Departement
Address : Zoological Laboratory
GR - 261 10 Patras
Head : Panayotis Kaspiris

Type of institute
University Department/Higher Education Institute
Regional/local

Research activities
Aquaculture
Reproduction
Growth, nutrition, feed composition
Upgrading of fishery products
Consumer response and demand studies

Facilities
Equipments for measuring the environmental conditions
(temperature, salinity, oxygene, pH), microscopes, stereoscopes, cages for catching the lavre

Staff: Scientists: 3 Others: 1

Research budget (ECU/year): 5.500

Description
We study the ichthyofauna in lagoons. Entering times, development and rowth of commercial interest species in comparison to physical and chemical parameters of the environment. Among different species, the biology of Ahtenia is our main interest.

28 Univ. Patras - Section of Animal Biology

Name : Section of Animal Biology
University of Patras
Address : GR - Patras
Head : I. Lykakis

Type of institute
University Department/Higher Education Institute
Regional / local

Research activities

Fisheries management
- Stock assessment

Aquaculture
- Reproduction
- Early stages of reared species
- Growth, nutrition, feed composition, live food production for aquaculture

Staff: Scientists: 4 Others: 1

29. Univ. Patras - Zoological Laboratory

Name : University of Patras
Department of Biology
Address : GR - 261 10 Patras
Head : Pr. Ioannis C. Ondrias
Tel. : 30.61.991917 **Fax**: 30.61.991909

Type of institute

Research centre
University Department/Higher Education Institute

Research activities

Fisheries management
- Multi-species biological models
- Environmental problems and their impact on resources

Aquaculture
- Disease, immunology, stress

Facilities

Rearing tank
Field site, electron microscope, specialized analytical apparatus
Photographic facilities, automatic photomicroscopy
Stereoscopes, microprocessors

Staff: Scientists: 7 Others: 4

Research budget (ECU/year): 14.800

Description

The Zoological Laboratory has a particularly wide range of interests and is equipped for research over a broad area. The research accomodation is currently being considerably enlarged and modernized. Current research interests of the Laboratory are as follows:

vertebrate biology, systematics, cytotaxonomy, evolution and ethology of fishes, amphibian, reptiles and mammals. Biology of fresh water fishes of Western Greece. Research investigation on the effects of heavy metals on shore animals. Studies on fish eggs and larval identification, growth rates and food of dominant fish larvae and assessment of fish spawning areas. Biology and ecological characteristics of grey mullets.

30. Univ. Thessaloniki - Aristotle - Dept. Genetics

Name	:	Aristotle University of Thessaloniki Department of Genetics Development and Molecular biology
Address	:	GR - Thessaloniki 54006
Head	:	Prof C. Triantaphyllidis
Tel.	:	031.998374/992813 **Fax**: 031.206138

Type of institute
University Department/Higher Education Institute National

Research activities
Genetic diversity of fish and mussel species
Environmental problems and their impact on genetic diversity
Cytogenetics of fish and mussel species
Fisheries management and improvement
Aquaculture
  Aquaculture/genetics and environment interaction
  Production-contaminants-genetic structure

Facilities
Specialized computer, software and hardware
Specialized analytical apparatus for enzyme and DNA work
Field sampling equipment

Staff: Scientists: 2 Others: 4

Research budget (ECU/year): 75.000

Description
The main research activities are:
To determine the genetíc structure, the growth rates, population density, population size, structure and seasonal condition index of commercialy important bivalve species. To make initial measurements of water quality in the Thermaikos Gulf (Thessaloniki) and of levels and sources of bioaccumulated contaminants.
To evaluate the changes in genetic diversity of natural populations of some European fish (*Salmo trutta, Leuciscus cephalus and Barbus meridionalis*) under human influence using

electrophoretical and molecular (DNA) methodology.
To examine any possible correlations between genetic composition and different financially important traits in order to develop improved fish and mussel strains in Greece, for example improved *Oncorhynchus myciss (Salmo gairdnerii)* strains.
The goals of the proposed research can be exemplified by the study of the genetic structure, identification, conservation and rehabilitation of *Silurus glanis* and *Silurus aristotelis*. The results obtained in the course of the project would provide methods of mass production of alevins for fish farming and restocking.
RECENT (1993) FINANCED PROJECTS
1. The effect of nutrient enrichment, industrial pollution and genetic composition on mussel culture and bivalve fisheries in the Thermaikos Gulf (N. Greece).
2. Genetic and reproduction of the European Silurid fishes: application to conservation and fish farming.
3. Genetic, management and improvement of the *Oncorhynchus myciss* (*Salmo gairdnerii*) in Greece.
4. Assessment of the two main and connected human influences, river disturbance and subsequent fish stocking, on the genetic diversity and stability of natural riverine fish populations.

31. Univ. Thessaloniki - Aristotle - Dept. of Veterinary Lab. Ichthyology

Name : Aristotle University of Thessaloniki
Department of Animal Production
Laboratory of Ichthyology and Fish Pathology
Address : P. O. Box 395
GR - 540 06 Thessaloniki
Head : M. Pneumatikatos
Tel. : 031.992748 **Fax**: 3031 200392

Type of institute
University Department/Higher Education Institute
National

Research activities
Aquaculture
Growth, nutrition, feed composition, live food production for aquaculture
Disease, immunology, stress
Aquaculture/environment interaction
Aquaculture structures, systems, techniques

Facilities
Specialised computer hardware and software
The laboratory is equiped with facilities for histological, microbiological and parasitological examinations as well as facilities for water analysis

Staff: Scientists: 2 Others: 2

Description
The laboratory is equiped with facilities for histological, microbiological and parasitological examinations as well as facilities for water analysis.

32. Univ. Thessaloniki - Aristotle - Dept. Zoology

Name : Aristotle University of Thessaloniki
Department of Zoology
Laboratory of Ichthyology
Address : Box 134
GR - 54006 Thessaloniki
Head : Prof. P S . Economidis/A.I. Sinis
Tel. : 031.992826 **Fax**: 031.206138 **Telex**: 412181 AUTH GR

Type of institute
University Department/Higher Education Institute
National

Research activities
Fisheries management
- Stock assessment
- Multi-species biological models
- Environmental problems and their impact on resources

Fishing techniques
- Selective gears
- Static gear

Aquaculture
- Aquaculture/environment interaction
- Aquaculture structures, systems, techniques

Facilities
Specialised computer, software and hardware
Specialised analytical apparatus
Controlled environmental chamber
National reference collection of fish
Field sampling equipment

Staff: Scientists: 2 Others: 5

Research budget (ECU/year): 75.000

Description
The main research activities are: - Systematics, Biology, Ecology and population structure of fish - Systematics and knowledge base systems - Comparisons using morphological and anatomical data - Determination and comparison of age with scales, opercula, vertebrae etc. - Determination of growth, reproduction and fecundity of fish - Measurement of physicochemical parameters of water - Recruitment, fisheries management.
Recent (1993) financed projects: - Threatened endemic freshwater fish - Common names of aquatic organisms of Mediterranean - Fish population dynamics of lakes and lagoons in nothern Greece - Heavy metals in Lake Koronia - Status of fisheries of Lakes Yliki and Paralimni - Elvers in estuaries of Nothern Greece.

33. Univ. Thessaloniki - Aristotle - Lab. of Fresh Water Fish and Wildlife

Name : Aristotle University of Thessaloniki
Laboratory of Freshwater Fish and Wildlife
Address : Box 248
GR - Thessaloniki 54006
Head : C. Neophitou
Tel. : 31.261506 **Fax**: 206138 **Telex**: 412181 AUTH GR

Type of institute
University Department/Higher Education Institute
Government Department/Regional or local authority
National

Research activities
Fisheries management
- Stock assessment
- Multi-disciplinary approach to fisheries management

Aquaculture
- Early stages of reared species
- Growth, nutrition, feed composition, live food production for aquaculture

Facilities
Data bases, extensive libraries
Two fishing boats, stereomicroscope, electronic fishing equipment, nets, macro-micro projector

Staff: Scientists: 2 Others: 3

Research budget (ECU/year): 9.400

Description

1. The stock assessment of the fish population in rivers at North Greece, including biological characteristics of stocks. It is leaded by me and financed by the Ministry of Agriculture. The duration is six years and 3 scientists are involved.
2. A pilot programme of the stock assessment of the eel elvers in river Strymer (State of Serres, N. Greece). It is leaded by me and financed by the Aristotelian University of Thessaloniki. The duration is two years and one scientist is involved.
3. A new programme will be started next June at the city of Volvos (Central Greece). The aim of the programme is ''the education and training of scientists on topics of aquaculture and the degree in the concrete subject gives them the possibility to be preoccupied with the water environment''. The project manager is the Institute of Information, Training and Development DIMITRA, with me as Scientific Consultant. The programme is financed by the C.E.C. (70%) and the General Secretary of Youth (30%). The duration is one year and 15 scientists are involved (Greek and English specialists).

State of Research

ICELAND

Fisheries related R&D is carried out by an increasing number of institutions and private companies in Iceland. In 1990 1.1% of GNP was devoted to R&D in Iceland. Of this 14,4% was spent on fisheries related marine research, 10,2% was devoted to fish processing R&D and 3.3% to aquaculture research. Thus nearly 28% of the total national R&D budget is spent on fisheries related R&D subjects.

The two principal actors are the Marine Research Institutes responsible for marine and fish biological research, and the Icelandic Fisheries Laboratories covering fish processing R&D. Both of these institutions report to the Ministry of Fisheries. Other important contributions come from the Agricultural Research Institute and the Institute of Freshwater Fisheries, both belonging to the Ministry of Agriculture and conducting research into aquaculture and fresh water fisheries. The University of Iceland and the University of Akureyri conduct research related to fisheries and aquaculture connected to their educational and training programs. The Institute of Experimental Pathology of the University of Iceland is responsible for all research and health care services in fish diseases.

The Icelandic Technological Institute conducts research and offers services to the food processing and environmental sector and reports to the Ministry of Industry.

The National Research Council, which reports to the Ministry of Culture and Education, carries the overall responsibility for research planning and coordination in all fields of applied R&D, including fisheries. It also finances the building of laboratory and equipment purchase for R&D and gives project financing based on annual processing of competing applications from institutions and industries. Over 50% of present funding of NRC goes to fisheries related projects.

The increasing interest by private industry in fisheries related R&D is mainly related to aquaculture, advanced fish processing and equipment development for product recovery and upgrading, as well as software development for various applications.

List of Institutes from ICELAND

1. Department of Physiology - University of Iceland
2. Icelandic Fisheries Laboratories
3. Institute of Experimental Pathology - University of Iceland - Department of Fish Disease
4. Marine Research Institute
5. The Agricultural Research Institute
6. The Fisheries Research Institute - University of Iceland
7. The Institute of Freshwater Fisheries
8. The Technological Institute of Iceland - IceTech Department of Environmental and Chemical Technology
9. The Technological Institute of Iceland - IceTech Department of Food Science
10. The University of Akureyri - Department of Fisheries

1. Department of Physiology - University of Iceland

Name	:	University of Iceland Department of Physiology
Address	:	Vatnsmyrarvegur 16, 101 Reykjavík
Head	:	Fish physiology section: Logi Jónsson
Tel.	:	354-1-694833 **Fax**: 354-1-694884

Type of institute
Research Centre/University Department

Research activities
Fish physiology

Facilities
Analytical laboratory
Library

Staff: Scientists: 4 Others: 1

Research budget (ECU/year): 0.2 M

Description
Research into the physiology of fish has for many years been a major part of the Physiology department. Research tasks: Rearing of supersmolts, effect of accelerated rearing on growth, sea maturing and sexual maturity, production of spawning stock and non-fertile stock in pisciculture, measurement of saline tolerance, effect on saline concentration on the growth of salmon, gill diseases in smolts.

2. Icelandic Fisheries Laboratories

Name	:	Icelandic Fisheries Laboratories
Address	:	Skúlagata 4, 101 Reykjavík
Head	:	Dr. Grímur Valdimarsson
Tel.	:	354-1-620240 **Fax**: 354-1-620740

Type of institute
Research Centre
Engineering/consulting

Research activities
Upgrading of fishery products. Methods for grading and quality assessment. New process methods for key fishery products. Analyses of process waste and discards. Integrated quality assurance and HACCP. Improved methods for handling, storing, processing and packing fish

Facilities
Analytical laboratory for bacteriological and chemical analyses. Laboratories and pilot plants for fish processing. Databases, library.

Staff: Scientists: 45 Others: 19

Research budget (ECU/year): 1.2 M

Description
The Institute falls under the Ministry of Fisheries. The laboratories are required by law to provide services and research for the fishing industry. The research projects of the laboratories are increasingly carried out in cooperation with companies and other institutions. The main research areas are: processing of new products, utilisation of the catch, pollution research, fishfat research, aquaculture feed, shelf life of fish products, biotechnology in the fishing industry.

3. Institute of Experimental Pathology - University of Iceland - Department of Fish Disease

Name	:	University of Iceland Institute of Experimental Pathology Department of Fish Disease
Address	:	Keldur v. Vesturlandsveg, 112 Reykjavik
Head	:	Dr. Sigurður Helgason
Tel.	:	354-1-674700 **Fax**: 354-1-673970

Type of institute
Research Centre

Research activities
Aquaculture
* bacteriological research
* immunological research
* vaccine development
* epidemology of fish diseases and health care in aquaculture

Facilities
Analytical and biological laboratory quarantine facilities

Staff: Scientists: 6 Others: 4

Research budget (ECU/year): 0.24 M

Description
The role of the fish disease department is research work on various aspects of fish diseases and official health control work and disease diagnosis.
Of research activity the following fields can be mentioned:
* *Aeromonas salmonicida subsp. achromogenes:*
  Pathogenesis, virulence and vaccine development
* *Renibacter salmoninarum:*
  Comparative studies of diagnostic methods
* *Immunology:*
  Research on cellular and humoral immunity, incl. the production of monoclonal antibodies against IgM in salmonids and marine species.

4. Marine Research Institute

Name : Marine Research Institute
Address : Skúlagata 4, 101 Reykjavík, Iceland
Head : Jakob Jakobsson
Tel. : 354-1-20240 **Fax**: 354-1-623790

Type of institute
Research Centre

Research activities
Fisheries management
* stock assessment
* multi species biological models
Oceanography
Plankton ecology
Aquaculture research

Facilities
Research vessels. Specialised research and analytical apparatus. Data bases/libraries. Rearing tanks for aquaculture and research

Staff: Scientists: 70 Others: 70

Research budget (ECU/year): 8.8 M

Description
The Institute falls under the Ministry of Fisheries. The primary objective of the Institute is to obtain knowledge on the sea around Iceland and its living resources.
The main research activities are oceanography and ecology - long term monitoring of hydrographic conditions in Icelandic waters - sediment formations in shallow waters - long term phyto- and zooplankton studies on primary production and distribution.
Fisheries - stock assessment - species interaction - whale research - marine aquaculture. Fishing gear testing and evaluation. Fisheries data analyses.

5. The Agricultural Research Institute

Name : The Agricultural Research Institute
Address : Keldnaholt, 112 Reykjavík
Head : Thorsteinn Tomasson
Tel. : 354-1-812230 **Fax**: 354-1-814604

Type of institute
Research centre
Experimental stations

Research activities
Agricultural research including horticulture, plant disease and pests, a pasture and range utilisation, farm technology, agricultural food, ecology, input control and animal husbandry including breeding and nutrition

Facilities
Analytical and biological laboratories. Experimental stations. Databases, libraries
Aquaculture

Staff: Scientists: 5 Others: 3

Research budget (ECU/year): 0.3 M

Description
The Agricultural Institute is an independent institute responsible to the Ministry of Agriculture. The institute has been running comparative study on salmon stocks and genetic research and husbandry on salmon, arctic char and trout. Nutrition research on salmon and arctic char is in progress.

6. The Fisheries Research Institute - University of Iceland

Name : University of Iceland
The Fisheries Research Institute
Address : Dunhagi 5, 107 Reykjavík
Head : Dr. Örn Jónsson
Tel. : 354-1-694724 **Fax**:354-1-694905

Type of institute
University Institute
Consulting

Research activities
Fisheries management and modelling. Multispecies research. Social aspects of fisheries. (Especially Benthic Invertibrates in Icelandic Waters).
Biotechnology, Aquaculture.

Facilities
Three research stations located in fisheries villages in different areas in Iceland. Access to the facilities at the various departments of the university. An object oriented simulation model for estimating various methods of fisheries regulation.

Staff: Scientists: 6 Others: 8

Research budget (ECU/year): 0.42 M

Description
The aim of the institute is to co-ordinate research in fisheries and aquaculture in which the University of Iceland participates and further cooperation with other institutes and the fishing industry in Iceland and abroad. To publish results, arrange courses and seminars in the field of fisheries and aquaculture. To establish a Masters degree in fisheries at the University of Iceland.

7. The Institute of Freshwater Fisheries

Name: : The Institute of Freshwater Fisheries
Address : Vagnhöfði 7, 112 Reykjavík
Head : Dr. Árni Ísaksson
Tel. : 354-1-676400 **Fax**: 354-1-676420

Type of institute
Research Centre
Experimental stations
Consulting

Research activities
Cultivation of juveniles. Fish genetics and husbandry. Ocean ranching of salmon. Cultivation of arctic char and trout. Assessment of stock in rivers and lakes

Facilities
Research laboratories. Experimental stations for juveniles and fish husbandry - stock improvement. Databases - libraries

Staff: Scientists: 5 Others: 13

Research budget (ECU/year): 1 M

Description
The Institute falls under the Ministry of Agriculture. The State Experimental Farm in Kollafjörður is linked to the Institute and the Fish farm is also a part owner of Stofnfiskur, a breeding and husbandry farm at Hafnir. The main activities of the Institute are: Collecting reports on aquaculture and fresh water fisheries, cultivation of arctic char, selective breeding for salmon ranching, genetics and husbandry of smolts, classification of rivers and lakes on ecological grounds, genetics of Icelandic salmon, release of salmon juveniles in rivers and lakes, ecology of arctic char and trout.

8. The Technological Institute of Iceland - IceTech Department of Environmental and Chemical Technology

Name : The Technological Institute of Iceland
IceTech Department of Environmental and Chemical Technology
Address : Keldnaholt, 112 Reykjavík
Head : Guðjón Jónsson
Tel. : 354-1-687000 **Fax**: 354-1-687409

Type of institute
Research Centre
Engineering/consultation

Research activities
Environmental technology, waste minimisation - clean technology, hydrology, chemical analysis

Facilities
Inorganic chemical analyses. Database, libraries.

Staff: Scientists: 7 Other: 1

Research budget (ECU/year): 0.32 M

Description
Process chemistry, R&D project in waste handling, pollution control in air, water and soil. One of the main activities in R&D projects is internal environment in closed land based fish farms, recirculation, hydrology and growth conditions.

9. The Technological Institute of Iceland - IceTech Department of Food Science

Name : The Technological Institute of Iceland
IceTech Department of Food Science
Address : Keldnaholti, 112 Reykjavík
Head : Hannes Hafsteinsson
Tel. : 354-1-687000 **Fax**:354-1-687409

Type of institute
Research Centre
Engineering/consulting

Research activities
Food science and technology
Biotechnology in food
Pet food

Facilities
Chemical and biochemical analyses. Pilot-plant.

Staff: Scientists: 5 Others: 2

Research budget (ECU/year): 0.4 M

Description
The Food Science Department is involved in product development, consulting and services to the food industry. The products include fish crisps, different aseptic fish products as well as pet food.

10. The University in Akureyri - Department of Fisheries

Name : The University in Akureyri
Department of Fisheries
Address : Glerárgata 36, 600 Akureyri
Head : Jón Pórðarson
Tel. : 354-6-11780 **Fax**: 354-6-25216

Type of institute
University Department

Research activities
Marine microbiology
Phycology research
Oceanography
Plankton ecology
Marine aquaculture

Staff: Scientists: 5 Others: 2

Research budget (ECU/year): 0.2 M

Description
The department is doing research in marine sciences as well as applied research in marine aquaculture. The department is also working with different fish processing factories.

State of Research

IRELAND

Two state organisations are involved in research and development concerning fishery:

* the Fisheries Research Center (FRC) of the Department of the Marine based at Abbotstown, Dublin
* Bord Iascaigh Mhara (BIM), a semi-state body, based at Dun Laoghaire, CO Dublin

The Fisheries Research Center pursues research projects which may have an influence and bearing on future National and EC legislation. In particular, it carries out biological investigations required for fishery management, which provide the basis for the setting of the TAC and Quotas by the EC Fisheries Council.

More precisely, the FRC conducts research programmes in five principal subject areas:

- stock evaluation
- resource evaluation
- aquaculture
- fish pathology
- environment

The overall objectives of these programmes are to provide scientific advice to resource and environmental management agencies (central government, regional government, fisheries board), development agencies and the industries themselves. Advice on fisheries management is formulated through participation in ICES working groups and ACFM, and the EC's STCF.

Other bodies conducting research concerning fishery management exist: The Central Fisheries Board is a state sponsored body funded but not controlled by the Department of the Marine. CFB is mainly concerned with freshwater fishery (generally leisure trout fishery), but is also studies some sea fish as well as salmon.

The Salmon Research Agency (SRA), which also operates under the aegis of the Department of the Marine, is responsible for scientific research for the purpose of improving, developing and extending salmon and sea-trout fisheries.

The Economics and Social Research Institute includes commercial and leisure fisheries in its research.

The other major centre with the FRC is An Bord Iascaigh Mhara (BIM). BIM, as a development agency, confines its activities to development projects designed to improve the catching efficiency and profitability of the Irish sea fishing industry, keeping in mind and respecting the constraints imposed by the Common Fishery Policy.

BIM is concerned with the overall development of the Irish fishing industry, including the processing and marketing of fish. It also takes part in the provision of credit finance for the fleet, having thus a strong element of market economics in its work. BIM is funded by the Department of Marine.

Universities participate occasionally in research projects related to different fishery topics, however these programmes are generally short term ones.

No other major organisation takes part in fisheries research and development. However, R&D in fishery products upgrading is undertaken by a number of third level institutions.

The work of these bodies is coordinated by the Department of Marine, which also tries to ensure that these researches are directed towards industry related problems.

Infrastructural problems exist for marine research in Ireland, both in terms of the number of people employed and in the facilities available ashore and at sea. The financial situation offers few possibilities of improvement of this problem. This creates a need for "new blood" among scientists, as restrictions on recruitment has created an age gap.

The Irish parliament has enacted legislation to create a Marine Research Institute to restructure marine research and combine the separate agencies into one organisation. This Institute will be more responsible for the development and coordination of all the research programmes realted to marine activities. It will also ensure the coordinaiton of the resources of the state used in this area.

Educational support is available at relevant departments of the different Irish Universities and Colleges of Technology. There is also BIM's National Fishery Training Centre, based in Greencastle, CO Donegal, which has its own 16m training vessel, educating students in fishery technology and techniques.

List of Institutes from IRELAND

1. Applied Aquaculture Research - Udaras Na Gaeltachta
2. BIM - Bord Iascaigh Mhara - Irish Sea Fisheries Board
3. Dublin Institute of Technology - School of Food Science
4. Eir Nor Teoranta
5. Environmental Management Services
6. EOLAS - Ocean Services Dept.
7. Fisheries Research Center - Abbotstown
8. Irish Hydrodata Ltd.
9. Irish Salmon Growers Association
10. Mari - Chem Ltd. - Galway
11. Mariculture Systems Ltd.
12. Regional Technical College - Cork - Dept. of Chemical and Biological Sciences
13. Regional Technical College - Dundalk - Dept. Experimental Science
14. Regional Technical College - Galway - Dept. Life Sciences
15. Regional Technical College - Galway - Dept. Physical Sciences
16. Regional Technical College - Letterkenny
17. Regional Technical College - Tralee
18. Salmon Research Agency of Ireland Inc.
19. Sherkin Island Marine Station
20. The National Food Centre - Department of Plant and Marine Foods
21. Univ. College Cork - Atlantic European Research
22. Univ. College Cork - Dept. of Civil Engineering
23. Univ. College Cork - Dept. of Zoology
24. Univ. College Dublin - Dept. of Zoology
25. Univ. College Galway - Dept. of Oceanography
26. Univ. College Galway - Dept. of Zoology
27. Univ. College Galway - Martin Ryan Marine Science Institute
28. Univ. College Galway - National Diagnostics Centre
29. Univ. College Galway - Shellfish Research Laboratory
30. Univ. Dublin - Trinity College - Dept. of Civil and Environmental Engineering
31. Univ. Dublin - Trinity College - Dept. of Microbiology
32. Univ. Dublin - Trinity College - Dept. of Physiology
33. Univ. Dublin - Trinity College - Dept. of Sociology
34. Univ. Dublin - Trinity College - Dept. of Statistics
35. Univ. Dublin - Trinity College - Dept. of Zoology
36. Univ. Dublin - Trinity College - Environmental Sciences Unit
37. Univ. Dublin - Trinity College - Dept. of Zoology - Aquaculture Research Group

1. Applied Aquaculture Research - Udaras Na Gaeltachta

Name : Applied Aquaculture Research - Udaras Na Gaeltachta Gaillimh
Address : IR - Furbo Co Galway
Head : John Lowry
Tel. : 091.92011 **Fax**: 091.92037

Type of institute
Government Department/Regional or local authority

Research activities
Aquaculture
Aquaculture economics

Research budget (ECU/year) : 194.500

2. BIM - Bord Iascaigh Mhara - Irish Sea Fisheries Board

Name : BIM - Bord Iascaigh Mhara
Address : P.O Box 12-Crofton Road
IR - Dun Laoghaire/Co Dublin
Head : A. Gannon
Tel. : 01.841544 **Fax**: 01.841123 **Telex**: 93237 EI

Type of institute
Government Department/Regional or local authority

Research activities
Fisheries management
Multi-disciplinary approach to fisheries management
Fishing techniques
Aquaculture
Upgrading of fishery products

Facilities
Data bases, extensive libraries
Specialised computer hardware and software
Training vessel

Staff: Scientists: 11 Others: 7

Research budget (ECU/year): 2.987.500

Description
Research activities:
Fisheries Management: market research on fish products, structure of primary producing sectors, feasibility studies of new ventures.
Fishing techniques: comparative studies of commercial trawls using scanner, new trawling techniques, introduction of new static gears, EC funded exploratory fishing voyages on non-quota species, investigation of the potential of automatic fishing robots (under EC FAR Programme).
Aquaculture: predesignation site surveys for aquaculture, economics, development of culture of oyster, mussels and scallops.
Upgrading: New products and process development assistance, processing advisory service.
Facilities: Modern 16 m steel training vessel fitted with sophisticated electronic equipment and can operate as a multipurpose fishery vessel.

3. Dublin Institute of Technology - School of Food Science

Name : Dublin Institute of Technology
College of Catering, School of Food Science and Environmental Health
Address : Cathal Brugha Street
IR - Dublin 1
Head : Marlene Proctor
Tel. : 353.1.747886 **Fax**: 353.1.743634

Type of institute
University Department/Higher Education Institute

Research activities
Upgrading of fishery products
Improvement of techniques for handling, storing, processing, packaging fish
Methods for grading and quality assessment of fish

Facilities
Laboratory Facilities/access to Fish Farms, Packing Stations,
Processing Plant and Consumer groups
Access to various online databases

Staff: Scientists: 2 Others: 1

Description
Research projects:
Studies on the onset and resolution of rigor mortis in skeletal muscle of fish incorporating biochemical, physicochemical and organoleptical aspects.
Quality implications of controling biochemical processes in skeletal muscle of fish post mortem.
Quality assurance in handling of fishery products.

4. Eir Nor Teoranta

Name : Eir Nor Teoranta
Address : Ballyhea
IR - Dingle, Co Kerry
Head : J. O'Driscoll
Tel. : 066.51139 **Fax**: 066.51139

Type of institute
Fish/Crustacean/mollusc farming

Research activities
Aquaculture

Facilities
Rearing tank

Staff: Scientists: 1 Others: 2

5. Environmental Management Services

Name : Environmental Management Services
Osborne House
Address : Seapoint avenue
IR - Monkstown Co Dublin
Head : Mr. Jack O'Sullivan
Tel. : 01.807867

Type of institute
Engineering/consulting firm
Fishing

Research activities
Fisheries management
  Environmental problems and their impact on resources
Aquaculture
  Aquaculture/environment interaction
  Aquaculture structures, systems, techniques

Facilities
Data bases, extensive libraries
Field site, electron microscope, specialized analytical apparatus

Staff: Scientists: 4 Others: 1

Description
Research expertise:
Site selection and evaluation: site survey and ranking of potential sites, hydrographic and water quality assessments, biological and chemical baseline data acquisition.
Feasibility and start-up: environmental impact assessment, design of shellfish and finfish cultivation systems, information and assistance with market development.
Aquaculture: biological and chemical monitoring, technical assistance in farming and quality control.
Coastal resources management.

6. EOLAS - Ocean Services Dept.

Name	:	EOLAS - Irish Science & Technology Agency Ocean Services Division
Address	:	Glasnevin IR - Dublin 9
Head	:	Mr. John O'Dea/Gerard Keane
Tel.	:	01.370101 **Fax**: 1.379620 **Telex**: 32501

Type of institute
Research centre
Engineering/consulting firm

Research activities
Fishing techniques.
  Other improvements in gears and/or fishing vessels
Aquaculture
  Aquaculture/environment interaction
  Aquaculture structures, systems, techniques

Facilities
Data bases, extensive libraries
Specialised computer hardware
Specialised software

Staff: Scientists: 5 Others: 1

Research budget (ECU/year): 300.000

Description
Research projects:
Energy efficiency and energy saving.
Facilities:
Structural monitoring equipment.
Specialised software: offshore engineering, offshore environment, offshore reliability.
Management of research projects.
Techno-economic studies of Marine Industry [all aspects].

7. Fisheries Research Center - Abbotstown

Name	:	Fisheries Research Center of Abbotstown
		Abbotstown
Address	:	Castleknock
		IR - Dublin 15
Head	:	David Griffith
Tel.	:	1.210111 **Fax**: 1.205078 **Telex**: 31236 EI

Type of institute
Research centre

Research activities
Fisheries management
- Stock assessment
- Multi-species biological models
- Multi-disciplinary approach to fisheries management
- Environmental problems and their impact on resources

Fishing techniques
Aquaculture

Facilities
Research vessel
Data bases, extensive libraries
Specialised computer hardware
Specialised software
Operation of multipurpose coastal marine research vessel, with two laboratories and accommodation for 8 scientists

Staff: Scientists: 14 Others: 14

Research budget (ECU/year): 298.300

Description
Research projects:
Management of salmonids, Nephrops, eels, demersal finfish, pelagic fisheries and development of non-commercial species.
Bioeconomics.
Selective fishing gear.
Aquaculture: Finfish and molluscs, phytoplancton bloom monitoring, fishfarm site monitoring. Fish and shellfish pathology/diseases.
Marine environmental monitoring and Marine and freshwater chemistry.

8. Irish Hydrodata Ltd.

Name	:	Irish Hydrodata Ltd.
Address	:	Rathmacuilig West
		IR - Ballygarvan/Cork
Head	:	T.N. Emerson
Tel.	:	021.311255 **Fax**: 021.968604

Type of institute
Engineering/consulting firm

Research activities
Fisheries management
 Multi-species biological models
Fishing techniques
 Electronics/acoustics
Aquaculture
 Aquaculture/environment interaction
 Aquaculture structures, systems, techniques
International level

Facilities
Research vessel
Flume tank
Simulated tidal tank
Data bases, extensive libraries
Specialised computer hardware
Specialised software

Staff: Scientists: 4 Others: 6

Research budget (ECU/year): 50.000

Description
Pollution problems and their consequences on the environment/other species.
Electronics/acoustics including detection of fish close to the sea bed.
Latest research project is the determination of sea bed type by acoustic methods, part funded by EC Hydrocarbons Projects (DG XVII) which will have direct relevance to bottom fishing/trawling.

9. Irish Salmon Growers Association

Name	:	The Irish Salmon Growers Association
Address	:	Crofton Road-P.O. Box 16 IR - Dun Laoghaire-Co Dublin
Head	:	Dr. J. Joyce
Tel.	:	353.1.800078 **Fax**: 353.1.800650 **Telex**: 93237

Type of institute
Fish/Crustacean/mollusc farming

Research activities
Aquaculture
- Genetics and conservation of gametes
- Disease, immunology, stress
- Aquaculture/environment interaction

Upgrading of fishery products
- Methods for grading and quality assessment of fish

Facilities
Data bases, extensive libraries

Staff: Scientists: 1 Others: 3

Research budget (ECU/year): 107.600

Description
Research projects:
Causes of pancreas diseases.
Quality control methodology on Irish salmon farm leading to creation of quality control manual for Irish Industry.

10. Mari - Chem Ltd. - Galway

Name	:	Mari - Chem Ltd.
		Unit No 8
Address	:	IDA Development Centre-Upper Newcastle
		IR - Galway
Head	:	J.F. Barrett
Tel.	:	091.2227

Type of institute
Chemical, Pharmaceutical Products

Research activities
Upgrading of fishery products
Novel processing methods for by-products, discards, and process wastes

Facilities
General Pilot-scale Processing Equipment and testing

Staff: Scientists: 1 Others: 2

Research budget (ECU/year): 25.900

Description
Mainly R&D unit investigating marine (fish+processing) wastes for useful, value-added products for the pharmaceutical, agri-chemical etc sectors.

11. Mariculture Systems Ltd.

Name	:	Mariculture Systems Ltd.
Address	:	Derryherbert
		IR - Letterfrack/Co Galway
Head	:	Niall Herriott
Tel.	:	95.43441

Type of institute
Engineering/consulting firm
Fishing
Distribution
Processing industry/mollusc

Research activities
Fisheries management
Environmental problems and their impact on resources
Aquaculture
Aquaculture/environment interaction
Aquaculture structures, systems, techniques
Upgrading of fishery products
Improvement of techniques for
handling, storing, processing, packaging fish
Novel processing methods for by-products, discards, and
process wastes

Facilities
Research vessel
Data bases, extensive libraries
Microscope, environmental monitoring equipment

Staff: Scientists: 1 Others: 1

Description
Facilities:
Research vessel, available for external users (to rent) with 34 foot catamaran and twin 60 HP diesels.
Mariculture systems is also part of a research and consultancy group called Environmental Management Associates, with its Head Office in Dublin (Jack O'Sullivan, Osborne House, Seapoint Av., Monkstown). The main areas of activity are Environmental Impact assessment, Rural Development (particularly in coastal areas), Shellfish Farming - both intensive and extensive, Pollution monitoring and control.
Projects have been carried out for numerous clients, including State and EC institutions, large industrial firms, co-operatives, Third Level educational establishments, small business and individual clients.

12. Regional Technical College - Cork - Dept. of Chemical and Biological Sciences

Name : Regional Technical College of Cork
Department of Chemical and Biological Sciences

Address : Rossa Avenue-Bishopstown
IR - Cork
Head : Michael Walsh/Michael Noonan (Head of Dept.)
Tel. : 021.545222 **Fax**: 021.545343

Type of institute
Higher Education Institute

Research activities
Upgrading of fishery products
Methods for grading and quality assessment of fish
Analysis of marine toxins

Facilities
Controlled environmental chamber
Field site, electron microscope, specialized analytical apparatus

Staff: Scientists: 3

Description
1. The laboratory undertakes microbiological analysis of shellfish (oysters, mussels, clams), fish (mackerel, salmon, trout, etc), processed fish products and water samples in connection with national /international environmental and sanitary regulations.
2. An additional topic of research is focussed on the occurrence of 'red tides' in the coastal waters of Ireland and their association with okadaic acid and dinophysis-toxin production.

13. Regional Technical College - Dundalk - Dept. Experimental Sciences

Name : Regional Technical College of Dundalk
Dept. of Experimental Sciences
Address : Dublin Road
IR - Dundalk, Co. Louth
Head : Dr. Dermot Douglas
Tel. : 042.34785 **Fax**: 042.33505

Type of institute
University Department/Higher Education Institute
Fish/Crustacean/mollusc farming

Research activities
Fisheries management
  Stock assessment
  Environmental problems and their impact on resources
Aquaculture

Facilities
Field sampling equipment
Field laboratory

Staff: Scientists: 5 Others: 4

Research budget (ECU year): 259.800

Description
Research projects:
Advice and consultancy to producers association.
Development of a protocol for management of a oyster nursery based on environmental monitoring.
Integrated research and development on shellfish production.

14. Regional Technical College - Galway - Dept. Life Sciences

Name : Regional Technical College of Galway
Departement of Life Sciences
Address : Dublin Road
IR - Galway
Head : Dr. Gerard Quinn
Tel. : 091.53161 **Fax**: 091.51107

Type of institute
University Department/Higher Education Institute
Institute for professional training/Secondary school

Research activities
Fisheries management
  Environmental problems and their impact on resources

Aquaculture
  Reproduction
  Growth, nutrition, feed composition, live food production for aquaculture
  Aquaculture/environment interaction
Freshwater ecology

Facilities
Rearing tank
Algae culture collection

Staff: Scientists: 5 Others: 2

Description
Research projects:
Marine macro and microalgae.
Freshwater microalgae.
Marine benthic ecology.
Freshwater ecology.
Physiology and genetics of marine molluscs.
Aquaculture.

15. Regional Technical College - Galway - Dept. Physical Sciences

Name	:	Regional Technical College of Galway Department of Physical Sciences
Address	:	Dublin Road IR - Galway
Head	:	Brian Ottway
Tel.	:	091.53161 **Fax**: 091.51107

Type of institute
University Department/Higher Education Institute

Research activities
Fisheries management
  Stock assessment

Aquaculture
  Genetics and conservation of gametes
  Growth, nutrition, feed composition, live food production for aquaculture
  Aquaculture/environment interaction
  Aquaculture structures, systems, techniques

Facilities
Rearing tank
Field site, electron microscope, specialized analytical apparatus
Shellfish ongrowing site

Staff: Scientists: 2 Others: 1

16. Regional Technical College - Letterkenny

Name : Regional Technical College Donagh O'Malley
Address : IR - Letterkenny Co Donegal
Head : Dr. W.J.W. Hines
Tel. : 074.624888 **Fax**: 074.24879

Type of institute
University Department/Higher Education Institute
Regional level

Research activities
Upgrading of fishery products
  Methods for grading and quality assessment of fish

Staff: Scientists: 10 Others: 4

17. Regional Technical College - Tralee

Name : Regional Technical College of Tralee
School of Science
Address : Clash
IR - Tralee-Co Kerry
Head : H. Lyons/J. Kelly
Tel. : 53.66.24666 **Fax**: 353.66.255711

Type of institute
University Department/Higher Education Institute
Institute for professional training/Secondary school

Research activities
Fisheries management
Environmental problems and their impact on resources
Aquaculture
Growth, nutrition, feed composition, live food production for aquaculture
Aquaculture/environment interaction
Upgrading of fishery products
Methods for grading and quality assessment of fish

Facilities
General biology, microbiology, chemistry, biochemistry labs etc.
Chemical & Biological Analytical Equipment

Staff: Scientists: 3 Others: 1

18. Salmon Research Agency of Ireland Inc.

Name	:	Salmon Research Agency of Ireland Inc. Farran Laboratory
Address	:	IR - Newport Co Mayo
Head	:	Dr. K. Whelan
Tel.	:	098.41107 **Fax**: 98.41107

Type of institute
Research centre

Research activities
Fisheries management
Stock assessment
Multi-species biological models
Environmental problems and their impact on resources
Aquaculture

Facilities
Rearing tank
Data bases, extensive libraries
Field site, microscopes, specialized analytical apparatus
Upstream and downstream traps for migratory salmonids

Staff: Scientists: 6 Others: 6

Research budget (ECU/year): 479.900

19. Sherkin Island Marine Station

Name : Sherkin Island Marine Station
Address : Sherkin Island-Baltimore
IR - West Cork
Head : Mr Matt Murphy
Tel. : 353.28.20187

Type of institute
Research centre
Institute for professional training/Secondary school
Field Study Centre

Research activities
Aquaculture
Aquaculture/environment interaction

Facilities
Research vessel
Data bases, extensive libraries
Field site, microscopes, specialized analytical apparatus

Staff: Scientists: 12

20. The National Food Centre - Department of Plant and Marine Foods

Name : The National Food Centre
Department of Plant and Marine Foods
Address : Dunsinea, Castleknock
IR - Dublin 15
Head : Dr. P.V. Tarrant/Dr. T.R. Gormley
Tel. : 1.383222 **Fax** : 1.383684

Type of institute
Semi-state

Research activities
Total quality and safety of fish and fish products
Quality evaluation and test processing of non-quota fish species
Evaluation of novel cryoprotectants
Studies on n.3 fatty acids
Developing added value fish products
Methods of dissemination to small industry

Facilities
Extensive laboratories and sensory facilities
Technology hall
Product development test kitchen

Staff: (full time on fish topics) Scientists: 1 Others: 2
(part-time on fish topics) 10

Description
Research projects:
Preparing added value herring products. Consumer preference of smoked salmon colour. Non-destructive measurement of the fat content of fish by NIR. Fatty acid (n.3) content of wild and farmed salmon. Evaluating the potential of non quota fish species for consumer products. Evaluation of novel cryoprotectants. Studies on the addition of n.3 encapsulated fish oils to food products. Evaluation of temperature profiles during the national and international distribution of chilled and frozen fish products. Evaluation (and use) of routes for the dissemination of information on food quality, safety and wholesomeness to the food industry including fish processors (the so-called FLAIR-FLOW EUROPE project).

21. Univ. College Cork - Atlantic European Research

Name : University College of Cork - Atlantic European Research
C/O Geography department
Address : IR - Cork
Head : Patrick O'Flanagan
Tel. : 021.276871 **Fax**: 021.275948 **Telex**: 76050 EI

Type of institute
University Department/Higher Education Institute

Research activities
Fisheries management
Multi-disciplinary approach to fisheries management
Environmental problems and their impact on resources
Aquaculture
Aquaculture economics

Facilities
Data bases, extensive libraries
Specialised software

22. Univ. College Cork - Dept. of Civil Engineering

Name : University College of Cork
Department of Civil Engineering
Address : Munster Institute Buildings
IR - University College Cork
Head : Dr. Tony Lewis
Tel. : 353.21.276871 ext. 2165 **Fax**: 353.21.343580

Type of institute
Research and Development Centre

Research activities
Coastal engineering, offshore engineering, aquaculture structures and equipment, marine technology, instrumentation and computing for marine activities, tank testing.

Facilities
Wave tank n1 15m*0.75m*0.7m. deep
Wave tank n2 26m*3.0m*2.0m deep
Wave basin n1 18m*6.8m*0.4m deep
Wave basin n2 25m*18m*2.5m deep
Electronics and mechanical workshops
Computer network and workstations

Staff: 10 technical staff including research engineers, support technicians and others. Postgrad. students work as research assistants.

Research budget (ECU/year): 150.000

Description
The H.M.R.C. is within the Department of Civil Engineering and has the function of carrying out repayment as well as fundamental research contracts. Courses taught within the Department, which include harbour and coastal engineering, basic oceanography and aquaculture structures use the facilities at the Centre. Research thesis are supervised. Courses in survival at sea will be offered soon.

23. Univ. College Cork - Dept. of Zoology

Name : University College Cork
Department of Zoology
Address : Lee Maltings-Prospect Row
IR - Cork
Head : Dr Richard FitzGerald/Gavin Burnell/Tom Cross
Tel. : 353 21 276871 **Fax**: 353 21 270562/277922

Type of institute
Research centre
University Department/Higher Education Institute
Fish/crustacean/mollusc farming

Research activities
Fisheries management
- Stock assessment
- Multi-species biological models
- Environmental problems and their impact on resources

Aquaculture
- Optimisation of production strategies
- New species: species and site selection; growth evaluation and enhancement
- Health protocols and disease control: development of vaccines; immunodiagnostic studies
- Genetics: stock discrimination; broodstock development and enhancement
- Recirculation system design
- Artificial reefs

Application of Image Processing and Analysis
Distance learning material and multi-media packages

Facilities

Immunology and Tissue culture (with ELISA) laboratory - Genetics (Electrophoresis and DNA probe) laboratory - Histology laboratory - Water analysis laboratory - Image Processing and Analysis laboratory - Stream systems - Aquatic isolation system (semi-static) - Extensive cold rooms and constant temperature rooms - S.E.M and T.M. are provided by a college service unit - Large recirculation facility on-campus which includes three groups of tanks (freshwater/marine) - Research Station located in West Cork - Marine pump-ashore unit with extensive tank farm and marine hatchery facilities - The unit also has a laboratory, offices and accommodation.

Staff: Scientists: 16 Others: 6

Research budget (ECU/year): 750.000

Description

Research projects:

Physiology, reproduction and genetics (salmon, trout, carp, tilapia and sea urchins). Early stages of turbot oysters and scallops; Growth nutrition and feed composition (salmon, trout, turbot, oysters and sea urchins). Disease and immunology of salmon and trout. *Bonamia* in oysters. Development of salmon vaccines for sea lice. Commercial Broodstock evaluation. Continuing (vocational) training programme for the aquaculture industry.

24. Univ. College Dublin - Dept. of Zoology

Name	:	University College Dublin Department of Zoology
Address	:	Stillorgan Road-Belfield IR - Dublin 4
Head	:	Prof. E. Duke/Dr. Yvette Kunz-Ramsay
Tel.	:	01.693244 **Fax**: 837009 **Telex**: 32693 EI

Type of institute

University Department/Higher Education Institute

Research activities

Fisheries management
- Stock assessment
- Multi-species biological models
- Environmental problems and their impact on resources

Aquaculture
- Aquaculture/environment interaction
- Aquaculture structures, systems, techniques

Facilities
Rearing tank
Data bases, extensive libraries
Large modern Aquaria. Full computer facilities. Research laboratory fully equipped for fishery science research projects.

Staff: Scientists: 3 Others: 10

Research budget (ECU/year): 120.000

Description
Research projects:
Investigation of fish productivity.
Impact of industrial pollution on rivers and streams.
Impact of cage fish farming.
Environmental studies on the edible mussel.
Distribution and importance of aquatic insects in relation to pollution and water quality monitoring.
Palaeolimnology and the use of lake sediments to assess changes in lake water acidification.
The use of satellite derived information to monitor lake water quality.
Sequence analysis of mitochondrial DNA to obtain genetic markers for fish populations.
Biological and statistical assessment of fish populations.

25. Univ. College Galway - Dept. of Oceanography

Name : University College Galway
Department of Oceanography
Address : IR - Galway
Head : Prof. Michael J. Orren
Tel. : 353.91.24411 Ext: 3201 **Fax**: 353.91.28302 **Telex**: 50023 UCG EI

Type of institute
Consulting activity
Research centre
University Department/Higher Education

Research activities
Fisheries management
Environmental problems and their impact on resources
Aquaculture
Aquaculture/environment interaction

Facilities
Research vessel
Data bases, extensive libraries
Field site, specialized analytical apparatus
Extensive facilities for physical and chemical oceanography

Staff: Scientists: 3 Others: 6

Research budget (ECU/year): 25.900

Description
Research projects:
Physical oceanography surveys, including measurements by dye dispersion, drogue and current meter measurement in order to estimate flushing times and to establish whether a given area is suitable for aquaculture activities.

26. Univ. College Galway - Dept. of Zoology

Name	:	University College Galway Department of Zoology
Address	:	IR - Galway
Head	:	P. O'Ceidigh/Dr. J. Fives c/o D. Dead(Bxl)
Tel.	:	24411 (2321) **Fax**: 25700 **Telex**: 50023 UCGEI

Type of institute
Research centre
University Department/Higher Education Institute
Fish/crustacean/mollusc farm

Research activities
Fisheries management
  Stock assessment
  Environmental problems and their impact on resources
Fishing techniques
  Testing techniques/instruments
Aquaculture

Facilities
Research vessel
Flume tank
Rearing tank
Mesocosm
Controlled environmental chamber
Field site, electron microscope, specialized analytical apparatus

Staff: Scientists: 4 Others: 4

27. Univ. College Galway - Martin Ryan Marine Science Institute

Name	:	The Martin Ryan Marine Science Institute University College Galway
Address	:	IR - Galway
Head	:	B.F. Keegan / P.B. Heffernan
Tel.	:	353.91.50351/50357 **Fax**: 353.91.25005 **Telex**: 50023 UCGEI

Type of institute
Higher Education/Research Institute
(Botany/zoology/microbiology/physical and chemical oceanography/meteorology/aquaculture).

Research activities
Morphology, taxonomy, life-histories and reproduction of algae; rhytmicity, photoperiodism and temperature control of cell division, growth and reproduction in marine algae; algal aquaculture for the production of fine chemicals; provision of seed stock phytoplankton for the shellfish industry; taxonomy of marine animals; zooplankton ecology; biology, ecology and behaviour of fish and benthic animals; fisheries technology; population/ecological genetics; environmental impact studies; sediment profile imagery; ecotoxicology; disease treatment in fish farming; phytoplankton ecology; effects of pollutants on microbially mediated processes; role of bacteria in biogeochemical fluxes in deep ocean waters; non-culture methods for the study of bacterial taxonomy and physiology; sediment chemistry; physical oceanography of the Irish slope; biogeochemical cycling of trace metals; remote sensing; air/sea interactions; distribution of metals in macro- and microalgae; sea level studies (Irish west coast); studies of particulate matter in Atlantic air; investigation of maritime sources of sulphur compounds; investigation of mercury and heavy metal concentrations in marine air; analysis of aerosols, precipitation and trace gases in Atlantic air; culture of marine invertebrates and finfish.

Facilities
Histo and cytological processing facilities - micro and macro algal clone library - fluorescent microscopy with photon accelerator - computer controlled light boxes - U/W still-, stereo-, digitising cameras, & video equipment - laser particle sizing - computer controlled flumes & incubaria - sediment profile imagery - image analysis - CHN analyser

- radiotracer suite - microscopy suite for bacterial counting/sizing - molecular probes for taxonomic/diagnostic purposes - autolab high sensitivity measurements and calibration for salinity - inductively coupled plasma spectrophotometry - furnace and flame atomic absorption measurement of heavy metals - alpkem autoanalysis of nutrients in sea water - UV/Vis spectrophotometer with kinetics facilities - on site continuous monitoring for O2, salinity and nutrients - clean room facilities for low level work - satellite image receiver - cloud condensation nuclei counter - TSI paticle classifier - hatcheries, nurseries and grow-out systems.

Staff: Scientists: 46 Others: 22

28. Univ. College Galway - National Diagnostics Centre

Name : University College Galway
National Diagnostics Centre/Bioresearch Ireland
Address : IR - Galway
Head : R. Smith/F. Gannon
Tel. : 353.091.66559 **Fax**: 091.66570

Type of institute
Research centre
University Department/Higher Education Institute

Research activities
Aquaculture
Genetics and conservation of gametes
Disease, immunology, stress
Smoltification. DNA probes. Immunodiagnostic tests

Facilities
Field site, electron microscope, specialized analytical apparatus
Fish holding facilities, molecular biology laboratory

Staff: Scientists: 12 Others: 1

Research budget (ECU/year): 260.000

Description

Research projects:

Development of DNA probes for organisms that cause disease in salmon.
Study of the molecular biology of smoltification in Atlantic Salmon.
Development of improved immunodiagnostic assays for salmon pathogens.
Study of the control of expression of genes in Atlantic Salmon [e.g. liver genes and globin].
Studies of Pancreas disease in salmon.
Diagnosis of disease in salmon.

Facilities:

Available fully equiped laboratory for work on bacteriology, virology, parasitology, histology and molecular biology techniques.
Skills in immunodiagnostics with the capacity to assay large number of samples.

29. Univ. College Galway - Shellfish Research Laboratory

Name : University College Galway
Shellfish Research Laboratory
Address : IR - Carna Co Galway
Head : Pr. P. O'Ceidigh/J. Mercer
Tel. : 091.24411 **Fax**: 091.25700 **Telex**: 500 23 EI UCG

30. Univ. Dublin - Trinity College - Dept. of Civil and Environmental Engineering

Name : University of Dublin
Department of Civil and Environmental Engineering
Address : Trinity College
IR - Dublin 2
Head : S. Perry/M. Hartnett
Tel. : 353.1.7022217 **Telex**: 25442 TCD EI

Type of institute

University Department/Higher Education Institute

Research activities
Fish farm management
Integrated environmental impact assessment
Carrying capacity studies
Water quality modelling
Marine engineering

Facilities
Hydraulics laboratory
Water quality modelling software
Hydrographic survey equipment
Geographical information system
Extensive computer hardware, PC's, workstations and mainframes

Staff: Engineers: 11 Others: 3

Description
1. Development and application of marine and freshwater eutrophication models.
2. Risk analysis on cage integrity
3. Solute and particulate pollutant prediction modelling
4. Regional zoning of aquaculture developments

31. Univ. Dublin - Trinity College - Dept. of Microbiology

Name	:	University of Dublin Department of Microbiology
Address	:	Trinity College IR - Dublin 2
Head	:	T.J. Foster/R.R. Russell
Tel.	:	353.1.7021990 **Fax**: 353.1.6799294 **Telex**: 25442 TCD EI

Type of institute
University Department/Higher Education Institute

Research activities
Molecular genetics
Vaccine development
Fish quality
Water quality
Rapid methods for the isolation and identification of pathogens in water, food and environment in general.

Facilities
BL2 and BL3 containment facilities
Conductance monitoring systems
Pathogen laboratories
Tissue culture facilities
Industry and Government linkages

Description
Research projects funded by FAR/AAIR:
An aromatic dependent mutant of *Aeromonas salmonicida* as a live vaccine for furunculosis in salmonid fish. Studies on the immune response of vaccinated fish and on the fate of bacteria after vaccination both in the vaccinated host and in the environment. (Project leader).

32. Univ. Dublin - Trinity College - Dept. of Physiology

Name : University of Dublin - Trinity College
Department of Physiology
Address : Trinity College
IR - Dublin 2
Head : J.V. McLoughlin
Tel. : 353.1.7021226 **Fax**: 353.1.793545 **Telex**: 93782 TCD EI

Type of institute
University Department/Higher Education Institute

Research activities
Aquaculture
Physiology
Upgrading of fishery products

Facilities
Rearing tank
Data bases, extensive libraries
Specialised computer hardware
Specialised software
Field site, electron microscope, specialized analytical apparatus

Staff: Scientists: 2 Others: 1

Description
Projects funded by the FAR/AAIR programme:
Quality implications of controlling nucleotide metabolism in skeletal muscle of fish (project partner).

33. Univ. Dublin - Trinity College - Dept. of Sociology

Name : University of Dublin
Department of Sociology
Address : Trinity College
IR - Dublin 2
Head : J. Wickham/H. Tovey
Tel. : 353.1.7021371 **Fax**: 353.1.771300 **Telex**: 25442 TCD EI

Type of institute
University Department/Higher Education Institute

Research activities
Aquaculture and rural development
Environmental issues in natural resource exploitation

Facilities
Direct access to university mainframe computer
Expertise in quantitative and qualitative research and data analysis.

34. Univ. Dublin - Trinity College - Dept. of Statistics

Name : University of Dublin
Department of Statistics
Address : Trinity College
IR - Dublin 2
Head : A.R. Unwin/J. Haslett
Tel. : 353.1.7021768 **Fax**: 353.1.615046 **Telex**: 25442 TCD EI

Type of institute
University Department/Higher Education Institute

Research activities
Interactive statistical and graphical analysis of large and complex environmental datasets including fishery and aquaculture applications.

Facilities
Specialised software for spatial analysis of data.

Staff: Scientists: 2 Others: 3

35. Univ. Dublin - Trinity College - Dept. of Zoology

Name	:	University of Dublin Department of Zoology
Address	:	Trinity College IR - Dublin 2
Head	:	D.I.D. Howie/J. Aldrich/J. Reynolds/O. Tully
Tel.	:	353.1.7021366 **Fax**: 353.1.772694 **Telex**: 25442 TCD EI

Type of institute
University Department/Higher Education Institute

Research activities
Freshwater and marine biology
Fisheries management
  Stock assessment
  Environmental problems and their impact on resources
Fishing techniques
  Selective gears
Aquaculture
  Reproduction, genetics, disease

Facilities
Rearing tank, incubation unit
Aquarium facilities for marine and freshwater species
Controlled environment chambers
Data bases, extensive libraries
Specialised software
Field site, minibus
Electron microscope, specialized analytical apparatus
Subaqua unit, two portable boats

Staff: Scientists: 13 Others: 4

Description
Research projects:
Biology of marine and freshwater species in relation to fisheries and aquaculture:
  Salmon, trout, wrasse, eels, lobsters, crayfish, amphipods, scallops, mussels oysters, echinoderms, marine annelids.
Fisheries management
  Impacts on environment and fishing on *Nephrops* population structure; population genetics of *Nephrops*; use of age pigment as index of age in crustaceans.

Environmental problems affecting fisheries management:
Parasites of sea-trout and freshwater fishes; impacts of sea-lice on wild salmonids; causes and effects of stream and lake acidification on salmonid stocks; impact of crayfish fungal plague on freshwater ecosystems.
Development of trapping methods for glass-eels, elvers and freshwater crayfish and of selective gears for Nephrops; impacts of fishing gear on other stocks.
Aquaculture
Effects of salmonid farming on the marine and freshwater environment; control of sea-lice on farmed salmon using wrasse as cleaner-fish; individual physiological variation in marine molluscs and crustaceans; interactions between habitat productivity and molluscs; experimental rearing; reproduction and growth of Astacid crayfish in relation to environmental parameters and feedstuffs.
Individual physiological variation in marine molluscs and crustaceans; reproduction and experimental rearing of Astacidae; experimental rearing of molluscan larvae; growth of Astacidae in relation to environmental parameters and development of feedstuffs; effects of salmonial farming on the environment; parasitic problems of farmed salmon and sea-lice; interactions between estuarine productivity and molluscs; effects of Aphanomycosis (crayfish fungal plague) on native stocks and potential for aquaculture.

36. Univ. Dublin - Trinity College - Environmental Sciences Unit

Name	:	University of Dublin Trinity College - Environmental Sciences Unit
Address	:	IR - Dublin 2
Head	:	C. Stillmann/M.J. Costello/J G. Wilson
Tel.	:	353.1.7021638/7022075 **Fax**: 353.1.718047

Type of institute
University Department/Higher Education Institute

Research activities
Fisheries management
Environmental problems and their impact on resources
Waste treatment from aquaculture and fish processing
Aquaculture/environment interaction

Facilities
Field site, electron microscope, specialized analytical apparatus
Range of analytical equipment for nutrient, metal and other pollutant analysis (including ion chromatograph, flow injection analyser, atomic absorption spectrophotometer (including graphite furnace).
Geographical information systems (GIS).

Staff: Scientists: 11 Others: 1

Description
The major research projects are carried out in the in the Environmental Science Unit in the following fields :
1) Biogeochemical cycling, including nutrients and organic matter.
2) Ecophysiology and ecotoxicology including effects of metals, sewage sludge on fish, molluscs and crustaceans.
3) Aquaculture/environment interactions including effluent treatment, chemicals (malachite green) and their impact on receiving waters.
4) Biological control (e.g. cleaner fish) diseases and parasites.
5) Fish pathology especially sea-lice control.
The ESU also offers an M.Sc environmental Sciences course with units in Aquaculture and Fisheries.
Projects funded by the FAR/AIR programme:
Effects of malachite green on the natural environment. (Project leader).
Cleaner fish technology: parasite control, alternative to pesticides and creating a new fishery. (Project leader).

37. Univ. Dublin - Trinity College - Dept. of Zoology - Aquaculture Research Group

Name : University of Dublin
Department of Zoology
Address : Trinity College
IR - Dublin 2
Head : J.D. Reynolds
Tel. : 353.1.7021063 **Fax**: 353.1.6793545

Type of institute
University Department/Higher Education Institute
Fish/Crustacean/mollusc farming
Consulting firm

Research activities
Fisheries management
  Stock assessment
  Environmental problems and their impact on resources
Fishing techniques
  Selective gears
Aquaculture
Upgrading of fishery products
  Consumer response and demand studies

Facilities
Rearing tanks
Aquarium facilities
Controlled environment chamber
Data bases, extensive libraries
Specialised software
Field site, electron microscope, specialized analytical apparatus
subaqua unit
2 portable boats

Staff: Scientists: 8 Others: 4

Description
Research projects:
Stock assessment studies on salmon, sea-trout, brown trout, eel, crayfish, mussels, scallop, other aquatic invertebrates.
Effects of pathogens and parasites on crayfish and salmonid stocks
Aquaculture: Ongoing programmes with culture of eels, mussels, scallops, crayfish and amphipods. Use of wrasse to clear fish-lice.

State of Research

ITALY

Italy has public (national or local authorities) and private research structures dedicated to research in the fishery sector. The main types of structures are:

Institutes of the National Research Council of Research (CNR)
Institute for Central Research Applied to the Sea (ICRAM) depending on the Ministry of the Merchant Navy (ICRAM)
University Institutes
Regional and provincial structures
Large private firms
Cooperatives

The National Research Council (CNR) and the Institute for Central Research Applied to the Sea (ICRAM) are the two main national institutions for research in the fishery sector.

Research on fisheries is also financed directly by the Ministry of the Merchant Navy, Direction for Fishery and by the National Research Council. Coordination at the national level is mainly performed by the Ministry of Merchant Navy and the Ministry of University and Scientific and Technical Research (MURST). It has been recently proposed that aquaculture activities should be coordinated within the Ministry of Agriculture.

Only a few University courses on fisheries and aquaculture are available. Graduates generally arrive in the field from biological, veterinary or agronomical faculties. Participation in international, and particularly EC programmes is increasing, which is allowing fruitful exchange of know-how and staff with other countries.

List of Institutes from ITALY

1. AGRITECO - Coop.
2. ANCP - Lega Pesca
3. AQUASTUDIO
4. A.S.A.L.M.
5. Ass. Italiana Cooperative Della Pesca
6. API - Ass. Piscicoltori Italiani
7. Ass. Produttori Ittici Veneto
8. BIOTECHNOLOGICAL srl
9. BIOTOP s.n.c.
10. Centro Interuniversitario di Biologia Marina - "G. Bacci" di Livorno
11. Centro Ricerca Interdip. Scienze del Mare
12. CIRSPE - Centro Ricerca Studi la Pesca
13. CETENA - Centro Tecnica Navale SpA
14. CNR - IRPEM - Istituto di Ricerche Sulla Pesca Marittima
15. CNR - Istituto di Biologia del Mare
16. CNR - Istituto di Cibernetica e Biofisica - Sezione Distaccata di Ecologia della Visione
17. CNR - Istituto di Tecnologia della Pesca e del Pescato
18. CNR - Istituto Italiano Idrobiologia
19. CNR - Istituto per L'Automazione Navale
20. CNR - Istituto per la Corrosione Marina dei Metalli
21. CNR - Istituto per lo Sfruttamento Biologico delle Lagune
22. CNR - Istituto per lo Studio della Dinamica delle Grandi Masse
23. CNR - Istituto Sperimentale Talassografico di Taranto
24. CNR - Istituto Sperimentale Talassografico di Trieste
25. CNR - Istituto Talassografico - Messina
26. CO.I.P.A. - Coop. Idrobiologia Pesca ed Acquacoltura
27. CEAC - Consorzio Ecologia e Acquacoltura Costiera
28. Consorzio Pesca Acquicoltura Trasimeno
29. CORISA - Consorzio Ricerche Sardegna
30. Coop. AGEI - Acquacoltura Gestione Ittica
31. Coop. BIOMAR
32. Coop. BIOSERVICE
33. Coop. BIOTECNO scrl
34. Coop. COISPA
35. Coop. TECNOMARE
36. Dr. Mura Graziella
37. ELMAR - Agency of Ship Design
38. ENEA - CASACCIA - Laboratorio Acquacoltura
39. ENEL SpA - DSR - CRTN
40. Environmental Marine Research Center
41. G.R.A.I.A. srl
42. HENDRIX SpA

43. HYDRORES s.a.s.
44. ICI - Istituto Culturale Italiano
45. IMC - International Marine Centre
46. IST - CSTA Istituto Nazionale per la Ricerca Sul Cancro
47. ICRAM Istituto Centrale per la Ricerca Applicata al Mare
48. Istituto Comunale di Biologia Marina di Nardo
49. Istituto di Zootecnica - Gallina
50. Istituto Idrobiologia e Acquacoltura G. Brunelli
51. Istituto Nazionale della Nutrizione
52. I.R.E.P.A. - Istituto Ricerche Economiche per la Pesca
53. Istituto Zooprofilattico delle Venezie
54. Istituto Zooprofilattico Sperimentale Piemonte Liguria Val d'Aosta
55. Italittica srl
56. KOBA srl (Biomass Provision and Conversion)
57. Lab. di Biologia Marina (LBM)
58. Lab. di Biologia Marina ed Acquario - Provincia di Bari
59. Lab. di Idrobiologia
60. M.A.R.E s.a.s.
61. Maricoltura Italia SpA
62. Polo Tecnologico Marino Marittimo
63. R.S.T.A.
64. Riserva Naturale Marina di Miramare
65. S.I.VAL.CO SpA - Societa Itticoltura Valli Comacchio
66. SSICA - Stazione Sperimentale Industria Conserve Alimentari
67. Stabilimento Ittiogenico - Regione Lazio
68. Stazione Zoologica - Anton Dohrn (SZN)
69. STM - Servizi Tecnici in Maricoltura
70. Univ. Ancona - Dipartimento Scienze Materiali e Terra
71. Univ. Bari - Dipartimento di Produzione Animale
72. Univ. Bari - Istituto di Chimica
73. Univ. Bari - Zoologia ed Anatomia Comparata
74. Univ. Basilicata - Dipt. Produzioni Animali - Lab. Acquacoltura
75. Univ. Cagliari - Ist. Zoologia
76. Univ. Genova - Ist. Anatomia Comparata
77. Univ. Genova - Scienze Ambientali Marine
78. Univ. Genova - Zoologia
79. Univ. Lecce - Dipartimento di Biologia
80. Univ. Milano - Inst. Pharmacological Science
81. Univ. Modena - Dipt. Biologia Animale
82. Univ. Palermo - Istituto Zoologia Giuseppe Reverberi
83. Univ. Pisa - Scienze dell Ambiente e del Territorio
84. Univ. Roma I - Ist. Parassitologia
85. Univ. Roma II - Dipt. Sanita'Pubblica e Biologia Cellulare
86. Univ. Sassari - Istituto di Zootecnica
87. Univ. Siena - Dipt. Biologia Ambientale
88. Univ. Trapani - Ist. Biologia Marina

89. Univ. Trieste - Dipt. Ingegneria Navale
90. Univ. Udine - Istituto di Scienze della Produzione Animale

1. AGRITECO - Coop.

Name : AGRITECO - Ambiente Progetto Territorio S.c.r.l
Address : Corso del Popolo, 96
I - 30171 Mestre - Venezia
Head : A. Bressan
Tel. : 41 5314467 **Fax**: 41 5313483

Type of institute
Research Centre
Engineering/consulting firm

Research activities
Fisheries management
Stock assessment
Multi-disciplinary approach to fisheries management
Environmental problems and their impact on resources

Facilities
Research vessels
Data bases
Digital Terrain model
CAD
Specialized analytical apparatus

Staff: Scientists: 6 Others: 5

Research budget (ECU/year): 228.000

Description
Environmental impact evaluation in Venice lagoon and Po Delta River. Control of coastal water quality in North Adriatic Sea. Projects on the commercial rearing of molluscs. Wild stock evaluation.

2. ANCP - Lega Pesca

Name : ANCP - Lega Pesca
Address : Via Guattani, 9/13
I - 00161 Roma
Head : Ettore Iani
Tel. : 06.844391 **Fax**: 06.84439247 **Telex**: 611346 LEGA I

Type of institute
Chamber of Commerce, of Industry and/or Agriculture

Description
Cooperative associations operating in fishing activities.
The main activity is representation and coordination of members.

3. AQUASTUDIO

Name	:	Aquastudio
Address	:	Via Trapani, 6 I - 98121 Messina
Head	:	Di Natale Antonio
Tel.	:	39.90 .91.189/55200 **Fax**: 39.90.391.189

Type of institute
Research institute (private firm)

Research activities
Commercial fisheries studies.
Assessment of fisheries impact.
Multi-disciplinary approach to fisheries management.
Marine biology and marine aquariums consultancies.
Coastal lagoons and marine parks management projecting.
Fisheries observers programmes.
Social and economic investigations in the fishery sector.
Artisanal tropical fisheries consultancies.

Facilities
Computers network, data bases, extensive libraries, specialised software, printing centre, service cars, microscopes, analytical balances, sediment sleves, underwater and surface cameras and lights, current-meters, bathythermographer, sampling bottles, dredges, sonar, GPS, diving equipments, marine mammals rescue equipment, board observers equipment.

Staff: Scientists: 9 Others: 10

Research budget (ECU/year): 520.000

Description
Aquastudio is a specialist research institute committed to advisory services for national and international public bodies involved in fisheries and marine management and for private institutions. Aquastudio is also a rescue unit for protected species.
Field of activities: fish stock structure, fishery statistics, commercial fisheries observers programmes, fisheries monitoring, marine environment studies, coastal lagoon studies, marine parks projecting, economic analysis applied to fisheries.
Research projects: studies on large pelagic fisheries, surface drifting gear fisheries analysis, fishermen's perception of EEC and national regulations, assessment of heavy oil

pollution on demersal resources, assessment of fishing gear impacts on protected species, cooperation assistance programmes for artisanal fishery in tropical developing countries, tuna fisheries interactions in the Pacific Ocean, scientific consultancy for new-building marine aquariums, studies on red shrimp fishing activities, socio-economical study of some Mediterranean fishing activities, monitoring system for techno-economic data, distributive channels and prices for fishery products.

4. A.S.A.L.M.

Name	:	A.S.A.L.M. - Azienda speciale per lo Sviluppo Ill'Acquacolturnella Laguna di Marano Lagunare
Address	:	Viale Ungheria, 15 I - 33100 UDINE
Head	:	S. Bozzetto
Tel.	:	432.503907/67970 **Fax**: 432.503919/67971 **Telex**: 450649 CFCE UDI

Type of institute
Chamber of Commerce, of Industry and Agriculture

Research activities
Aquaculture
- Reproduction and mollusc farming
- Economic studies
- Disease of molluscs
- Aquaculture/environment interaction
- Aquaculture structures, systems, techniques

Facilities
Research vessels (GPS, ECHOSOUNDER, RADAR).
Field sites, pilot plants, specialized analytical apparatus
Laboratory with standard equipment for sea water analysis
Rearing tanks

Staff: Scientists: 3 Others: 5

Research budget (ECU/year): 374.100

Description
A.S.A.L.M. is a scientific operation established in 1985 for the development of the Marano Lagoon, and for the management of coastal resources in the nothern Adriatic Sea. In particular A.S.A.L.M. has developed:
- culture techniques for mussels, clams, oysters and scallops;

- monitors and collects bivalve spat;
- evaluates the status of natural mollusc beds;
- carries out general monitoring of the marine and brackish environments;
- economic studies on aquaculture products.

5. Ass. Italiana Cooperative Della Pesca

Name : Associazione Italiana Cooperative Della Pesca A.G.C.I.
Address : Via Salaria 222
I - 00198 Roma
Head : Cono Bruno (Presidente)
Tel. : 06.8413586 **Fax**: 06.8848258 **Telex**: 622285

Type of institute
Research centre
Association of fish cooperatives

Research activities
Upgrading of fishery products
Consumer response and demand studies

Facilities
Central structure and several offices at the more important Italian' fishery harbours.

Description
The "Associazione Italiane Cooperative della Pesca" is a national association which joins together the fishery, aquaculture, service and research cooperatives.
It has been juridically recognized in representative and development activities. It gives them associated fiscal, technical, economical and management assistance, and it participates in aquaculture projects, marketing structures and feasibility studies. It represents the fishery and research cooperatives in Health, Finances, Ministero Marina Mercantile and Environmental Ministries.

6. API - Ass. Piscicoltori Italiani

Name : API - Associazione Piscicoltori Italiani
Address : c/o Agricenter
Viale del Lavoro, 8
I - 37135 Verona
Head : Dr. A. Trincanato

Tel. : 045.580978 **Fax**: 045.582741 **Telex**: 480538

Research activities
Aquaculture
Aquaculture economics, statistics and marketing

Description
The API, a trade-union association, juridically recognized with D.P.R. n.1011 on the 31st of March 1970, operates within the sphere of aquaculture activities development and economic protection. The association promotes and carries out studies relevant to this field, with particular reference to economic, statistical and marketing themes, giving the results of the surveys carried out to all its members through specific working documents, and to all the fish farmers through its own publications.

7. Ass. Produttori Ittici Veneto

Name : Associazione tra i Produttori Ittici del Veneto
Address : Via C. Monteverdi, 15
I - 30170 Mestre (VE)
Head : Lucia Scudeller
Tel. : 39.041.987400 **Fax**: 39.41.989034

Type of institute
Processing Industry/ agro-food / diets producer

Research activities
Upgrading of fishery products

Staff: Scientists: Others: 60

Description
Introduction of fishery products on market.
Representation of Producers associations.
Promotion of research projects directed to the upgrading of fishery products.

8. BIOTECNOLOGICA srl

Name : BIOTECNOLOGICA srl
Address : Campo Viola
I - 27043 Broni (PV)
Head : Dssa B. Zuppini
Tel. : 0385.44015 **Fax**: 385.44270

Type of institute
Research Centre

Research activities
Aquaculture
  Growth, nutrition, feed composition, live food production for aquaculture
Aquaculture economics
  Upgrading of fishery products
  Novel processing methods for by-products, discards, and process wastes

Facilities
Field site, pilot plants, specialized analytical apparatus

Staff: Scientists: 9 Others: 15

Research budget (ECU/year): 1.070.000

Description
Aquaculture: research,setting up technologies for food production for aquaculture. Feed production and formulation utilizing specific microalgae biomasses. Feed production and formulation utilizing proteins recovered from organic waste.
Upgrading: Direct and indirect energy recovery from fish processing industry. Production of fish meals.
FACILITIES: Lab. equipment for microalgae research and production 50 Lt. bioreactor for algae production.
Chemical-physical lab. with: gas chromatographs HPLC Atomic adsorption, IR, UV spectrophotometry. Microbiological lab with: scanning electron microscope (Available to external users). Anaerobic chamber. A series of fermenters (up to 250 Lt) for aerobic and anaerobic tests.

9. BIOTOP s.n.c.

Name : BIOTOP s.n.c.
Address : Via Benedetto Croce, 19
I - 80134 Napoli
Head : S. Abbadessa
Tel. : 081.5517131

Type of institute
Research Centre
Engineering/consulting firm

Research activities

Fisheries management
- Multi-species biological models
- Multi-disciplinary approach to fisheries management
- Environmental problems and their impact on resources

Aquaculture
- Disease, immunology, stress
- Aquaculture structures, systems, techniques
- Aquaculture economics

Facilities

Specialised computer hardware
Specialised software
Equipped laboratories of histology, microbiology and chemistry.
Possibility of the use of the structures of a diving centre belonging to one of the company members.

Staff: Scientists: 5 Others: 1

Description

The Biotop Society was established in 1989 by a group of scientists.
Several activities on aquaculture research have been carried out: Publications of biological and ecological models, based on taxonomical and physiological data.
Studies on the management and nutrition of Decapod Crustaceans.
A project for the economical utilization of Anguilla in a region of Southern Italy.
Publication of a text on ichthyopathology.
A project on the commercial rearing of edible Molluscs.
Collaboration of other researchers on production and demand studies.

10. Centro Interuniversitario di Biologia Marina - "G. Bacci" di Livorno

Name : Centro Interuniversitario di Biologia Marina "G. Bacci"
Address : P. le Mascagni 1
I - 57127 Livorno
Head : Dr. Stefano De Ranieri_
Tel. : (39-586)805504 **Fax**: (39-586)807287

Type of institute

Research Centre (University of Pisa, Firenze, Siena, Modena, Bologna, Torino and Municipality of Livorno).

Research activities
Fisheries management
  Stock assessment
  Multi species biological models
  Multi-disciplinary approach to fisheries management
  Environmental problems and their impact on resources
Fishing techniques
  Selective gears
  Static gears
Aquaculture
  Growth, nutrition, feed composition
  Disease, immunology, stress

Facilities
Rearing tank
Specialised software
Equipped laboratories for oceanography, biochemistry and histology
A motor boat (25 CV, 5m)
Equipment for diving

Staff: Scientists: 12 Others: 3

Research budget (ECU/year): 150.000

Description
Research projects:
Demersal resources evaluation.
Evaluation and analysis of the interaction of fishing gears in the demersal fisheries of the western Mediterranean.
Settlement and mortality patterns of Mediterranean littoral fish juveniles, in protected and Unprotected areas.

11. Centro Ricerca Interdip. Scienze del Mare

Name : Centro di Ricerca Interdipartimentale
Scienze del Mare
Address : Via Zanolini, 3
I - 40126 Bologna
Head : R. Viviani
Tel. : 051.243019/53 **Fax**: 051.243053

Type of institute
Research Centre
University Department/Higher Education Institute

Research activities

Fisheries management
  Environmental problems and their impact on resources
Aquaculture
  Growth, nutrition, feed composition, live food production for aquaculture
  Aquaculture/environment interaction
  Biochemistry of marine organisms (Phytoplankton, Molluscs, Crustaceas and Fishes)
  Anaerobic metabolism in molluscs
  Trace metals metabolism
  Marine biotoxins

Facilities

Rearing tank
NHR
HPLC
Gas chromatography
Specialized analytical apparatus

Staff: Scientists: 14 Others: 10

Research budget (ECU/year): 300.000

12. CIRSPE - Centro Ricerca Studi la Pesca

Name : CIRSPE - Centro Italiano Ricerche e Studi per la Pesca
Address : Via Benedetto Croce, 68/116
I - 00144 Roma
Head : Massimo Coccia
Tel. : 0039 65406991 **Fax**: 0039 65425093

Type of institute

Research Centre
Institute for professional training/Secondary school
Engineering/consulting firm

Research activities

Fisheries management
  Stock assessment
  Multi-disciplinary approach to fisheries management
  Fishing techniques

Aquaculture
Upgrading of fishery products

Facilities
Research vessel

Staff: Scientists: 11 Others: 2

Research budget (ECU/year): 151.500

Description
Research projects:
Stock evaluation, fishing techniques (mesh size effects, purse seining in coast for white fish, technologycal aspects of reconversion of small trawlers).
Aquaculture: Mussel production in sea, clam culture.
Environmental studies: Bentonic anoxic phenomena in some Adriatic area.
Study, project, realisation and experimentation of a ship specialised in mitiliculture in open sea.
Study and experimentation of molluscs hatching in open sea as a alternative for mitili.

13. CETENA - Centro Tecnica Navale SpA

Name : CETENA SpA
Centro per gli Studi di Tecnica Navale
Address : Via al Molo Giano (Calata Grazie)
I - 16126 Genova
Head : C. Castelli-President/G. Siffredi-Director
Tel. : 010.5995460 **Fax**: 010.5995790 **Telex**: 271559 CETENA I

Type of institute
Research Centre
Engineering/consulting firm

Research activities
Fishing techniques
Other improvements in gears and/or fishing vessels
New type of fishing vessels.

Facilities
Data bases, extensive libraries
Specialised software

Staff: Scientists: 65 Others: 21
Dedicated to fisheries Scientists: 3 Others: 3

Research budget (ECU/year): 5.600.000
dedicated to fisheries : 0.5

14. CNR - IRPEM - Istituto di Ricerche Sulla Pesca Marittima

Name : CNR - IRPEM-Istituto di Ricerche sulla Pesca marittima
Address : Molo Mandracchio
I - 60100 Ancona
Head : Prof. G. Bombace
Tel. : 071.55313-4 **Fax**: 071.55313

Type of institute
Research Centre

Research activities
Fisheries management
Stock assessment
Multi-disciplinary approach to fisheries management
Artificial reefs and open sea mariculture
Environmental problems and their impact on resources
Fishing techniques

Facilities
Research vessel
Data bases, extensive libraries
Specialised computer hardware
Specialised software
Research vessels : 'S. Lo Bianco', 30m, 660 HP;
'Tecnopesca', 15m, 360 HP. Data bases: Data series of acoustic survey and population dynamic of pelagic fish from 1976 to 1993 Library.
Hardware: Measurement of geometrical and hydrodynamical parameters for trawl gears.
Software: Library of programmes for biological character data.

Staff: Scientists: 30 Others: 6

Research budget (ECU/year): 700.000

Description
Major research activities:
Valuation of the pelagic population in Italian seas using acoustic methodology.
Study of the population dynamics of sardine and anchovy stocks in Adriatic sea.
Bivalve molluscs fishing in Adriatic sea.

Feeding and growing of the fish spcecies from the trawling bottoms.
Fishing effort distribution in the Italian seamanships.
Coastal oceanography and ecology. Dynamic of the Adriatic water masses. Artificial reef mariculture. Improvements in gears and fishing vessels. Fish marketing.

15. CNR - Istituto di Biologia del Mare

Name : CNR - Istituto di Biologia del Mare (Reparto Acquacultura)
Address : Castello 1364/a
I - 30122 Venezia
Head : Dr. Paolo Franco
Tel. : 041.5207622 **Fax**: 39.41 5204126

Type of institute
Research Centre

Research activities
Aquaculture
Reproduction
Genetics and conservation of gametes
Early stages of reared species
Growth, nutrition, feed composition, live food production for aquaculture

Facilities
Data bases, extensive libraries
Field site, specialized analytical apparatus

Staff: Scientists: 4 Others: 1

16. CNR - Istituto di Cibernetica e Biofisica - Sezione Distaccata di Ecologia della Visione

Name : CNR - Istituto di Cibernetica e Biofisica
Sezione distaccata di Ecologia della Visione
Address : Lungomare Eleonora d'Arborea, 22
I - 09072 Torregrande - Oristano
Head : Franco Gambale (Director of the Institute)
Silvana Vallerga (Responsible of the section)
Tel. : (39)783.22032/22027 **Fax**: (39)783.22002

Type of institute
Research Centre
Higher Education Institute

Research activities
Fish vision
Aquaculture
  Physiology
  Aquaculture/environment interaction

Facilities
Electrophysiology set up
Microscopes and image analysis system
Specialized analytical apparatus

Staff: Scientists: 2 Others: 2

Research budget (ECU/year): 50.000

Description
The centre has been established in 1992, hosted on premises of the IMC - International Marine Centre to allow CNR researchers to do research on blue water marine organisms on a field site. The current research going on are eel and sea bass vision at different stages of development.

17. CNR - Istituto di Tecnologia della Pesca e del Pescato

Name	:	CNR - Istituto di Tecnologia della pesca e del pescato Consiglio Nazionale delle Ricerche
Address	:	Via L.Vaccara 61 I - 91026 Mazara del Vallo (TP)
Head	:	Dr. D. Levi
Tel.	:	0923.948723 **Fax**: 923.906634

Type of institute
Research Centre

Research activities
Fisheries management
  Stock assessment
  Multi-disciplinary approach to fisheries management

Facilities
Data bases, extensive libraries
Specialised computer hardware and software

Staff: Scientists: 7 Others: 2

Research budget (ECU/year): 329.700

Description
The institute will be involved in the next three years in the strenthening action of the National Research Council for research in the 'Mezzogiorno'. That means doubling the research team and provision of a research vessel.
The areas of research can be summarized as follows:
A) Assessment and modelling of demersal stocks in the Sicilian channel.
B) Artificial reefs-Monitoring and modelling; study areas;
Gulf of Castellamare; Capo Granitola.
C) Operational oceanography: interacting with POEM, with particular reference to pelagic resources.

18. CNR - Istituto Italiano Idrobiologia

Name	:	CNR - Istituto Italiano Idrobiologia
Address	:	Largo Vittorio Tonolli, 50/52
		I - 28048 Verbania Pallanza(NO)
Head	:	R. de Bernardi
Tel.	:	0323.556571 **Fax**: 556513

Type of institute
Research Centre

Research activities
Fisheries management
- Stock assessment
- Environmental problems and their impact on resources

Aquaculture
- Growth, nutrition, feed composition, live food production for aquaculture
- Ecology and Biology of freshwater fish, aquatic food chain.

Facilities
Research vessel
Field site, electron microscope, specialized analytical apparatus

Staff: Scientists: 2 Others: 5

Description
Research projects:
Biomanipulation of aquatic food chain as a tool for recovery of eutrophication process.

19. CNR - Istituto per L'Automazione Navale

Name	:	CNR - Istituto per l'Automazione Navale
Address	:	Via de Marini, 6 - Torre Francia I - 16149 Genova
Head	:	Ing. Filippo Aldo Grassia
Tel.	:	39.10.64751 **Fax**: 39.10.6475600

Type of institute
Public research organisation belonging to Italian National Research Council (CNR).

Research activities
The Institute operates in the marine-maritime research field, specializing in the study of systems for the enhancement of navigational systems and the protection of coasts and the marine environment.

Facilities
HP 9000 computer, tests tank, oceanographic buoy in open sea, oceanographic vessel provided by C.N.R.

Staff: Scientists: 9 Others: 11
(Technicians+Administr.)
External : Scientists: 3

Research budget (ECU/year): 433.000

Description
Research projects:
1. Navigational safety and efficiency in maritime transport
2. Technology for monitoring the marine environment
3. Marine robotics
4. Telematic support systems for marine transport
5. Detection on the earth surface of possible emissions in the ULF/ELF band by the TSS1 satellite
6. Ship noise evaluating in coastal waters (CEC - MAST I, SNECOW)
7. Advanced manipulators for deep underwater systems (CEC - MAST II, Amadeus)
8. Design and assessment of a regional traffic management system (CEC - EURET, 1.3 RTIS).

20. CNR - Istituto per la Corrosione Marina dei Metalli

Name : CNR - Istituto per la Corrosione Marina dei Metalli
Address : Centre San Benigno, Via de Marini 6
I - 16149 Genova
Tel. : 39.10.64751 **Fax**: 39.10.6475400

Type of institute
Research Institution

Research activities
The institute's activities are thus based on interdisciplinary approaches directed to define the chemical, physical and biological aspects of corrosion in sea water, in order to develop a protection system which employed in natural environment, might respect the marine ecosystem.

Facilities
Library, data bank, offshore laboratory for field test in sea, analytical facilities (GC, GC/MS, HPLC, SEM, XPS, X-ray-microdiffractometry).

Staff: Scientists: 11 researchers Others: 7
Students: 11 technical

Research budget (ECU/year): 600.000

Description
The Istituto per la Corrosione Marina dei Metalli is the only organisation of the Consiglio Nazionale delle Ricerche devoted to the study of corrosion behaviour of materials in marine environments.
Research projects:
The Institute's main research projects concern the understanding of the corrosion mechanisms and the study of modalities for the prevention in marine environment. The Institute in particular attaches great interest to the interferences of marine organisms on the life and the efficiency of metallic structures or plants using natural seawater.

21. CNR - Istituto per lo Sfruttamento Biologico delle Lagune

Name : CNR - Istituto per lo Sfruttamento Biologico delle Lagune

Address : Via Pola, 4
I - 71010 Lesina (FG)
Head : Dott. Robero Purini
Tel. : 0882.91166/91483 **Fax**: 0882.91352

Type of institute
Research Centre
National Research Council
Government Department

Research activities
Fisheries management
Stock assessment
Multi-disciplinary approach to fisheries management
Environmental problems and their impact on biological resources
Continuous monitoring of lagoon through automatic system
Aquaculture
Growth, nutrition, feed composition, live food production for aquaculture
Advanced system for algae-culture
Physiology, reproduction of captivity broodstocks
Live algae collection
Early stages of warmwater species
Disease and stress of fingerling fish

Facilities
Rearing tank, special conditioning rooms
Field site, microscope, analytical apparatus
Computer, gas chromatograph, spectrophotometers, etc.

Staff: Scientists: 4 Others: 22

Research budget (ECU/year): 309.600

Description
Research projects:
Management aspects in shrimp culture of Japanese prawn PENAEUS JAPONICUS Bate and Mediterranean prawn PENAEUS KERATHURUS Foskal.
Larval breeding and growing out tests in sea bass (DICENTRARCHUS LABRAX L.).
Algal cultures and live food chains. Lesina lagoon field survey aimed at the phytoepuration acted by GRACILARIA VERRUCOSA prairie enhancement. Husbandry and tank breeding of VENUS VERRUCOSA, MODIOLUS BARBATUS and other bivalves. High nutrition value diet compounds in live food chains for mariculture purposes.

Spawning areas and ascent of fingerling finfish into the Lesina and Varano lagoons (Southern Italy).

22. CNR - Istituto per lo Studio della Dinamica delle Grandi Masse

Name	:	CNR-Istituto per le Studio della Dinamice delle Grandi Masse
Address	:	San Polo 1364 I - 30125 Venezia
Head	:	Dr. Gianfranco Dallaporta (Director)
Tel.	:	39(0)41.5216811 **Fax**: 39(0)41.5216871

Type of institute
Parastatal

Research activities
Short and long term evolution of the morphology of landforms and subsoil (subsidence, hydrogeology of fresh and brackish waters, bottom sedimentation and erosion, evolution of the littorals and coastal erosion, geodetic levellings, groundwater), marine and lagoon dynamics (flood and tides, air-sea and sea bottom interactions, models thermohaline currents, pollution, remote sensing); transfer of methodologies and prototypes for environmental control and planning.

Facilities
Offshore oceanographic platform.

Staff: Scientists: 26 Technical: 33 Others: 10

Research budget (ECU/year): 1.600.000

Description
Oceanography Division: Research in physical oceanography, mathematical models, experimental studies and observational methods regarding the propagation of tide from Adriatic, the contribution of water and pollutants from the mainland and balances of internal processes in the Venice Lagoon. studies on the general circulation and tides, in the Adriatic and Mediterranea seas.
Geology Division: Description, numerical modelling, geodetic and hydrologic control of land subsidence in the Venetian area.
Environmental Division: Development of techniques to measure physical and chemical parameters of lagoon and marine waters and interpretation of in situ collected data.
Research projects:
EC MAST contracts, POEM, ERS-1, project on the Venice Lagoon system, Adriatic Sea Project.

23. CNR - Istituto Sperimentale Talassografico di Taranto

Name : CNR - Istituto Sperimentale Talassografico di Taranto
Address : Via Roma, 3
I - 74100 Taranto
Head : Dr. Sebastiano Geraci (Director)
Tel. : 099.4525434/490627 **Fax**: 099.49481

Type of institute
Research Institute (of the CNR) - Marine Sector

Research activities
Oceanographical, chemical and biological research

Staff: Scientists: 6 Others: 14

Research budget (ECU/year): 264.000

Description
Aquaculture/environment interaction
Fisheries/multispecies studies
Research projects:
Oceanography and environmental impact
Algology and marine biology
Toxicology and microbiology

24. CNR - Istituto Sperimentale Talassografico di Trieste

Name : CNR - Istituto Sperimentale Talassografico di Trieste
Address : Viale Romolo Gessi, 2
I - 34123 Trieste
Head : Dott. Davide Bregant
Tel. : 040.305312 **Fax**: 040.308941

Type of institute
Governmental

Research activities
Physico-chemical oceanography

Staff: Scientists: 7 Others: 8

Research budget (ECU/year): 40.000

Description
This budget is comprensive for ordinary and extraordinary research
Research projects:
Air/sea interaction
Physico-chemical conditions in South Adriatic sea, distribution of chemical parameters in Ross sea.

25. CNR - Istituto Talassografico - Messina

Name : CNR - Istituto Sperimentale Talassografico
Address : Spianata S. Ranieri, 86
I - 98100 Messina
Head : Dr. Ermanno Crisati
Tel. : 090.669003 **Fax**: 090.669007

Type of institute
Research Centre

Research activities
Fisheries management
- Stock assessment
- Multi-species biological models
- Multi-disciplinary approach to fisheries management
- Environmental problems and their impact on resources
- Upgrading of fishery products

Aquaculture
- Growth, nutrition and feed composition
- Reproductive physiology
- Larval development
- Live food production
- Microbiology of fish

Facilities
Research vessels
Rearing tanks
Hatchery
Water filtration apparatus
Deep-freezers
Light-controlled rearing tanks
Lyophylisator

Refrigerated centrifuge
Laboratories (microbiology, histology, chemical and physical oceanography, phytoplankton).

Staff: Scientists: 18 Others: 12

Research budget (ECU/year): 141.700

Description
Some projects carried out :
A trawl survey in the area from Capo Calava to Capo Gallo in Sicily Thyrrhenian Sea (1984-1988). Collection of fingerlings and rearing of Seriola Demerili in tanks, survey of little pelagic fish.

26. CO.I.P.A. - Coop. Idrobiologia Pesca ed Acquacoltura

Name : CO.I.P.A.-Cooperativa Idrobiologia Pesca ed Acquacoltura
Address : Viale Mazzini, 55
I - 00195 Roma
Head : Dr. Roberto Minervini
Tel. : 6.3217546 **Fax**: 6.3217582 **Telex:** 623658 SCPA I

Type of institute
Research Centre
Engineering/consulting firm

Research activities
Hydrobiology
Aquaculture
Environmental study, pollution
Fisheries management
Fishing techniques
Stock assessment
Feasibility studies and projects
Multi-species biological models
Multi-disciplinary approach to fisheries management
Environmental problems and their impact on resources
Upgrading of fishery products

Facilities
Data bases
Extensive libraries

Specialised software
Field site, microscopes, specialized analytical apparatus
Sea-water hatchery
Fresh-water hatchery
Lagoon of 400ha
Phyto and zooplankton cultures centre

Staff: Scientists: 11 Others: 5

Research budget (ECU/year): 225.000

Description
Created in 1978, the CO.I.P.A. company gathers biologists and engineers specialized in various fields such as fisheries, aquaculture, ecology, limnology, marine biology and zoology. CO.I.P.A. thus relies on a group of outstanding researchers and well adjourned technicians who can:
- Participate in the formulation, implementation and management of breeding plants for fish and shellfish, in salt, brackish and fresh waters.
- Formulate and carry out environmental research, aiming at various problems both in marine and fresh waters.
- Organize research campaign on sea fishing, either for off-shore or on-shore fisheries.
- Formulate and carry out research on intensive farming of fish, crustaceans, molluscs and plankton, including problems of feeding, pathology and ecology.
- Elaborate and manage development plans, for industrial or small-scale fisheries and for aquaculture, in developing countries, with emphasis on the assessment of resource potential.
- Organize training and vocational courses for technicans and staff at various levels, already active in different sectors of fishery and aquaculture.

27. CEAC - Consorzio Ecologia e Acquacoltura Costiera

Name : CEAC - Consorzio per l'Ecologia e l'Acquacoltura Costiera
Address : Via Trieste, 26/Loc. Capisani
Contarina-Rovigo
I - 35141 Padova
Head : Lino Zanella/Ettore Grimaldi
Tel. : 049.661144/380438 **Fax**: 049.655532

Type of institute
Research Centre
Institute for professional training/Secondary school

Research activities
Aquaculture
 Genetics and conservation of gametes
 Early stages of reared species
 Growth, nutrition, feed composition, live food production for aquaculture
 Aquaculture structures, systems, techniques

Facilities
Research vessel
Rearing tank
Specialised computer hardware
Field site, electron microscope, specialized analytical apparatus

Staff: Scientists: 1 Others: 4

Research budget (ECU/year): 123.800

Description
The major research subjects are Sparus aurata wintering and Dicentrarchus Labrax diet.

28. Consorzio Pesca Acquicoltura Trasimeno

Name	:	Consorzio Pesca ed Acquicoltura del Trasimeno Centro Ittiogenico
Address	:	Via della Palazzetta I - 06060 S.Arcangelo di Magione (PG)
Head	:	R.M. Gallo/M. Natali
Tel.	:	075.848041 **Fax**: 075.848041

Type of institute
Fishing
Fish/Crustacean/mollusc farming

Research activities
Fisheries management
Stock assessment
Fishing techniques
 Testing techniques: instruments
 Selective gears
 Static gear
 Aquaculture
Upgrading of fishery products

Facilities
Research vessel
Rearing tank
Controlled environmental chamber

Staff: Scientists: 1 Others: 6

29. CORISA - Consorzio Ricerche Sardegna

Name : Università degli Studi di Sassari
CORISA - Consorzio Ricerche Sardegna
Address : Legale: Piazza Università, 21,
presso l'Università degli Studi
I - 07100 Sassari
Operativa: S.P.55 "Porto Conte Capo Caccia
Km 8,400 - Loc. Tramariglio
I - 07041 Alghero (SS)
Postale: Casella Postale
I - 07040 Santa Maria La Palma (SS)
Head : A. Milella
Tel. : 079.998400 **Fax**: 079.946692

Type of institute
Research Centre

Research activities
Fisheries management
- Stock assessment
- Multi-disciplinary approach to fisheries management
- Environmental problems and their impact on resources
- Fishing techniques
- Selective gears
- Static gear

Aquaculture
- Reproduction
- Early stages of reared species
- Growth, nutrition, feed composition, live food production for aquaculture
- Aquaculture/environment interaction
- Aquaculture structures, systems, techniques

Facilities
Research vessel
Rearing tank

Specialised computer hardware
Specialised software
Field site, electron microscope, specialized analytical apparatus
Equipped laboratories for oceanography, biochemistry and histology.

Staff: Scientists: 10 Others: 5

30. Coop. AGEI - Acquacoltura Gestione Ittica

Name : Coop. AGEI - Acquacoltura Gestione Ittica
Address : Via Ugo de Carolis, 100
I - 00136 Roma
Head : Dr. Giampaolo Buonfiglio
Tel. : 06.343149/3451741 **Fax**: 06.492174/3453552

Type of institute
Research Centre
Engineering/consulting firm

Research activities
Fisheries management
Stock assessment
Multi-disciplinary approach to fisheries management
Environmental problems and their impact on resources
Aquaculture

Facilities
Data bases, extensive libraries
Free admittance to specific instrumentation such as electronic microscope, spectrophotometer etc.

Staff: Scientists: 15 Others: 4

Description
Research projects:
Carried out in the last years: Plan for the reorganization and development of the professional lake fishery.
Study on the mounting of fish juveniles in the Salerno area.
Study on the glass-eel mounting in the Tiber river and on elvers and adults migrations downstream and upstream the castel Giubeleo dam.
Preparation and processing of the data relevant to the I,II,III,IV experimental fishing campaign. Study on the ichthyfaune of the waters bodies of Provincies di Campobasso and project for the exploitation of fish resources. Stock evaluation of demersal fish resources

in the middle and lower Tyrrhenian.
Stock evaluation and population dynamics of natural fry suitable for fish farming, with a study of the correlations with environment for fishery management purpose.
Technical assistance to the corporation of the development agency of the central region in the fisheries sector (Venezuela).
Feasibility study for an aquaculture plant in an area of naturalistic interest.
Rearing trials of marine coryaline species in low salinity waters.

31. Coop. BIOMAR

Name : Coop. BIOMAR
Address : Via XXIV Maggio 60
I - 19100 La Spezia
Head : Marcello Boni
Tel. : 0187.24404

Type of institute
Research Centre

Research activities
Fisheries management
Environmental problems and their impact on resources
Aquaculture
Reproduction
Aquaculture/environment interaction
Aquaculture structures, systems, techniques

Facilities
Specialised computer hardware
Field site, electron microscope, specialized analytical apparatus

Staff: Scientists: 2 Others: 2

32. Coop. BIOSERVICE

Name : BIOSERVICE s.c. a.r.l. -
Studio e Gestione di risorse ittiche
ed ambientali

Address : Via Belvedere 212,
I - 80127 Napoli
LABORATORY: Vico S. Domenico Maggiore 9
I-80134 Napoli

Head : M. Scardi
Tel. : 081.5517064 **Fax**: 081.5528922

Type of institute
Research Centre
Engineering/consulting firm

Research activities
Marine benthos and plankton ecology
Fisheries management
Mathematical modelling

Facilities
Oceanographic research instruments (CTD probe, current meters, dredges, plankton nets, etc;); phase contrast, inverted and sterio microscopes; oven and sieve shaker for granulometrical analysis; Scuba diving equipment and underwater camera; specialised software, literature and species data bases (Tyrrenian Sea and Posidonia oceanica prairies), extensive libraries; computers, printers, plotter, digitizer, etc; outboard engine pneumatic boat; generator set; etc.

Staff: Scientists: 8 Others: 3

Description
Intervention fields:
Marine, brackfish and freshwater environmental monitoring; numerical analysis of mono- and multidimensional ecological data sets; thematic mapping; compatibility analysis and environmental impact assessment; mathematical models of ecological processes; stock assessment; fisheries and natural resources management; site evaluation for natural parks, aquaculture plants, etc.; personnel training in ecology, aquaculture and fishery sciences; data base development and management; vertical software development.

33. Coop. BIOTECNO scrl

Name : BIOTECNO scrl
Address : C.da Birgi S. Teodoro
I - 91020 Marsala (TP)
Site address : P.zza S. Agostino, 12
I - 91100 Trapani
Head : G. Basciano
Tel. : 0923.966733 **Fax**: 0923.966733

Type of institute
Research Centre
Fish/Crustaceans/molluscs farming

Engineering/consulting firm
Professional training

Research activities
Aquaculture
Fishery management
Environmental problems and their impact on resources
Biotechnology

Facilities
Research laboratory
Rearing tank
Specialised software

Staff: Scientists: 4 Others: 10

Research budget (ECU/year): 266.600

Description
Artificial reproduction, larval rearing of fish (Puntazzo puntazzo, Diplodus sargus) crustaceans (Homarus gammarus) and molluscs (Tapes and Haliotis); growing of Seriola dumerilii and Epinephelus guaza.
Biotechnology in aquaculture:
Transgenic fish production.
Triploidy induction in molluscs (Tapes and Haliotis).
Environmental impact.

34. Coop. COISPA

Name	:	COISPA Tecnologia, Ricerca Scrl
Address	:	Statale 16 Km.818,8 - Complanare Est Strada Padovano-Casella postale 62 I - 70042 Mola di Bari (BA)
Head	:	D.ssa Maria Teresa Spedicato
Tel.	:	080.8732077 **Fax**: 080.8733440

Type of institute
Research Centre
Engineering/consulting firm

Research activities
Fisheries management
Stock assessment

Aquaculture
Reproduction
Early stages of reared species
Live food production for aquaculture

Facilities
Rearing tank
Data bases, extensive libraries
Specialised software
Experimental units with tanks ranging from 150L to 40 tons
Controlled environmental chamber for phyto- and zooplankton cultures

Staff: Scientists: 6 Others: 3

Research budget (ECU/year): 230.000

Description
Title of major projects :
Fisheries management :
"Assessment of groundfish resources in Central and Southern Thyrrenean Sea by seasonal trawl-surveys".
Duration of programme: 1990-1995.
Aquaculture :
Marine aquaculture. "Larval and post-larval rearing of PARALICHTHYS OLIVACEUS (japanese flounder)". Duration of programme: 3 years.
Reproduction of EPINEPHELUS GUAZA oriented to develop the Restocking techniques". Duration of programme: 1990-1996.

35. Coop. TECNOMARE

Name : Cooperativa Tecnomare soc. coop. a.r.l.
Servizi e Assistenza Pesca e Acquacoltura
Address : Via dei Marrucini, 21
I - 65127 Pescara
Head : Viola G.
Tel. : 085.693360 **Fax**: 085.693360

Type of institute
Research Centre
Engineering/consulting firm

Research activities
Fisheries management
Stock assessment

Multi-species biological models
Multi-disciplinary approach to fisheries management
Environmental problems and their impact on resources
Aquaculture
Upgrading of fishery products

Facilities
Physical, chemical and biological laboratories for water analysis and aquatic organisms. Possibility to use a boat for sampling.

Staff: Scientists: 4

Description
Tecnomare Cooperative is active in technical, biological and administrative services and assistance to fishing and aquaculture. It is a very young cooperative. Research activities will be carried out on fishing, aquaculture, up-grading and environmental problems of the waters.

36. Dr. Mura Graziella

Name : Mura Graziella
Address : Viale Universita 32
I- Roma
Head : Mura Graziella
Tel. : 6.4958259 **Fax**: 0039.6.49912812

Type of institute
University Department/Higher Education Institute

Research activities
Upgrading of fishery products
Consumer response and demand studies

Facilities
Field site, electron microscope, specialized analytical apparatus

Staff: Scientists: 1 Others: 7

Research budget (ECU/year): 11.900

37. ELMAR - Agency of Ship Design

Name : ELMAR di Giuliano Pagliarani
Address : Via dello Squero, 29
I - 47042 Cesenatico (FO)
Head : Dr. Giuliano Pagliarani
Tel. : 0547.80331 **Fax**: (0)547.80673

Type of institute
Engineering/consulting firm
Fishing
Equipment manufacturer

Research activities
Fishing techniques
- Selective gears
- Static gear
- Hydrodynamics of gears
- Other improvements in gears and/or fishing vessels
- Upgrading of fishery products
- Consumer response and demand studies
- Specific characteristics of vessels

Facilities
Research vessel
Specialised software

Staff: Scientists: 1 Others: 2

Research budget (ECU/year): 140.200

Description
Planning of fishing boats adaptable for each type of fishing in the Adriatic sea and Mediterranean sea.

38. ENEA - CASACCIA - Laboratorio Acquacoltura

Name : ENEA - CASACCIA - Unità a Progetto
Acquacoltura
Dipartimento Agrobiotecnologia
Address : Via Anguillarese, 301
I - 00060 S.M. di Galeria (Roma)
Head : Dr. A. Lazzari/R. Ceccarelli/F. Barbato
Tel. : 39.6.30484716 **Fax**: 39.30484995 **Telex**: 613296 ENEACA I

Type of institute
Government Department
Research Centre

Research activities
Fisheries management
Multi-disciplinary approach to fisheries management
Environmental problems and their impact on resources
Aquaculture

Facilities
Data bases, extensive libraries
Specialised software

Staff: Scientists: 4 Others: 4

Research budget (ECU/year): 230.600

Description
Major projects:
Artificial reproduction of the Mediterranean yellowtail.
Offshore cages rearing.
Floating cages in coastal lagoon.

39. ENEL SpA - DSR - CRTN

Name : ENEL SpA - DSR - CRTN (Nuclear and Thermal Research Center) - Aquaculture section
Address : Via Montfalcone, 15
I - 20132 Milano
Head : Ing. A. Taschini/Dr. P. Bronzi
Tel. : 39.2.7224.3452 **Fax**: 39.2.7224.3496

Type of institute
Research Centre
National Electricity Producer

Research activities
Aquaculture
Thermal aquaculture: optimization of residual heat utilization.
Technology
Physiology

Mariculture
Coastal area management

Facilities
Experimental plant for freshwater thermal aquaculture
Experimental plant for sea water thermal aquaculture
Rearing tank and basins
Laboratories equipped for studies on physiology, microbiology, chemistry

Staff: Scientists: 3 Others: 6

Description
Since 1980 Enel has been involved in research for the optimization of technologies of utilization of warm water discharged by power stations in aquaculture both in sea water and fresh water. Research is conducted in collaboration with several Universities and Institutions, and external assistence (24 persons) is employed for the management of the plants. The main fields of research are nutrition and fish quality improvement, biological control of animals, physiology and stress on sturgeon, eel, carp, tilapia and sea bass.

40. Environmental Marine Research Center

Name : Environmental Marine Research Center
Address : P.O. Box 316
I - 19100 La Spezia
Head : Enea Cram
Tel. : 0039(0)187.5336111 **Fax**: 0039(0)187.536273

Type of institute
Marine Research Center

Research activities
Biology, chemistry, oceanography, sedimentology.

Facilities
Laboratories

Staff: Scientists: 35 Others: 13

Research budget(ECU/year): 368.968,69
(total budget: 632.517,75)

Description
Research projects:
Global climate-change programme, marine data bank, coastal ecotopologies, sea

monitoring and advanced technologies, marine information system, benthos and sediments mapping.

41. G.R.A.I.A. srl

Name : G.R.A.I.A srl - Gestione e Ricerca Ambientale Ittica Acque
Address : Via Milano, 14
I - 21028 Travedona Monate (VA)
Head : Prof. E. Grimaldi
Tel. : 0332.977494 **Fax**: 0332.977494

Type of institute
Research and consulting firm

Research activities
Freshwater ecology
Fisheries management
Stock assessment
Population dynamics
Multi-disciplinary approach to fisheries management
Habitat quality and habitat suitability indices
Models and field applications
IFIM - Instream Flow Incremental Methodology
Eutrophicated lakes restoration
Coastal lagoons productivity
Environmental impact assessment
Aquaculture

Facilities
Data bases, extensive libraries
Field site
Biological laboratory
Electrofishing equipment
Hatchery

Staff: Scientists: 5 Others: 1

Description
Research projects:
Coregonid biology in lake Maggiore. Ecological characterization of Italian rivers in order to point out the effects of hydroelectric development projects. Restoration of northern Italy eutrophicated lakes through phosphorus precipitation.

42. HENDRIX SpA

Name : HENDRIX SpA
Address : Frazione S. Zeno
I - 37060 Mozzecane-Verano
Head : G. Bauce
Tel. : 45.6340120 **Fax**: 45.6340058 **Telex**: 480367

Type of institute
Processing Industry/agro-food/diets producer
Fish/Crustacean/mollusc farming
Industry/Feed production for fish farming: trouts, eels, sea bass, sea bream, sturgeons, carps, cat fish
Distribution

Research activities
Aquaculture
Growth, nutrition, feed composition, live food production for aquaculture
Disease, immunology, stress, feed conversion, reduction of environment pollution

Facilities
Rearing tank n.44

Staff: Scientists: 3 Others: 2

Research budget (ECU/year): 263.500

43. HYDRORES s.a.s.

Name : HYDRORES s.a.s.
di Mario Bussani & C.
Address : ACRG - CNR
I - Genova
Head : Dott. Giuseppe Cotugno
Tel. : 040.34131/302527

Type of institute
Research Centre
Fish/Crustacean/mollusc farming
Research activities
Aquaculture/environment interaction

Aquaculture structures, systems, techniques
Fisheries techniques
Artificial reefs
Macro and micro structures

Staff: Scientists: 2 Others: 5

Research budget (ECU/year): 166.000

44. ICI - Istituto Culturale Italiano

Name	:	ICI-Istituto culturale italiano
Address	:	Via Campania 57-P.O. Box 91 I - 63039 S. Benedetto del Tronto (AP)
Head	:	P. Scipioni Fraticelli
Tel.	:	0735.594029/594410 **Fax**: 0735.584863/83021 **Telex**: 561268 I

Type of institute
Research Centre
Institute for professional training/Secondary school

Research activities
Fishing techniques
Aquaculture
Upgrading of fishery products
Improvement of techniques for handling, storing, processing, packaging fish
Consumer response and demand studies

Facilities
Data bases, extensive libraries
Specialised software
Electromechanical, cold and chemical laboratories. Nautical room. (Facilities available for external users)

Staff: Scientists: 6 Others: 2

Research budget (ECU/year): 50.000

Description
The principal lines in research field are the knowledge about fisheries management in the S. Benedetto del Tronto area; the study of the effect of different kinds of rigging in the trawl nets on resources, environment and fish quality; the test and comparison of various cage patterns for fish culture in open sea; experimentation of fixed nets and their

selectivity; survey and trails of a stern towed mechanized hydraulic dredge for striped venus.

45. IMC - International Marine Centre

Name : IMC - International Marine Centre
Address : Lungomare Eleonora d'Arbores, 22
I - 09072 Torregrande - Oristano
Head : Silvana Vallerga (Scientific Director)
Renato Covacci(General Manager)
Tel. : 39.783.22032/22027 **Fax**: 39.783.22002

Type of institute
Research Centre
Higher Education Institute

Research activities
Fish sensory systems
Genetics
Basic research on fish ecology
Aquaculture
 Physiology
 Aquaculture/environment interaction
Fishery management
 Multi-species biological models
 Environmental problems and their impact on resources
Upgrading of fishery products
 Novel processing methods for by-products, discards, and process wastes
Water quality and ecosystem management

Facilities
Laboratories equipped for studies on marine biology, molecular biology, marine ecology, neuromorphology and neurophysiology.
Laboratories to rent for visiting scientists - Lodging for visiting scientists.
Advanced summer schools - International conferences.
Data base.
Field site and specialized analytical apparatus
Rearing tanks
Research boat

Staff: Scientists: 10 Others: 16

Research budget (ECU/year): 2.500.000

Description

Research on aquatic organisms and their environment is carried out by four transnational teams, a further group is devoted to diffusion and transfer of research results and of innovation. Funding comes mainly from EC programmes.

Research projects:

1. "*Study of the eel biology and the development of the use of eel in aquaculture*", organized into 4 subprojects "Comparison of North Atlantic and Sardinian eel", "Artificial sexual maturation of the Sardinian eel", "The visual pigments of Sardinian and North Atlantic eels", "The natural environment of the eel".
2. "*Fish vision and feeding*". Morpho-physiological studies on fish visual system, characterization of underwater photic environment, study of the relation between vision and feeding in zooplanctivorous larvae.
3. "*Fresh, brackish and marine water environment*". Water quality in river, lagoons and coastal waters is studied as a global model, analysing atmospheric dust, sediment, fresh and marine water pollution.
4. "*Algal ecology and water productivity*". The algal populations are studied for applications of integrated aquaculture and environmental protections.

46. IST - CSTA - Istituto Nazionale per la Ricerca Sul Cancro

Name : National Institute for Cancer Research
Centre for the Study of Tumors of
Environmental Origin
Istituto Nazionale per la Ricerca sul Cancro
Centro per lo Studio dei Tumori di Origine
Ambientale (IST - CSTA)
Address : 10 Viale Benedetto XV
I - 16132 Genova
Head : Prof. Dr. Leonardo Santi
Tel. : 39.1035341 **Fax**: 39.10352999

47. ICRAM - Istituto Centrale per la Ricerca Applicata al Mare

Name : ICRAM - Istituto Centrale per la Ricerca
Scientifica e Tecnologica Applicata al Mare
Address : Via L. Respighi, 5
I - 00197 Roma
Head : Dr. Paolo Arata
Tel. : 06.8077592 **Fax**: 06.8870326/8088326 **Telex**: 623227 I

Type of institute
Research Centre

Research activities
Fisheries management
  Stock assessment
  Multi-species biological models
  Multi-disciplinary approach to fisheries management
  Fishing techniques
Aquaculture
  Upgrading of fishery products
Marine environment

Facilities
Field site, specialized analytical apparatus.
Equipped laboratories for chemistry and biology.
Instruments for oceanography studies.

Staff: Scientists: 47 Others: 20

Research budget (ECU/year): 3.875.500

Description
Research projects:
Demersal and pelagic fish stocks evaluation, life-cycle and population dynamics of Antarctic species, environmental studies (heavy metals determination, environmental impact), development of production methods for several commercial species in aquaculture, pathology in intensive and semi-intensive cultures. Fixed gears and fishing boat technology.

48. Istituto Comunale di Biologia Marina di Nardo

Name	:	Istituto Comunale di Biologia Marina di Nardo
Address	:	S. Caterina di Nardo'
		I - 73048 Nardo' (LE)
Head	:	Prof. G. De Metrio
Tel.	:	0833.823261

Type of institute
Research Centre

Research activities
Fisheries management

Stock assessment
Multi-disciplinary approach to fisheries management
Aquaculture
Early stages of reared species
Aquaculture/environment interaction

Facilities
Research vessel
Data bases, extensive libraries

Staff: Scientists: 3 Others: 4

Research budget (ECU/year): 52.800

Description
Research projects:
Evaluation of Scombridae stocks in Mediterranean, including cooperation with Greece.

49. Istituto di Zootecnica - Gallina

Name : Prof. Scerra Vincenzo
Address : P.zza S. Francesco 7
I - 89061 Gallina (RC)
Head : Prof. Scerra Vincenzo
Tel. : 0965.682816 **Fax**: 682453

Type of institute
University Department/Higher Education Institute

Research activities
Fisheries management
Stock assessment
Aquaculture
Growth, nutrition, feed composition, live food production for aquaculture
Aquaculture/environment interaction
Upgrading of fishery products discards,and process wastes

Staff: Scientists: 6

Description
Istituto Zootecnica is a very new research centre so facilities have just been bought, which means the Institute has been until recently unable to conduct its researches.

50. Istituto Idrobiologia e Acquacoltura G. Brunelli

Name : Istituto di Idrobiologia e Acquacoltura "G. Brunelli"
Address : Casali di Paola
I - 04016 Sabaudia (LT)
Head : Dr. Claudio Costa
Tel. 39.773596703 **Fax**: 39.63217582

Type of institute
Research Centre

Research activities
Fisheries management
- Stock assessment
- Multi-species biological models
- Multi-disciplinary approach to fisheries management

Aquaculture

Facilities
Rearing tank
Data bases, extensive libraries
Specialised software
Field site, microscopes, specialized analytical apparatus
Scientific journal

Staff: Scientists: 4 Others: 2

Research budget (ECU/year): 80.000

Description
The institute was established in 1978 to promote, manage and spread out studies and researches in the field of hydrobiology, fisheries and aquaculture. Among the most relevant studies carried out by the Institute are: the survey on bivalve beds of the Tyrrhenian coast; the experimental feasibility and project for a salt-water hatchery, supported with non-conventional energies; the integrated proposal for the establishment of the marine park of Portofino; the evaluation of the economic and biological impact of the compulsory fishing halt; artificial feeding of larval stages of marine species.
The Institute is also interested in R&D studies in and for the LCD's.
Its major achievements are: the feasibility and project for a 400ha shrimp farm in Andra Pradesh (India); the field survey and operative proposal for the development of the fisheries in Uganda; the various courses for foreign aquaculture workers.
Besides specific studies, the Institute aims at monitoring the environment of Sabaudia Lake through regular sampling and detailed fisheries records. The Institute is publishing an international scientific journal, the "Quaderni dell'Istituto Brunelli". At present it is

carrying out a research program "Elaboration of sword-fish fishing methods compatible with the safeguard either of juvenile sword-fish or of protected species", EEC contract MED/91/005.

51. Istituto Nazionale della Nutrizione

Name : Istituto Nazionale della Nutrizione
Address : Via Ardeatina, 546
I - 00178 Roma
Head : Dr. Giovanni Battista Quaglia
Tel. : 06.5032412 **Fax**: 06.5031592

Type of institute
Research Centre

Research activities
Aquaculture
Growth, nutrition, feed composition, live food production for aquaculture
Upgrading of fishery products
Novel processing methods for by-products, discards, and process wastes

Staff: Scientists: 4 Others: 8

Description
Major projects:
AQUACULTURE:
Whitefish eggs biochemical composition as an approach to the nutritional requirements of fry, for the improvement of fisheries management in the lake of Bracciano.
Biometrical and biochemical analysis of freshwater Anostracans for their valorization in aquaculture.
Biochemical evaluation of the sardinian population of Artemia salina from S. Antioco for its valorization in aquaculture.
UPGRADING:
Enzymatic modification of the chemical and functional properties of proteins from underutilized fish. Valorization of small pelagic fish by identification and trial of processing lines for obtaining innovative products. Suitable technologies for developing countries for the valorization of small pelagic species of fish.

52. I.R.E.P.A. - Istituto Ricerche Economiche per la Pesca

Name : I.R.E.P.A.
Istituto Ricerche Economiche per la Pesca
e l'Acquacoltura
Address : Via B. Croce, 35
I - 84100 Salerno
Head : Prof. Massimo Spagnolo
Tel. : 089.241503 **Fax**: 89.241145

Type of institute
Research Centre
Consulting firm

Research activities
Multi-disciplinary approach to fisheries management
Aquaculture economics
Consumer response and demand studies

Facilities
Data bases, extensive libraries
Specialised software
Both available to external users

Staff: Scientists: 8 Others: 20

Research budget (ECU/year): 625.300

Description
I.R.E.P.A. is a specialist research institute committed in advisory services to national and international public bodies involved in fisheries and aquaculture management.
Field of activity: mathematics, statistics, econometrics, economic analysis and political economy applied to fisheries and aquaculture.
Research projects:
Demand and price formation studies for fish products; bio-economic modelling for Italy and Mediterranean; monitoring system for techno-economic data in fisheries; artisanal fisheries; technical and economic standard parameters, fishing industry input-output; marketing outlets and distributive channels for fishery products; coastal lagoon management; development and markets for aquaculture products; countertrade in the fishing sector.

53. Istituto Zooprofilattico delle Venezie

Name : Istituto Zooprofilattico delle Venezie
Fish Diseases Dept.
Address : Via della Roggia 92
I - 33030 Basaldella di Campoformido (UD)
Head : Prof. G. Giorgetti
Tel. : 0432.561196 **Fax**: 0432.561532

Type of institute
Research Centre

Research activities
Fisheries management
  Environmental problems and their impact on resources
Aquaculture
  Physiology
  Disease, immunology, stress
  Aquaculture/environment interaction
  Aquaculture structures, systems, techniques

Facilities
Flume tank
Rearing tank
Data bases, extensive libraries
Specialised software
Field site, electron microscope, specialized analytical apparatus

Staff: Scientists: 3 Others: 4

Description
Diagnostic support for the fish farmers.
Research on parasites, bacteria, virus of the fish species farmed in Italy.
Organisation of eradication plans against V.H.S. and I.H.N.
Test of efficacy on bacterial vaccines (Vibrio anguillarum - Yersinia ruckeri).
Efficacy's test of associated vaccines.
Efficacy's test of vaccines administred orally. Efficacy's test of new antibiotics also for the registration purposes.
Serologic identification of various Yersinia's strains.

54. Istituto Zooprofilattico Sperimentale Piemonte Liguria Val d'Aosta

Name : Ist. Zooprofilattico. Sperim Piemonte, Liguria
Valle d'Aosta I.Z.S.P.L.V
(Experimental Institute for zooprophylaxis in
Piedmont, Ligura and Aosta Valley)
Centro per lo Studio delle Malattie dei Pesci
(Fish Disease Laboratory)
Address : Via Bologna, 148
I - 10154 Torino
Head : Dr. Claudio Ghittino
Tel. : 011.2686251 **Fax**: 011.851093

Type of institute
Diagnostic and Research Centre

Research activities
Aquaculture
Aquaculture techniques
Fish diseases

Facilities
Laboratory Equipment for fish pathology
Experimental rearing tank

Staff: Scientists: 5 Others: 6

Description
In the Fish Disease Laboratory of Experimental Institute for Zooprophylaxis, Institute of Turin, we do a preeminent diagnostic activity on diseases affecting the main reared species of freshwater and marine fish. We have histological, virological, bacteriological, toxicological units and a wet laboratory at our disposal. Research on fish diseases is also done in the field, by private fishfarms which provide facilities.

55. Italittica srl

Name : Italittica srl
Acquacoltura
Address : Via Vincenzo Florio, 23
I - 91025 Marsala (TP)
Head : Prof. V. Nania
Tel. : 0923.981588 **Fax**: 0923.981294 **Telex**: 910632

Type of institute
Fish/Crustacean/mollusc farming
Processing Industry/agro-food/diets producer
Engineering/consulting firm
Research Centre

Research activities
Aquaculture

Facilities
Research vessel

Staff: Scientists: 4 Others: 4

Research budget (ECU/year): 310.600

Description
A research programme on Seriola dumerilii has been achieved:
Possibility of breeding in particular pools at an aquaculture plant. Study of Seriola reproduction in his natural environment. During this year, we are setting up a programme of research for a possible reproduction of "Pagrus-Pagrus" and we are also developing the study of a type of feeding for every kind of fish that we reproduce during all the different stages of growing.

56. KOBA srl (Biomass Provision and Conversion)

Name : KOBA srl (Biomass Provision and Conversion)
Address : Via dei Bollo, 4
I - 20123 Milano
Head : Dr. Virginia Bombelli
Tel. : 39.2.26410069 **Fax**: 39.2.26410069

Type of institute
Engineering and consulting company

Research activities
Live food production for aquaculture: Study, research and development of aquatic micro-macrophytes cultures as food for aquaculture.
Aquaculture structures, systems, techniques:
Study, research and development of:
a) integrated systems of aquaculture combining fish breeding with aquatic biomass.
b) phytodepuration systems for the treatment of effluents from intensive aquaculture.
c) micro-macroalgae culture for industry, energy and commercial purposes.
d) study of lagoons and aquatic environments (mapping, phytosociological studies).

Facilities
Specialized software
Analytical and culture apparatus

Staff: Scientists: 3 senior biologist Others: junior, technicians and collaborators including engineering

Description
The activity of KOBA set up in 1983 covers the study, development and optimization both at research stage and in the realization of Biomass Integrated Systems for industrial, energy and nutritional purposes.

57. Lab. di Biologia Marina (LBM)

Name : Laboratory of Marine Biology
Laboratorio di Biologia Marina (LBM)
Address : Strada Costiera, 336-S. Croce
I - 34010 Trieste-Friuli Venezia-Giulia
Head : Pr. Serena Fonda Umani (Scientific Director)
Tel. : (39) 40 224400/224464/224620/224621 **Fax**: (39) 40 224437

Type of institute
Consortium inter-university

Research activities
Physical oceanography, marine chemistry and geochemistry (nutrients, sediments, waters, suspended matters); marine geology and geophysics (sedimentology); marine biology (bacterioplankton, microphytoplankton, phytoplankton, microzooplankton, mesozooplankton, benthos, algae, eutrophication, molluscs, crustaceans); marine physiology; marine biochemistry; marine microbiology; toxicology; taxonomy and systematics; marine ecology (community and population studies); aquaculture, marine environmental management; monitoring; assessment; pollution; feeding behaviour of fish larvae.
Scientific support to the WWF-Riserva Parco di Miramare (Trieste).

Facilities
Publications of "Nova Thalassia" journal.
Library with 10.000 volumes/reprints and 50 periodicals.
Computers and various printers.
Laboratory equipment: spectrophotometer, spectrofluometer, IPLC, autoanalyser, coulter counter, CHNS, CTD, salinometer, centrifuges, thermostas, autoclave, rearing cells, microscopes, fume hoods.
Research crafts RV "Noctiluca" (1983).

Facilities for 6 visiting scientists, 12 students.
Conference/seminar facilities for 40 participants.

Staff: : Scientists: 24 Others: 9

Research budget (ECU/year): 1.000.000

Description

Research projects:

-Eutrophication control and water quality evaluation in the mussel farms of the Friuli Venezia-Giulia Region coastal waters.

-Study of coastal phytoplankton and evaluation of blooms and mucilaginous exudates and their impact on the North Adriatic ecosystem, compared to physical, chemical and biological properties of the water column and sediments during the year.

-Study of biology and ecology of potentially toxic algae and detection of toxine in molluscs.

-Impact on the mucous aggregates formation of zeolite Na-A and polycarboxylic acids, using as detergents.

-Control on mucous aggregates development along the water column in the North Adriatic Sea.

-New artificial reef in the Gulf of Trieste: co-ordination of research in progress or planned.

58. Lab. di Biologia Marina ed Acquario - Provincia di Bari

Name	:	Laboratorio Provinciale di Biologia Marina ed Acquario
Address	:	Molo Pizzoli (Porto) I - 70123 Bari
Head	:	Prof. G. Marano
Tel.	:	080.5211200 **Fax**: 080.5213486

Type of institute

Research Centre

Research activities

Fisheries management
Fishing techniques
Testing techniques/instruments
Aquaculture

Facilities

Research vessel
Data bases, extensive libraries

Field sites, optical microscopes, specialized analytical apparatus

Staff: Scientists: 9 Others: 10

Research budget (ECU/year): 331.000

Description
Research projects:
Research concerns the fishing resources assessment in the South Adriatic (trawl fishing, small and large pelagic fish, clams) and reproduction cycles and growth of molluscs and teleosteans. Another research project is carried out regarding essentially the degrading of coastline and investigations of biocenosis damaged by pollution.

59. Lab. di Idrobiologia

Name	:	Laboratorio Centrale di Idrobiologia Ministero dell'Agricoltura e delle Foreste
Address	:	Viale del Caravaggio, 107 I - 00147 Roma
Head	:	S. Panella
Tel.	:	06.5140296

Type of institute
Research Centre

Research activities
Fisheries management
Environmental problems and their impact on resources
Aquaculture

Facilities
Data bases, extensive libraries
Specialised computer hardware
Specialised software
Field site, electron microscope, specialized analytical apparatus

Staff: Scientists: 6 Others: 10

Research budget (ECU/year): 98.900

Description
Research projects:
Environmental impact for industrial waifs, hydrobiological research in salty waters, studies on yields of extensive aquaculture in salty lakes.

Facilities:
Specialized library available for external use.
Hewlett-Packard, NCR and IBM computers with printers and plotter. Computer programmes for statistical analyses, graphics and data bases.
Chemical and biological laboratories.
Instruments of laboratory for field work.

60. M.A.R.E. s.a.s.

Name	:	M.A.R.E. s.a.s. (Marinescience Application in Research and Education)
Address	:	Seat: Via Antica Romana di Quinto, 37 I - 16166 Genova Agency: Viale Tommaseo, 19 I - 09072 Torregrande-Oristano
Head	:	Renato Covacci
Tel.	:	(39)10.335626/783.22200/22027/22032 **Fax**: (39)783.22002

Type of institute
Engineering/consulting firm

Research activities
Aquaculture
Aquaculture/environment interaction
Aquaculture structures, systems, techniques
Upgrading of fishery products
Consumer response and demand studies

Facilities
Specialised computer hardware
Specialised software

Staff: Scientists: 4 Others: 10

Description
M.A.R.E. acts at transnational level organizing services for research centres and enterprises, disseminating research's results, planning procedures, designing techniques, organizing joint projects to promote and improve the sustainable exploitation of aquatic organisms and to preserve coastal marine environment. Expertise includes training and editing. The main outcome of M.A.R.E.'s activity has been the establishment of the IMC - International Marine Centre in Sardinia. The research projects are channelled to the IMC, while M.A.R.E. deals with the diffusion and the application of results and custom tailored projects.

61. Maricoltura Italia SpA

Name : Maricoltura Italia SpA
Address : Via Timavo, 76-Gorizia CP 84
I - 34074 Monfalcone (GO)
Head : S.Bozzetto/I. di Maria
Tel. : 0481.410035 **Fax**: 0481.791385 **Telex**: 461220 MARIT

Type of institute
Fish/Crustacean/mollusc farming

Research activities
Aquaculture
- Early stages of reared species
- Disease, immunology, stress
- Aquaculture structures, systems, techniques

Facilities
Rearing tank
Field site, electron microscope, specialized analytical apparatus

Staff: Scientists: 2 Others: 7

Research budget (ECU/year): 79.400

Description
Research programmes:
Setting up of a technique for swim bladder inflation.
Broodstocks'conditioning to thermo-photo period.

62. Polo Tecnologico Marino Marittimo

Name : Polo Tecnologico Marino Marittimo
Address : Via al Molo Vecchio, Catala Gadda
I - 16126 Genova
Head : G. Pattarini
Tel. : 010.2471020 **Fax**: 010.2471097

Type of institute
Semi-private institute regarding maritime and marine biological aspects.

Research activities
Management of the sea
Research network

Aquaculture and mariculture
Education updating project

Facilities
Automatic analyser for nutrients, spectrophotometer, gas chromatograph/MS, multiparameter probe with fluormeter and transmissionmeter, geographic information system for satellite images and data elaboration.
In 1994 we foresee the buying of one boat for marine monitoring and buoys for continuous monitoring.

Staff: 5 employees plus 20 to be hired

Research budget (ECU/year): 1.600.000

Description
The PTMM is a s.r.l.c. issued in Genova in June 1992 with the support of the E.E.C. and the Ligurian Region as a centre for the development of those problems related to the management of marine environment and of maritime activities.
The projects advanced presently (with reference to the above-mentioned directives) are:
- Management of the sea: systems for integrated monitoring and for the control of navigation.
- Research network: concentration of knowledge regarding marine-maritime issues and its integration with other projects.
- Aquaculture and mariculture: an integrated system that aims to increase areas that are already productive (restoking) and to create high-technology aquaculture and mariculture sites.
- Education updating project: the formation of marine technologies and regarding the marine professions.

The expected activities for the next years not only include the above projects but will also encompass:
- Development and enhancement of a system for the control of navigation (V.T.S.).
- Development of monitoring activities (sensor and tele-monitoring).
- Control system of freshwater inputs to the sea.
- Control system of mercantile goods.

63. R.S.T.A.

Name	:	R.S.T.A.-Ricerca e Sviluppo Tecnologie Appropriate S.c.r.l.
Address	:	Via Al Molo Vecchio Calata Gadda I - 16126 Genova
Head	:	Dott. ssa Claudia Revelli
Tel.	:	(010)20921

Research activities
Fisheries management
Stock assessment
Aquaculture

Facilities
Field site, electron microscope, specialized analytical apparatus

Staff: Scientists: 9

Research budget (ECU/year): ± 90.000

64. Riserva Naturale Marina di Miramare

Name	:	Riserva Naturale Marina di Miramare
Address	:	Viale Miramare, 349
		I - 34014 Grignano (Trieste)
Head	:	G. Benedetti/M. Spoto
Tel.	:	0039.40224147 **Fax**: 0039.40224147

Type of institute
Natural marine reserve

Facilities
Research vessel
Simulated tidal tank
Field site

Staff: Scientists: 4 Others: 1

Research budget (ECU/year): 10.600

Description
Stock assessment, particulary of shellfish in a protected area. Dynamics of seagrass beds. Dismass migrations. Seasonality of benthic populations. The main activities concern research and education, which are managed in cooperation with the Laboratory of Marine Biology of Aurisina (Trieste).

65. S.I.VAL.CO SpA - Societa Itticoltura Valli Comacchio

Name : S.I.VAL.CO SpA
Centro Ricerche di Valle Campo
Address : Via Mazzini, 200
I - 44022 Comacchio (FE)
Head : F. Ghion
Tel. : 0533.330271 **Fax**: 0533.313053

Type of institute
Research Centre
Fish/Crustacean/mollusc farming

Research activities
Aquaculture
- Reproduction
- Early stages of reared species
- Growth, nutrition, feed composition, live food production for aquaculture
- Disease, immunology, stress
- Aquaculture/environment interaction
- Aquaculture structures, systems, techniques

Facilities
Rearing tank
Field site, specialized analytical apparatus

Staff: Scientists: 3

Description
The research centre operates in strong connection with fish farming plants of S.I.VAL.CO. Company.

66. SSICA - Stazione Sperimentale Industria Conserve Alimentari

Name : Stazione Sperimentale per l'Industria
delle Conserve Alimentari
Reparto Conserve Ittiche
Address : Viale F. Tanara, 31/A
I - 43100 Parma

Head : Gianfranco Baldrati
Tel. : 521.7951 **Fax**: 521.771829 **Telex**: 532259 SSICA I

Type of institute
Research Centre

Research activities
Upgrading of fishery products
Improvement of techniques for handling, storing, processing, packaging fish
Methods for grading and quality assessment of fish
Novel processing methods for by-products, discards, and process wastes

Facilities
Electron microscope, specialized analytical apparatus and processing equipments
Cooperation with the scientists working in other departments of the Institute on different sectors.
Data bases, long-term series, extensive libraries.
Microbiology, packages inspection, heat processing, waste-waters processing.

Staff: Scientists: 2 Others: 2

Research budget (ECU/year): 126.000

Description
Research projects:
UPGRADING:
Modified atmosphere storage of fishery products. Use of D-Glucono-8-Lactone as acidifier to increase the shelf-life of fresh fish and to enhance the acidic effect of acetic acid in marinated products. Tests on shelf-life of several frozen fish products. Development of novel products from fish minces. Upgrading of sardines: investigation of mince preparation and new texturisation ways using mince, surimi and mixed materials. Preparation and evaluation of surimi from different fish species.

67. Stabilimento Ittiogenico - Regione Lazio

Name : Stabilimento Ittiogenico di Roma
Regione Lazio
Address : Via della Stazione Tiburtina, 11
I - 00162 Roma
Head : Dr. E. Gelosi
Tel. : 06.4462414 **Fax**: 06.4456689

Type of institute
Research Centre
Government Department/Regional or local authority

Research activities
Fisheries management
  Stock assessment
  Multi-species biological models
  Multi-disciplinary approach to fisheries management
  Environmental problems and their impact on resources
  Fishing techniques
Aquaculture

Facilities
Rearing tank
Data bases, extensive libraries

Staff: Scientists: 3 Others: 7

Research budget (ECU/year): 131.800

68. Stazione Zoologica - Anton Dohrn (SZN)

Name : Stazione Zoologica "Anton Dohrn"
Naples Zoological
Address : Villa Communale
I - 80121 Napoli
Head : Dr. Lucio Cariello (Director General)
Tel. (39)81/5833111 **Fax**: (39)81/7641355 **Telex**: 7200256 SZN I

Type of institute
Research Centre

Research activities
Biological and marine environmental research:
Water masses and circulation, analytical marine biochemistry, modelling, community and population studies in marine organisms, coastal and marine management, water quality study.

Facilities
Laboratory provided with controlled ambient chambers and continuous seawater supply.
Equipments for underwater sampling and measurements, aquariums.
Analytical instruments, electron microscopes, computer network, remote access for E-mail, library, conference/seminar facilities.

Staff: Technicians: 46 Others: 16

Research budget (ECU/year): 10.000.000

Description
Research projects:
Fertilization and development in marine organisms
Molecular biology
Comparative physiology of marine organisms
Comparative neurobiology
Behavioural aspects in marine invertebrates
Ecophysiology of marine organisms
Biological oceanography
Benthic ecology
Marine botany
Aquariology
CEE: Box and flux of nitrogen and carbon in *Posidonia oceanica*

69. STM - Servizi Tecnici in Maricoltura

Name : STM - Servizi Tecnici in Maricoltura
Address : Via Durer, 38
I - 35132 Padova
Head : Alessandro Moretti
Tel. : 39.49.617171 **Fax**: 39.49.8640866

Type of institute
Engineering/consulting firm for fish/crustacean/mollusc farming

Research activities
Aquaculture
- Reproduction
- Early stages and growing of reared species
- Nutrition, feed composition, live food production
- Disease, immunology, stress
- Aquaculture-environment interaction
- Aquaculture structures, systems, techniques
- Economics

Staff: Scientists: 8 Others: 2

Description
Research projects:
Reproduction techniques for seabream, seabass, prawns, manila clam. Live food

production. On-growing techniques of fish under intensive, semi-intensive and extensive rearing conditions. Semi-intensive farming of shrimp with integrated diets. STM has carried out fish, shrimp and clam farming projects in Italy (3), Portugal (1), Spain (1), Turkey (1), Greece (3), Tunisia(1).
It has cooperated as aquaculture consultant with FAO, GTZ (German Agency for Technical Cooperation) and NATO.

70. Univ. Ancona - Dipartimento Scienze Materiali e Terra

Name : Università degli studi di Ancona
Dipartimento scienze materiali e terra
Address : Via Brecce Bianche
I - 60131 Ancona
Head : Prof. P. V. Curzi
Tel. : 071.2204712 **Fax**: 071.2204714 **Telex**: 561838 UNIVAN I

Type of institute
Research Centre
University Department

Research activities
Fishing techniques
Electronics/acoustics
Upgrading of fishery products
Consumer response and demand studies
Grain size, geotechnical, mineralogical, chemical, biochemical, geochemical, sedimentological analysis.
Elaboration and interpretation of high resolution profiles acoustic and seismic profiles (bathymetry, Subbottom profiles UNIBOOM, Sparker, Side Scan Sonar)

Facilities
Data bases, extensive libraries
Field site, electron microscope, specialized analytical apparatus

Staff: Scientists: 12 Others: 8

Research budget (ECU/year): 35.000

Description
Geochemical, mineralogical, geotechnical and sedimentological studies are carried out on the bottom sediments with benthonic population.
The aim of the above studies coupled with biological and oceanographic analyses carried out by other Institutes, is to define the environmental parameters which determine the distribution and density of benthic population.

It must be noted that during 1991-1992 academic year, a new faculty of Marine and Oceanographic biology will start in Ancona University.
This kind of Faculty will be unique and the first in Italy in the next years. As a consequence, during the next years, the Ancona University will get more appropriate laboratories, facilities and scientific staff.

71. Univ. Bari - Dipartimento di Produzione Animale

Name : Università degli Studi di Bari, Facolta
Medicina Veterinaria
Dipartimento de Produzione Animale, Biologia
della Pesca
Address : Strada per casamassina, KH 2
I - 70010 Valenzano
Head : G. De Metrio
Tel. : 080.8779076 **Fax**: 080.8779077

Type of institute
Research Centre
University Department/Higher Education Institute

Research activities
Fisheries management
  Stock assessment
  Multi-disciplinary approach to fisheries management
Aquaculture

Facilities
Research vessel
Controlled environmental chamber
Data bases, extensive libraries
Specialised computer hardware
Specialised software
Field site, electron microscope, specialized analytical apparatus

Staff: Scientists: 5 Others: 10

Research budget (ECU/year): 80.000

Description
Research projects:
Studies on evaluation of Scombridae stocks in Mediterranean, including cooperation with other countries (Greece-Turkey) and international organisations (FAO, CGPM).

72. Univ. Bari - Istituto di Chimica

Name : Università degli Studi di Bari
Facolta di Medicina Veterinaria
Istituto di Chimica
Address : Via Gentile, 182
I - 70126 Bari
Head : Prof. G.C. Marcotrigiano
Tel. : 080.583593 **Fax**: 080.334906

Type of institute
University Department

Research activities
Fisheries management
Environmental problems and their impact on resources
Aquaculture
Aquaculture/environment interaction
Upgrading of fishery products
Methods for grading and quality assessment of fish
Novel processing methods for by-products, discards, and process wastes

Facilities
Specialised software

Staff: Scientists: 10

Research budget (ECU/year): 57.000 ECU for 3 years,
100.000 ECU for 2 years.

Description
Research projects:
Evaluation of large pelagic stocks (swordfish and tuna) in Italian and Greek seas and amount of pesticides and heavy metals in their meats.
Influence of several pollution agents (PAHs, PCBs, heavy metals) in marine and commercial species (bivalves, crustaceans and fishes).
OBJECTIVES :
Control methods to ensure the safety and enhance the quality and commercial value of products;
Methods for grading and quality assessment of fish and fishery products;
Consumer response and demand studies.

73. Univ. Bari - Zoologia ed Anatomia Comparata

Name : Università degli Studi di Bari
Istituto di Zoologia ed Anatomia Comparata
Address : Via Amendola, 165 A
I - 70126 Bari
Head : Prof. A. Tursi
Tel. : 080.243350 **Fax**: 080.243358

Type of institute
University Department/Higher Education Institute

Research activities
Fisheries management
Stock assessment
Environmental problems and their impact on resources

Facilities
Data bases, extensive libraries
Specialised software

Staff: Scientists: 6 Others: 4

Research budget (ECU/year): 125.400

Description
At present, we are carrying out research projects as follows:
1) Stock assessment of demersal resources of the Ionian sea (Italy) financed by the Ministry of Merchant Navy (project leader: Prof. A. Tursi). 2) Investigation of the abundance and distribution of demersal stocks of primary importance to the Greek fishery in the north Aegean sea (Greece) founded by the EEC (project leader: Dr. C. Papaconstantinou-National Centre for Marine Research of Athens). The objective of these studies is to contribute to the knowledge of the biology and population dynamics of some demersal fish stock in the mentioned areas and to propose new or supplementary measures for their management.

74. Univ. Basilicata - Dipt. Produzioni Animali - Lab. Acquacoltura

Name : Università della Basilicata
Facoltà di Agraria
Dipartamento di Produzioni Animali

Address : Via Nazario Sauro, 85
I - 85100 Potenza
Head : Prof. Marco Saroglia
Tel. : (0)971 474295 **Fax**: (0)971.470719

Type of institute
Research Centre
University Department/Higher Education Institute
National level

Research activities
Aquaculture

Facilities
Research vessel
Rearing tank
Field site, electron microscope, specialized analytical apparatus
A new laboratory with research vessels and rearing tanks is in progress

Staff: Scientists: 3 Others: 2

Research budget (ECU/year): 100.000

Description
Research projects: (project leader: Prof. M. Saroglia)
The staff has been involved on studies related with fish physiology, sea bass reproduction and early stages development, then on growth, Vit C on nutrition, nitrogenous catabolites, aquaculture/environmental interaction, stress.
At the moment, two main projects are on progress:
1) Minimisation of ammonia excretion peaks through "on-field" assessment of the alimentation schedule. The physiology of nitrogen secretion is on study, with the goal of clarifying the role of environmental factors on the ratio of excreted ammonia-N and urea-N. The research is going to be supported by the EEC and will be developed in cooperation with Heriot-Watt Univ. (Edinburgh, UK).
2) A study is in progress, on the environmental impact of coastal-marine aquaculture. The main goal of the study is to evidentiate the most appropriate analytical index for a correct forecast and post operational evaluation of the impact on the receiving water body.

75. Univ. Cagliari - Ist. Zoologia

Name : Università degli Studi di Cagliari
Dipartimento Biologia Animale ed Ecologia
Address : Viale Poetto, 1
I - 09100 Cagliari

Head : Prof. A. Cau
Tel. : 070.370263 **Fax**: 070.380285

Type of institute
University Department/Higher Education Institute

Research activities
Fisheries management
- Stock assessment
- Multi-species biological models
- Fishing techniques
- Selective gears
- Static gear

Facilities
Research vessel
Rearing tank
Controlled environmental chamber
Data bases, extensive libraries
Specialised software
Field site, electron microscope, specialized analytical apparatus
Basic biology

Staff: Scientists: 8 Others: 14

Research budget (ECU/year): 800.000

Description
Research projects:
Population dynamics and evaluation of stocks of fry useful to farming and studies of correlations with environmental parameters. Evaluation and biology of crustacean stocks in Western Mediterranean. Uniqueness of genetic population.

76. Univ. Genova - Ist. Anatomia Comparata

Name : Università degli Studi di Genova
Istituto di Anatomia comparata
Address : Viale San Benedetto XV, 5
I - 16100 Genova
Head : A. Morescalchi/M. Wurtz
Tel. : 010.3538050 **Fax**: 3538047

Type of institute
University Department/Higher Education Institute

Research activities
Fisheries management
  Stock assessment
Aquaculture
  Physiology
  Reproduction
  Genetics and conservation of gametes
  Growth, nutrition, feed composition, live food production
  for aquaculture

Facilities
Field site, electron microscope, specialized analytical apparatus

Staff: Scientists: 15 Others: 7

Research budget (ECU/year): 131.700

Description
Field of research:
- Evolutive convergence between aquatic vertebrates and invertebrates: growth; mathematical models of growth; age determination; morphology and morphometrics; population dynamics; population genetics; competition and feeding interactions.
- Biology of aquatic vertebrates and invertebrates: reproduction; distribution; stock assessment; fishery resources biology; plant-animal interactions.

77. Univ. Genova - Scienze Ambientali Marine

Name	:	Università di Genova Istituto di Scienze Ambientali Marine Laboratorio di Biologia Marina ed Ecologia Animale
Address	:	Corso Rainusso, 14 I - 16038 S. Margherita Ligure (Genova)
Head	:	Prof. N. Della Croce
Tel.	:	0185.286195 **Fax**: 0185.281089 **Telex**: 271114 UNIVEGEI

Type of institute
Research Centre

Research activities
Fisheries management
  Stock assessment
  Environmental problems and their impact on resources
Aquaculture
  Aquaculture/environment interaction

Facilities
Data bases, extensive libraries
Specialised computer hardware
Specialised software
Field site, electron microscope, specialized analytical apparatus
The institute offers housing capacity for 5 visitors.

Staff: Scientists: 8 Others: 15

Description
The following research programmes are being developed:
- Macrobenthic communities and pelagic-benthic coupling
- Plankton distribution, ichtyoplankton stock assessment, hydrology in the Ligurian sea
- Energetical exchanges in the planktonic ecosystem
- Copepoda and Bivalves in the Antarctic sea
- Seston and organic matter in the sediments used for nourishing the superior trophic zones.

78. Univ. Genova - Zoologia

Name	:	Università degli Studi di Genova Istituto di Zoologia
Address	:	Via Balbi, 5 I - 16126 GENOVA
Head	:	Prof. A. Arillo / Prof. G. Relini
Tel.	:	39.10.2039379 **Fax**: 39.10.20260

Type of institute
University Department/Higher Education Institute

Research activities
Fisheries management
  Environmental problems and their impact on resources, stock assessment
Upgrading of fishery products
  Consumer response and demand studies

Facilities
Field site, electron microscope, specialized analytical apparatus

Staff: Scientists: 6 Others: 4

Research budget (ECU/year): 131.700

Description
Several research programmes are carried out :
Valuation of fish resources, particularly demersal resources.
Fishing efficiency.
Biology of the more interesting species for the market, definition of biological parameters. Artificial reef. Submerged aquaculture.
Contamination of the food chain of the deep water bottoms.
SPECIALITIES :
Taxonomy of the Mediterranean fishes, decapod crustaceans and cephalopod molluscs.
Determination of the age, growth, maturity-breeding, diet in interesting species for the market of Ligurian Sea.

79. Univ. Lecce - Dipartimento di Biologia

Name : Università degli Studi di Lecce
Dipartimento di Biologia
Address : Via Provinciale Monteroni
I - 73100 Lecce
Head : Prof. Carlo Storelli
Tel. : 0832.6201 **Fax**: 0832.351504

Type of institute
University Department/Higher Education Institute

Research activities
Fisheries management
Aquaculture

Facilities
Field site (aquatina lagoon)
Rearing tank
Electron microscope, specialized analytical apparatus
Fully equipped greenhouse for fish holding

Staff: Scientists: 15 Others: 10

Research budget (ECU/year): 50.000

Description
Research projects:
1) Repopulation of lagoonal environments: study on ecology and economic aspects.
2) Physiology of teleost digestive tract: study on sugar, amino acid, vitamin and ion transport systems.
3) environmental adaptation of teleosts of great interest for aquaculture: physiology of kidney and osmoregulation in freshwater and seawater fishes.
4) Metabolism of fatty acid in the liver of seawater teleosts.
5) Dietology and feed composition for aquaculture: phytoplancton and zooplancton production.
6) Application of molecular biology techniques in obtaining rapidly growing and/or disease resistant transgenic varieties of fishes.
7) Aquaculture/environment interaction.

80. Univ. Milano - Inst. Pharmacological Science

Name	:	University of Milano Laboratory of Nutrition and Comparative Biology
Address	:	Via Balzaretti, 9 I - 20133 Milano
Head	:	R. Paoletti/E. Agradi
Tel.	:	39.2.20488356 **Fax**: 39.2.29404961

Type of institute
Research Centre
University Department/Higher Education Institute

Research activities
Aquaculture
Upgrading of fishery products

Facilities
Rearing tank
Specialized analytical apparatus

Staff: Scientists: 3 Others: 3

Description
Research projects:.
Development of nutritional protocols for the improvement of growth survival and nutritional quality of aquacultured fish. Role of lipid supplements on biochemical responses to stressors in intensivety of fish meat enriched in polyunsaturated fatty acids (PUFA) with beneficial effects on human health.

81. Univ. Modena - Dipt. Biologia Animale

Name : Università degli Studi di Modena
Dipartimento di Biologia Animale
Address : Via Università, 4 - Via Bescugario, 14
I - 41100 Modena
Head : Milena Marini
Tel. : 059.225067/243566 **Fax**: 059.581069

Type of institute
University Department/Higher Education Institute

Research activities
Fisheries management
Stock assessment
Aquaculture
Physiology
Early stages of reared species

Staff: Scientists: 4 Others: 1

Research budget (ECU/year): 700.000

Description
Research projects:
Evaluation of stocks and dynamics of natural population of fry useful to farming.
Migration patterns of glass eel.
Cartography of fish resources in Emilia-Romagna region. Chemoreception of glass eel.

82. Univ. Palermo - Istituto Zoologia Giuseppe Reverberi

Name : Università degli Studi di Palermo
Istituto di Zoologia
Address : Via Archirafi, 18
I - 90123 Palermo
Head : Prof. Nicolo Parrinello
Tel. : 091.6161603 **Fax**: 091.6172009

Type of institute
University Department/Higher Education Institute

Research activities
Fisheries management
Stock assessment
Multi-disciplinary approach to fisheries management
Aquaculture
Early stages of reared species
Disease, immunology, stress
Aquaculture structures, systems, techniques

Staff: Scientists: 3

Description
Research projects:
Evaluation of demersal resources by trawl survey
(NW Sicily)
Evaluation of fishing yields of small scale fishing (W. Sicily) Biology and feeding habits of some high prized benthic fishes and prawns.
Artificial reefs (restocking practices and mariculture technologies).
Aquaculture and rearing of marine fish.
Consequences of stress on the biology and immunitary response of fish.
Stock assessment, including biochemical characteristics of stocks.

83. Univ. Pisa - Scienze dell Ambiente e del Territorio

Name	:	Università degli Studi di Pisa - Dip. Scienze Ambiente e Territorio Renewable Resource Management Section
Address	:	Via A. Volta, 6 I - 56127 Pisa
Head	:	St. De Ranieri
Tel.	:	(39.50)500018 **Fax**: (39.50)49694

Type of institute
University Department/Higher Education Institute

Research activities
Fisheries management
Stock assessment
Multi-species biological models
Multi-disciplinary approach to fisheries management
Environmental problems and their impact on resources
Fishing techniques
Selective gears
Static gear

Facilities
Rearing tank
Data bases, extensive libraries
Specialised software
Field site, electron microscope, specialized analytical apparatus

Staff: Scientists: 10 Others: 2

Research budget (ECU/year): 105.000

Description
The research group works on stock assessment projects in the Tuscany Archipelago. It is also involved in monitoring of the mercury contents in fisheries products. Another ongoing project deals with the fishery and stock assessment in the Ethiopian Red Sea, which is in cooperation with Asmara University in Ethiopia.
Major research projects :
Demersal resources evaluation.
Mercury contents in fishery products.
Study for the assessment and management of fisheries in the Western Mediterranean.
Evaluation and analysis of the interaction of fishing gears in the demersal fisheries of western Mediterranean. Settlement and mortality patterns of Mediterranean littoral fish juveniles, in protected and unprotected areas.

84. Univ. Roma I - Ist. Parassitologia

Name	:	Università degli Studi di Roma "La Sapienza" Instituto di Parassitologia
Address	:	P. le Aldo Moro, 5 I - 00185 Roma
Head	:	Prof. Lia Paggi
Tel.	:	39 6 4463868 **Fax**: 39 6 49914644 **Telex**: UNISAP I 620564

Type of institute
Research Centre
University Department/Higher Education Institute

Research activities
Fisheries management
Environmental problems (pollution, diseases, parasites) and their impact on resources)
Upgrading of fishery products
Consumer response and demand studies

Facilities
Specialised hardware
Several software programmes specialized on population genetics analysis, statistical numerical taxonomy and population dynamics. Specialized equipment for starch gel electrophoresis.

Staff: Scientists: 5 Others: 10

Research budget (ECU/year): 100,000

Description
Research projects:
- Genetic structure, biochemical taxonomy, life cycle and biological distribution of various genera of anisakid nematodes whose larvae affect seriously the market quality of some commercially valuable fish species and have public health importance.
- Morphology and systematics of helminth parasites of fish; use of helminth parasites as biological indicators of fish stocks population dynamics, of fish breeding habits, of changes in a predator or a prey population, of pollution effects, etc.

85. Univ. Roma II - Dipt. Sanita'Pubblica e Biologia Cellulare

Name : Università degli Studi di Roma "Tor Vergata"
Dipartimento di Sanità Pubblica e Biologia
Cellulare
Address : Via Orazio Raimondo, 8
I - 00173 Roma
Head : Prof. Paola Orecchia
Tel. : 739 6 72592025 **Fax**: 39 6 7233063

Type of institute
Research Centre
University Department

Research activities
Parasitic diseases in Fisheries and Aquaculture

Facilities
Specialised software
Field site, specialized analytical apparatus

Staff: Scientists: 3 Others: 3

Research budget (ECU/year): 58.000

Description
Research projects:
- Morphology and systematics of parasites of marine, brackish water and freshwater fish.
- Statistical evaluation of morphometric data using univariate and multivariate.
- Parasitic diseases, control and treatment of parasite borne diseases, control of parasites of wild fry or wild breeder fish in aquaculture.

86. Univ. Sassari - Istituto di Zootecnica

Name : Università degli Studi di Sassari
Istituto di Zootecnica - Facoltà di Agraria
Address : Via E. De Nicola
I - 07100 Sassari (I) O
Head : Prof. Paolo Brandano/Prof. Lorenzo A. Chessa
Tel. : 079.229306 **Fax**: 079.229302

Type of institute
University Department/Higher Education Institute

Research activities
Fisheries management
Environmental problems and their impact on resources
Lagoon resource assessment
Stock assessment
Aquaculture
Environmental impact
Early stages of reared species
Nutrition, growth and diet composition of commercial species

Facilities
Research vessel
Research laboratories
Rearing tanks and aquaria
Controlled environmental chamber
Specialized computer hardware and software
Diving equipment

Staff: Scientists: 2 Others: 3

Research budget (ECU/year): 20.000

Description
Research projects:
Artificial feeding of marine fish
Investigation on red coral
population dynamics of shellfish in lagoon

87. Univ. Siena - Dipt. Biologia Ambientale

Name	:	Università degli Studi di Siena Dipartimento di Biologia Ambientale
Address	:	Via della Cerchia, 3 I - 53100 Siena
Head	:	Prof. Silvano Focardi
Tel.	:	0577.298833 **Fax**: 0577.298860/298806 **Telex**: 572459 I

Type of institute
Research Centre
University Department/Higher Education Institute

Research activities
Aquaculture
- Reproduction
- Early stages of reared species
- Aquaculture/environment interaction
- Aquaculture structures, systems, techniques

Aquatic toxicology

Facilities
Studies of new methods of mollusc rearing

Staff: Scientists: 4 Others: 4

Research budget (ECU/year): 100.000

Description
This department has developed studies on aquaculture for 15 years. During the last 5 years, we have developed several techniques concerning the aquaculture of mussels and oysters in Italian lagoons. Particular attention was focussed on the techniques of reproduction and rearing of new species introduced in Italy for marketing interests; this species is a clam, TAPES PHILIPPINARUM, introduced into the Italian ecosystem for

the first time in 1982. In 1991, our main project concerning mollusc culture involves a study on recruitment of juvenile stage of scallops in the Adriatic sea. This research is based on the experimentation of artificial surfaces to collect the benthonic stage of scallops and in particular the species: PECTEN JACOBAEUS, CHLAMIS (AQUIPECTEN) OPERCULARIS, and CHLAMIS VARIA. In parallel, this study has the aim to develop several techniques of artificial reproduction of PECTEN JACOBAEUS and culture in an artificial marine system.

88. Univ. Trapani - Ist. Biologia Marina

Name : Università Libera di Trapani
Istituto di Biologia Marina
Address : Villa Nasi
Viale del Lazzaretto
I - 91100 Trapani
Head : V. D'Amelio
Tel. : 0923.24771 **Fax**: 0923.568300

Type of institute
Research centre

Research activities
Aquaculture
- Physiology
- Genetics and conservation of gametes
- Growth, nutrition, feed composition, live food production for aquaculture
- Aquaculture/environment interaction

Facilities
Research vessel
Rearing tank
Field site, electron microscope, specialized analytical apparatus.
Laboratory equipped for biochemical analysis (electrophoresys, chromatography, ...)
A motor boat (20 CV diesel, 6,5 m).
Equipment for diving.
Portable equipment for measuring physical-chemical parameters.
Aquaculture plant with 100 l, 300 l, 500 l, 1.000 l and 7.000 l tanks.

Description
Research projects:
Rationalization of aquaculture techniques in salt ponds.
Lipid absortion and transport in fishes.

Carnitine use in aquaculture.
Biochemical genetic in aquaculture.

89. Univ. Trieste - Dipt. Ingegneria Navale

Name	:	Università degli Studi di Trieste Dipt. Ingegneria Navale, Del Mare e per l'Ambiente
Address	:	Via Valerio, 10 I - 34127 Trieste
Head	:	Prof. Antonio Cardo
Tel.	:	39.40.6763500 **Fax**: 39.40.6763443 **Telex**: 460865 UNIVTS I

Type of institute
University Department

Research activities
Fishing techniques
Other improvements in gears and/or fishing vessels

Facilities
Specialised software
Hull fairing - Hydrostatics - Stability - Seakeeping-
Structural computations
The hydrodynamic laboratories of the Institute include:
a rectilinear 50m long towing tank, with wavemaker and the possibility of performing seakeeping, resistance, self propulsion and captive test.
A cavitation tunnel for propeller testing.
A stability balance.

Staff: Scientists: 3 Others: 1

Research budget (ECU/year): 50.000

Description
Research projects:
Fishing vessels improvement (design optimisation, stability, seakeeping, operational capability, manoeuvering, non conventional vessels).

90. Univ. Udine - Istituto di Scienze della Produzione Animale

Name : Università degli Studi di Udine
Dipartimento di Science delle Produzione Animale
Facoltà di Agraria
Address : Via San Mauro 2
I - 33010 Pagnacco (UD)
Head : Prof. Piero Susmel
Tel. : 432.650110 **Fax**: 432.660614

Type of institute
University Department

Research activities
Section Aquaculture (Head: Prof Domenico Lanari)
Early stages of reared fish species
Growth, nutrition, feed composition, live feed production for aquaculture
Aquaculture/environment interaction
Aquaculture systems, structures and techniques

Facilities
Field sites, research vessels rearing tank, recirculating systems
Feed mill, mixers and pelletting machine
Laboratory and portable equipment for water analysis
Data bases, extensive libraries
Chemical and Biochemical Laboratories
Equipped laboratories for histology and pathology
Data bases, extensive libraries
Statistical software

Staff: Scientists: 7 Others: 8

Research budget (ECU/year): Aquaculture 110.000

Description
Research projects:
- Live feed substitution for dry feed for weaning marine fish larvae;
- Assessment of essential amino acids requirement of sea bass fingerlings;
- Assessment of n.3 PUFA requirements of sea bass fingerlings;
- Development of dry pelletted diet for *Dentex dentex*

fingerling and juveniles:
- Optimal protein to lipid dietary levels for finishing sea bass (100 g to market size);
- Development of low pollution diets and of water recirculating systems for rainbow trout and sea bass;
- Culture systems for Crassostrea gigas and Pecten jacobaeus in the Northern Adriatic regions.

State of Research

THE NETHERLANDS

The Ministry of Agriculture and Fisheries supervises a central research organisation, DLO, of which 5 research institutes are involved in fisheries and aquaculture. The other research institutes, involved in fisheries and aquaculture, are public or semi public institutes, funded by the government. Their research projects and programmes, as far as they are funded by the government, are coordinated and are approved on a national governmental level.

The industry has its influence through the Advisory Committees of the DLO institutes, the industry funds a part of the research.

There are some institutes particularly relevant for fishery:

* Rijksinstituut voor Visserij Onderzoek - IJmuiden
* Landbouweconomisch Instituut afdeling Visserij - The Hague
* Vakgroep Vissteelt en Visserij, Landbouw Universiteit Wageningen - Wageningen.

Other institutes from different universities can also participate in fisheries and aquaculture research.

Funding of these institutes originates both from the Government and from industry.

One of the main problems arising from the organisational structure is that - though the research fields of the different institutes are very much interrelated - it is difficult to organise integrated research programmes which result in coherent overall research results.

Educational support exists in related faculties; agricultural (Wageningen), veterinary, biological, and specific fishery research training is normally given by the institutes themselves.

List of Institutes from THE NETHERLANDS

1. Dienst Getijdewateren - Den Haag (RWS-DGW)
2. Dienst Getijdewateren - Middelburg (RWS-DGW)
3. DLO - Instituut voor Bos- en Natuuronderzoek (IBN-DLO)
4. DLO - Dept. Herd Health, Pathology and Epidemiology - Section Fish Diseases (CDI-DLO)
5. DLO - Landbouw Economisch Instituut, LEI-DLO Afd. Visserij
6. DLO - Rijksinstituut voor Visserijonderzoek (RIVO-DLO)
7. Netherlands Institute for Sea Research (NIOZ)
8. Netherlands Institute of Ecology - Centre for Estuarine and Coastal Ecology (NIOO - CEMO)
9. Organization for the Improvement of Inland Fisheries (OVE)
10. RIKILT - Rijks Kwaliteitsinst. voor Land- en Tuinbouwprodukten
11. TNO Lab. for Applied Marine Research - Den Helder
12. Univ. Utrecht - Dept. Experimental Zoology
13. Univ. Wageningen - Landbouw - Exp. Animal Morph. Cell Biol.
14. Univ. Wageningen - Landbouw - Fish Culture and Fisheries

1. Dienst Getijdewateren - Den Haag (RWS - DGW)

Name : RWS/DGW Dienst Getijdewateren
Address : Postbus 20904
NL - 2500 EX Den Haag
Head : Drs. A.D. Vethaak
Tel. : 70.889373 **Fax**: 70.805199 **Telex**: 33566

Type of institute
Research centre
Government Department/regional or local authority

Research activities
Fisheries management
Environmental problems and their impact on resources

Facilities
Research vessel
Simulated tidal tank
Mesocosm
Controlled environmental chamber
Data bases, extensive libraries
Specialised software
Field site, electron microscope, specialized analytical apparatus

Staff: Scientists: 3 Others: 2

Research budget (ECU/year) : 130.000

Description
The research focusses on cause-effect relationships between pollution factors and the occurence of fish diseases. Methods:
- Controlled chronic exposure of fish in sediments/water systems in which the pollution dose is included in the sediment
- Monitoring of changes in disease occurence etc.
- Linkage of experiment data to field data.

2. Dienst Getijdewateren - Middelburg (RWS - DGW)

Name : RWS/DGW-Dienst Getijdewateren
Address : Postbus 8039
NL - 4330 EA Middelburg
Head : J.V.D. Meer/R.J. Leewis/A.C. Smaal/M. Knoest
Tel. : 1180.11851 **Fax**: 1180.16500 **Telex**: 37777

Type of institute
Research centre
Government Department/Regional or local authority

Research activities
Fisheries management
Environmental problems and their impact on resources
Aquaculture
Physiology

Facilities
Research vessel
Mesocosm
Field site, electron microscope, specialized analytical apparatus

Staff: Scientists: 2 Others: 2

Research budget (ECU/year) : 5.188.400

3. DLO - Instituut voor Bos- en Natuuronderzoek (IBN-DLO)

Name : DLO - Instituut voor Bos- en Natuuronderzoek (IBN)
Afdeling Aquatische Ecosystemen - Texel
Address : Postbus 167
NL - 1790 AB Den Burg (Texel)
Head : Prof. Dr. W.J. Wolff/Dr. N. Dankers
Tel. : 02220.69700 **Fax**: 02220.19235

Type of institute
Research centre

Research activities
Fisheries management
  Stock assessment
  Environmental problems and their impact on resources
Aquaculture
  Aquaculture/environment interaction

Facilities
Flume tank
Simulated tidal tank
Mesocosm
Specialised software
Portable flume which can be used successfully on densely agregated organisms such as mussel or oyster beds and for mesocosm studies.
8 tidal tanks which simulate intertidal flats.

Staff: Scientists: 4 Others: 2

Research budget (ECU/year): 430.000

Description
Research projects:
Impact of mussel culture and cockle fisheries on the Wadden Sea ecosystem.
In collaboration with other institutes the department works with a comprehensive mathematical ecosystem model of the western Wadden Sea.
Impact of fisheries on the marine ecosystem. Impact of eutrophication on marine ecosystems.

4. DLO - Dept. of Herd Health, Pathology and Epidemiology - Section Fish Diseases (CDI - DLO)

Name : CDI-DLO (Centraal Diergeneeskundig Instituut)
Dept. of Herd Health, Pathology and Epidemiology -
Section Fish Diseases
Address : Postbus 65
NL - 8200 AB Lelystad
Head : Dr. D .van Zaane/Ms Ir. O.L.M. Haenen
Tel. : 3200.73911 **Fax**: 3200.73473 **Telex**: 40227

Type of institute
Research centre
Government Department / regional or local authority

Research activities
Fisheries management
  Environmental problems and their impact on resources
Aquaculture
  Diseases, immunology, stress, diagnostics, prophylaxis

Facilities
Research vessel
Field site, electron microscope, specialized analytical apparatus
Aquarium facilities with 32 aquaria for experiments with fish pathogens
A full equipped diagnostic laboratory including necropsy, parasitology, bacteriology, virology, histopathology, and waterquality
Research of fish diseases. Other disciplines (immunology, molecular biology, and toxicology) are available for support.

Staff: Scientists: 1 Others: 3

Research budget (ECU/year): 250.000

Description
Research projects:
The Section Fish Diseases is mainly focussing on diagnostic work as well as the development of fish virological, bacteriological and certification work.
Diagnostic of fish diseases as a support to veterinarians and fishfarmers. Certification of fish farms and fish related to import and export of alive fresh water fish.
Research on fish diseases based on problems out of daily practice.

5. DLO - Landbouw Economisch Instituut, LEI - DLO Afd. Visserij

Name	:	Landbouw Economisch Instituut, LEI-DLO Afdeling Visserij
Address	:	Conradkade, 175 NL - 2517 CL Den Haag
Head	:	Drs. W. Smit
Tel.	:	31.70.3308330 **Fax**: 70.615624

Type of institute
Research centre

Research activities
Fisheries management
Multi-disciplinary approach to fisheries management
Estimation of the economic impact of proposed fisheries management; General fisheries economics
Aquaculture
Aquaculture economics
Upgrading of fishery products
Consumer response and demand studies

Facilities
Data bases, extensive libraries
Specialised software

Staff: Scientists: 8 Others: 6

Research budget (ECU/year): 1.067.500

Description
Research programme: (project leader=[])
- Financial data from fishing enterprise [H. Harmsma]
- Fisheries statistics and database [J. de Jager]
- Costs and earnings, productivity, incomes [W.P. Davidse]
Economics of:
- Changing technology and decision surroundings [J.P. de Wilde]
- Fish processing, marketing and promotion [J.G.P. Smit]
- National fisheries management [W. Smit]
- EC and international fishery and trade [P. Salz]
- Fishery in developing countries [W. Smit]
- Development of methodology and expertise [H. Harmsma]

6. DLO - Rijksinstituut voor Visserijonderzoek (RIVO-DLO)

Name : Rijksinstituut voor Visserijonderzoek (RIVO-DLO)
Address : Postbus 68-Haringkade 1
NL - 1970 AB IJmuiden
Head : Jan Willem Henfling Ph. D.
Tel. : 31.255064646 **Fax**: 31.255064644

Type of institute
Research centre
Regional or local fish/crustacean/mollusc farming

DEPENDANCE

Name : Dependance Rijksinstituut voor Visserijonderzoek (RIVO-DLO)
Address : Postbus 77-Korringaweg 5
NL - 4400 AB Yerseke
Head : R. Dijkema/M.R. van Stralen
Tel. : 31.11312781 **Fax**: 31.11313477

Type of institute
Field laboratory of RIVO-DLO, IJmuiden

Research activities
Fisheries management
  Stock assessment
  Multi-species interaction
  Ecosystem effects of fishing activities
Environmental problems and their impact on resources
Fishing techniques
Quality of fisheries products
Fishery technology
Aquaculture

In dependance:
Molluscan fisheries research
  Stock assessments, population studies
  Fisheries management advice
  Interactions of molluscan fishery and ecosystems
Culture of molluscan shellfish
  Culture techniques
  Storage of live molluscs
Water quality in molluscan culture and fishing areas
  Bacteriological monitoring and studies
  Monitoring and studies of toxic phytoplankton species

Facilities
Controlled environmental chamber
Data bases, extensive libraries
Specialized analytical apparatus
Specialized software
Seawater aquarium

Research vessel: One of 28 meters beamtrawler with a draught of 2,50 meters built for research work in the near coastal water. One of 73 meter long, type sterntrawler, also fitted out for beamtrawling, 4000HP.
Two vessels, equipped for beamtrawling, about 21 nets for working in inshore waters.

Staff: Scientists: 26 Others: 67

Research budget (ECU/year) : 7.500.000

Description
The activities of the institute covers:
Biological research. Technical and technological research; Environmental research. Aquaculture research. On request specialized courses in fish processing and quality control can be designed. Organoleptic panel.

7. Netherlands Institute for Sea Research (NIOZ)

Address : P.O. Box 59
NL - 1790 AB Den Burg (Texel)
Head : Dr. W.G. Hock
Tel. : 31.2220.69300 **Fax**: 31.2220.19674

Type of institute
Research centre

Research activities
Fisheries management
Stock assessment
Impact of fisheries on the marine ecosystem and seabird populations

Facilities
Research vessels
Extensive set-up for experimental work (aquaria, recirculation-systems, controlled environmental chambers, mesocosm, etc.)
Data bases, extensive libraries

Staff: Scientists: 120 Others: 150 (Scientists involved in fish and fisheries: 4 Others: 4)

Research budget (ECU/year) : 12.000.000 (in relation of fish and fisheries: 650.000)

Description

Research projects:

Impact of fishing-activities (beamtrawl) on the marine ecosystem (including seabirds) and benthos specifically (Arctica islandica). Survival of bycatch of the commercial fisheries. Importance of the Wadden Sea for the flatfish-stocks in the Northsea.

8. Netherlands Institute of Ecology - Centre for Estuarine and Coastal Ecology (NIOO-CEMO)

Name	:	Netherlands Institute of Ecology Centre for Estuarine and Coastal Ecology
Address	:	Vierstraat 28 NL - 4401 EA Yerseke
Head	:	Prof. Dr. C.H.R. Heip
Tel.	:	01131.1920 **Fax**: 01131 3616

Type of institute

Research centre

Research activities

Fisheries management
- Stock assessment
- Environmental problems and their impact on resources

Facilities

Research vessel
Simulated tidal tank
Rearing tank
Mesocosm
Controlled environmental chambers
Data bases, extensive libraries

Staff: Scientists: 25 Others: 55

Research budget (ECU/year) : 3.100.000

Description
Research projects:
Impact of the storm surge barrier in the Oosterschelde on fish population. The fish population of the Westerschelde. The importance of fish and crustaceans on intervidal flats. Impact of beam-trawl fisheries on benthos.

9. Organization for the Improvement of Inland Fisheries (OVE)

Name : Organization for the improvement of Inland Fisheries
Research Department
Address : Postbus 433
NL - 3430 AK Nieuwegein
Head : A.J.P. Raat
Tel. : 3402.58411 **Fax**: 3402.39874

Type of institute
Fishing
Fish farming
Fishery management

Research activities
Fisheries management
- Stock assessment
- Multi-species biological models
- Environmental problems and their impact on resources
- Fishing techniques

Aquaculture
- Early stages of reared species
- Upgrading of fishery products
- Improvement of techniques for handling, storing, processing, packaging fish

Facilities
Data bases, extensive libraries
Specialised software
Experimental Station with 80 drainable ponds and laboratory facilities
Monitoring facilities for fisheries research in stagnant and flowing waters

Staff: Scientists: 4 Others: 8

Research budget (ECU/year) : 432.000

Description
Major research projects:
- restauration of salmonids in Dutch waters
- biomanipulation studies (effects of depletion of fisheries, effects of benthivorous fish, effects of o+ cyprinids, bioregulation of o+ cyprinids, with northern pike and pike perch).
- optimalisation of research fisheries
- habitat evaluation programme-hydro acoustics

10. RIKILT - Rijks Kwaliteitsinst. voor Land- en Tuinbouwprodukten

Name	:	RIKILT - Rijks Kwaliteitsinstituut voor Land- en Tuinbouwprodukten (RIKILT - DLO)
Address	:	Postbus 230 NL - 6700 AE Wageningen
Head	:	Prof. Dr. W. de Wit
Tel.	:	08370.19110 **Fax**: 08370.17717 **Telex**: 75180

Type of institute
Research and quality control centre

Research activities
Fundamental and applied research focussed on the advancement of safe food production and food products.
Disciplines: analytical chemistry, instrumental analysis, biochemistry, microbiology, toxicology, microscopy, histology.

Staff: Scientists: 25

Research budget (ECU/year): 2.000.000

Description
Fish quality control is one item in a wide range of RIKILT research topics.
The main tasks of the Institute are:
- Fundamental and applied research on food quality and safety.
- Acting as a supervising and reference laboratory for the inspection of the quality and safety of agricultural products.
- Support to the legal inspection under the responsibility of the Ministry of Agriculture, Nature Management and Fishery.

11. TNO Lab. for Applied Marine Research - Den Helder

Name : TNO Lab. for Applied Marine Research
Address : P.O. Box 57
NL - 1780 AB Den Helder
Head : Dr. Martin Scholten
Tel. : 02230.32924 **Fax**: 2230.30687

Type of institute
Research centre

Research activities
Fisheries management
Multi-species biological models
Environmental problems and their impact on resources
Aquaculture
Growth, nutrition, feed composition, live food production for aquaculture
Aquaculture/environment interaction
Aquaculture structures, systems, techniques of materials prevention from fouling and corrosion

Facilities
Research vessel
Simulated tidal tank
Mesocosm
Controlled environmental chamber
Specialised software
Field site

Staff: Scientists: 12 Others: 9

Research budget (ECU/year): 1.570.000

Description
The TNO Laboratory for Applied Marine Reseach in Den Helder (The Netherlands) is part of the Netherlands Organisation for Applied Scientific Research TNO, an independent organisation.
Research is directed towards solving practical problems, and is commissioned by government, industry and overseas organisations.
The laboratory in Den Helder is specialised in environmental research and mariculture research, related to marine conditions.

The laboratory possesses extensive facilities to study the corrosion behaviour of materials in the marine environment, and to test or develop corrosion prevention and antifouling systems.
The environmental research includes field studies; field experimentation (with the help of 1-75 cubic metre mesocosms); and ecological risk evaluation (with HAZARD and REFEREE).
The laboratory is also involved in studies on the economic, technical and biological feasibility of mariculture operations with fish (viz.cod), shellfish (viz.mussels) and algae, as well as in environmental monitoring of mariculture operations.

12. Univ. Utrecht - Dept. Experimental Zoology

Name	:	Universiteit Utrecht Dept. of Experimental Zoology/Faculty of Biology
Address	:	Research Group for General & Comp. Endocrinology Padualaan 8 NL - 3584 CH Utrecht
Head	:	Dr H.J.Th. Goos
Tel.	:	30.30533664 **Fax**: 31.30521877 **Telex**: 70353

Type of institute
University Department

Research activities
Aquaculture
 Physiology
 Reproduction
 Endocrine regulation of reproduction in teleost fish
 Regulation of gonadal cycles and sex differentiation

Facilities
Field site, electron microscope, specialized analytical apparatus for hormone measurement, hormone receptor assay, steroid biochemistry
Instruments for thin layer chromatography, gas chromatography, mass spectrometry, HPLC, radioinmunoassays, inmunocytochemistry, light microscopy, electron microscopy, cell and tissue culture.

Staff: Scientists: 6 Others: 6

Research budget (ECU/year): 22.000

Description
The main research project is the RBA (Reproductive Biology in Aquaculture).
The aim of RBA is the world-wide promotion of application directed fundamental research dealing with reproductive biology in invertebrates and vertebrates of aquacultural importance.
In this formulation, reproductive biology includes reproductive physiology, physiology of gametes and fertilization, as well as genetic engineering and early development.

13. Univ. Wageningen - Landbouw - Exp. Animal Morph. Cell Biol.

Name : Agricultural University
Dept. Exp. Anim. Morph. & Cell Biology
Address : P O Box 338
NL - 6700 Wageningen
Head : Prof. W.B. van Muiswinkel
Tel. : 8370.83336 **Fax**: 8370.83962 **Telex**: 45854

Type of institute
University Department/Higher Education Institute

Research activities
Aquaculture
Genetics and conservation of gametes
Disease, immunology, stress

Facilities
Rearing tank
Electron microscope - flowcytometer - Inbred lines of fish - tissue culture
Disease challenge unit for testing vaccines.

Staff: Scientists: 8 Others: 6

Research budget: 30% of salaries : 160.000-consumables : 45.000-equipment : 35.000

Description
Research projects:
All projects fit in a general theme: "Immunology and Health", and the three main areas are:
- Local immunity: Administration of antigens through the surrounding waters (bath vaccination).

- Immunogenetics and disease resistance: Biochemistry and the biological role of Major Histocompatibility Complex (MHC) in carp and trout.
- Functional and morphological characterization of white blood cells.

14. Univ. Wageningen - Landbouw - Fish Culture and Fisheries

Name : Universiteit Wageningen/Landbouw
Vakgroep Visteelt & Visserij
Address : Postbus 338
NL - 6700 AH Wageningen
Head : Dr. E. A. Huisman
Tel. : 08370.83307 **Fax**: 8370.83962 **Telex**: 45015

Type of institute
University Department/Higher Education Institute

Research activities
Fisheries management
Stock assessment
Multi-species biological models
Multi-disciplinary approach to fisheries management
Fishing techniques
Selective gears
Aquaculture
Physiology

Facilities
Controlled environmental chamber
Data bases, extensive libraries
Field site, electron microscope, specialized analytical apparatus
Specialised computer hardware and software
1000 m^2 experimental fish hatchery

Staff: Scientists: 11 Others: 9

Research budget (ECU/year): 550.000

State of Research

NORWAY

The Research Council of Norway which is publicly funded, has its main responsibilities in planning, promoting, evaluating and funding research, also within the fisheries- and aquaculture- sector. As the Research Council, however, is not engaged in research itself, its objective will in a sense be to function as a national "market-place" for science activities in question.

Most of the research funding goes into programmes, leaving a lesser part to single projects.

Additionally, an important means for carrying out the Councils strategy, is through supporting certain research institutions with a basic financial foundation.

Distribution of different fellowships, including both such for training of scientists and different travel grants, is another essential task for the Council.

An important duty is also to coordinate the national participation in different topical international bodies, including international research cooperation.

The following is a brief description of the most important institutions which carry out research within the fisheries- and aquaculture sector:

The universities are vital elements in the research system since they assume the main responsibility for basic research and the education of scientists. They also have the main responsibility for dissemination of research results. These institutions receive their basic appropriations through the budget of the Ministry of Education and Research, but also consume quite an amount of strategic funding from the Research Council.
At the University of Tromsø an important part of the subjects are related to the fisheries.

The university is, as the only university in Norway, teaching law related to fisheries-legislation.

There is also close integration with other research establishments in the region, which helps enhance the quality of both the research and teaching. The Norwegian College of Fishery Science is the main institution for higher education in the fisheries in Norway.

The universities of Oslo and Bergen also have considerable fisheries- and aquaculture research.

The public research institutes receive their basic grants either directly from the Ministry of Fisheries or through the Research Council. The financial support from the

Council will be either basic- or project-/programme support. The Institute of Marine Research in Bergen has a central position in Norwegian fisheries research and is placed under the Ministry of Fisheries. The institute also receives most of its funding from the budget of this ministry. Apart from research tasks concerned with the fundamental role of the marine environment and marine biology for the fisheries, catch technology and aquaculture, the institute has special advisory tasks for the Director of the Fisheries, the Ministry of Fisheries, other authorities, the fishing industry and other industries.
The institute also takes active part in ICES - activities.

The Directorate of Fisheries Institute of Nutrition has its responsibility within research on marine resources as a source of human nutrition together with nutrition of fish in fish farms.

The Norwegian College of Veterinary Medicine together with The National Veterinary Institute in Oslo, are doing research in relation to fish diseases, as well as genetics.

The contract research institutes finance their activities through contract work from industry and business, together with public sector grants through the Research Council. These institutions have considerable expertise in applied research. The Norwegian Institute of Fisheries and Aquaculture is an independent research institute which main objectives are to promote fishery technology research and development in areas that are important for the Norwegian fisheries.
Institute of Aquaculture Research is the prime Norwegian institution for research on i.a. breeding and feeding of fish.

The Foundation for Scientific and Industrial Research at the Norwegian Institute of Technology in Trondheim is a large institution with a considerable amount of research on processing and upgrading of fish, in addition to a strong Marine Technology section.

One should also mention research institutions operating on behalf of the fisheries industry. The most important are the research institutions for the Norwegian herring oil and -meal industry together with that for the canning industry.

List of Institutes from NORWAY

1. Agricultural University of Norway - Dept. of Agricultural Engineering
2. Agricultural University of Norway - Dept. of Dairy and Food Industries/Biochemistry and Molecular Biology
3. Akvaforsk - Institute for Aquaculture Research Ltd.
4. Allforsk, Social Research Centre - SESAM
5. Central Laboratory, Department of Quality Control, Directorate of Fisheries
6. Centre for Fisheries Economics, Norwegian School of Economics and Business Administration
7. Department of Biotechnology, the Norwegian Institute of Technology, University of Trondheim
8. Department of Fisheries and Marine Biology (IFM)
9. Division of General Physiology, Department of Biology, University of Oslo
10 Institute of Marine Research
11. Institute of Nutrition, Directorate of Fisheries
12. Institute of Zoology - University of Trondheim
13. Laboratory for Biotechnology, University of Bergen
14. Nordland College and Nordland Research Institute
15. Norwegian College of Veterinary Medicine
16. Norwegian Institute for Nature Research (NINA)
17. Norwegian Institute of Fisheries and Aquaculture
18. Norwegian Marine Technology Research Institute - MARINTEK
19. Research Institute Norconserv
20. SINTEF Applied Chemistry
21. SINTEF Group
22. State Veterinary Laboratory for North Norway - Harstad
23. State Veterinary Laboratory - Bergen
24. State Veterinary Laboratory - Sandnes
25. State Veterinary Laboratory - Trondheim
26. The Norwegian College of Fisheries, University of Tromsø
27. The Norwegian Herring Oil and Meal Industry Research Institute
28. The Norwegian Institute of Technology - Division of Refrigeration Engineering
29. Trondheim Biological Station
30. VESO
31. Veterinarinstituttet (Central Veterinary Laboratory)
32. Vetmest - Centre of Veterinary Medicine - Tromsø
33. Zoological Institute - University of Bergen
34. Zoological Museum - University of Oslo

1. Agricultural University of Norway - Dept. of Agricultural Engineering

Name : Department of Agricultural Engineering
Agricultural University of Norway
Address : P.b. 5065
N - 1430 Aas
Head : Prof. Herman Qvam
Tel. : 47.9.948700 **Fax**: 47.9.948810

Type of institute
Research and development, testing and certification, institution for the university.

Research activities
Aquacultural engineering
Building of on-shore fish farms, handling of fish, feeding of fish, feediing equipment, electronics and instruments, management.
Environmental control
Combating of parasites, equipment for reduced pollution, techniques for treatment of waste water.
Cultivating
Measuring techniques for quality of fish.

Facilities
Fish laboratory, water laboratory, design and construction office, different types of workshops, metal, wood, electronic, hydraulics, data etc.

Staff: Scientists: 10 Others: 4

Research budget (ECU/year): 500.000

Description
Dept. of Agricultural Engineering's main activities are concerned with traditional agricultural engineering. However, two new areas have been built up, "aquacultural engineering and environmental technology". In aquaculture several projects considering fish handling such as grading and transport, have been educated. Also planning and building of on-shore fish farms. In the area of feeding equipment several developing works have been executed. In the area of environmental control much work has been done in the last few years, such as development of new types of parasite control, composting of waste, experiments and systems which take care of feed waste in aquaculture.

2. Agricultural University of Norway - Dept. of Dairy and Food Industries/Biochemistry and Molecular Biology

Name : Dept. of Dairy and Food Industries/Biochemistry and Molecular Biology
The Agricultural University of Norway
Address : P.O.Box 5036
N - 1432 Aas
Head : Peter Aleström
Tel. : 47.9.948544/948550 **Fax**: 47.9.943789

Type of institute
University institute

Research activities
Molecular biology of fish reproduction.
- Characterization of GnRH genes from salmonid fish and zebrafish.
- Gene expression control mechanisms of GnRH promoter.

Transgenic model fish (zebrafish).
- Improvement of gene transfer methods.
- Sterile fish with non natural anti-fertility genes.

Facilities
Laboratories for high level molecular biology research

Staff: Scientists: 3 Others: 3

Research budget (ECU/year: 156.000

Description
In 1988 a major Biotechnology Laboratory at the Agricultural University of Norway was established. This lab. is located at the Dept. of Dairy and Food Industries and houses approximately 30 researchers working with problems at the level of molecular biology, giving a reasonable "critical mass". Also at the university campus is The Institute of Aquacultural Research (AKVAFORSK), with which our group has close collaboration. AKVAFORSK has among other activities developed the breeding program for the farmed Norwegian salmon which recently and successfully has been applied to farmed tilapia in the Philippines.

3. Akvaforsk - Institute for Aquaculture Research Ltd.

Name : Akvaforsk - Institute for Aquaculture Research Ltd.
Address : P.O. Box 5010
N - 1432 Ås
Head : Erling Fimland(Director)
Tel. : 64.94.95.00 **Fax**: 64.94.95.02

Type of institute
Research institute

Research activities
Selective breeding/genetics, nutrition/fish feed, prophylactic health care, production technology, quality, pollution, marine fry production and aquaculture in developing countries.

Facilities
AKVAFORSK is located at three geographically separated places. The main office in Ås, at the Agricultural University of Norway, an on-shore research station in Sunndalsøra and a marine research station in Averøy at the west coast. The institute has at its disposal tanks, ponds and seacages of different sizes, chemical/biological laboratories, laboratories for physiological studies of fish, slaughterhouse, engineering workshop etc. At the research station in Sunndalsøra, both fresh and seawater, free from pathogenic organisms and with various temperatures are available.

Staff: Scientists: 37 Others: 53

Research budget (ECU/year): 4.000.000

Description
The aim of the institute is to produce and distribute knowledge about aquaculture that will contribute to a sustainable, economical and socially beneficial industry. AKVAFORSK, with its roots at the Agricultural University of Norway, was founded as an independent research institute in 1984. But research regarding aquaculture has been carried out at the Agricultural University of Norway as early as the 1860's, and in particular from 1970, at the beginning of the success of the Norwegian salmon industry. AKVAFORSK was established by The Agricultural Research Council of Norway, but was in 1992 converted to a stock company, owned with a majority by the Agricultural University of Norway and the Veterinary College of Norway. AKVAFORSK receives a regular grant from the Agricultural Research Council of Norway corresponding to one quarter of the annual budget. The rest of the income includes grants from Norwegian research councils for research projects, other governmental funding and research projects for the industry in Norway and abroad.

4. Allforsk, Social Research Centre - SESAM

Name : ALLFORSK, Social Research Centre - SESAM
Address : The University of Trondheim
N - 7055 Dragvoll
Head : Oddmund Otterstad
Tel. : 07.59.18.35 **Fax**: 07.59.17.08

Type of institute
Social Research Centre, part of the Research Foundation at the College of Arts and Science, the University of Trondheim.

Research activities
All activities applied to social research. Important projects at the moment: a quantitative database of fishing communities, process evaluation, social work projects, organizational studies, teaching-at-a-distance and action research.

Facilities
Part of the infrastructure at the University, closely affiliated with all institutes.

Staff: Scientists: 10 Others: 1

Research budget (ECU/year): 798.900

Description
Social Research Centre

5. Central Laboratory, Department of Quality Control, Directorate of Fisheries

Name : Central laboratory, Department of quality Control,
Directorate of Fisheries
Address : P. O. Box 185
N - 5002 Bergen
Head : Bjarne Boe
Tel. : 05.23.80.00 **Fax**: 05.23.80.90

Type of institute
University Institute

Research activities
Molecular genetics of marine micro-organisms (virus, bacteria and microalgae). Molecular genetics of fish.

Facilities
Fully equipped for organic and inorganic analytical chemistry, microbiology and sensory analyses including test panel.

Staff: Scientists: 5 Others: 15

Research budget (ECU/year): 79.900(total budget 639.800).

Description
The Central Laboratory is a part of the Departement of Quality Control in the Directorate of Fisheries. The main activities concern the quality of fish and fish products. Contents of naturally occuring substances as well as contaminants are measured by means of chemical, sensoric and microbiological methods.

6. Centre for Fisheries Economics, Norwegian School of Economics and Business Administration

Name : Centre for Fisheries Economics, Norwegian School of Economics and Business Administration
Address : Helleveien, 30
N - 5035 Bergen-Sandviken
Head : Dr. Trond Bjørndal/Dr. Røgnvaldur Hannesson
Tel. : 05.95.92.50 **Fax**: 05.95.95.43

Type of institute
Research Institute

Research activities
1. Bioeconomics analysis and modelling of fish stocks.
2. The industrial organization of fishing and aquaculture.
3. Market research

Staff: Scientists: 6 Others: 7

Research budget (ECU/year): 568.900

Description
The Centre for Fisheries Economics is a joint institute between the Norwegian School of Economics and Business Administration and the Foundation for Research in Economics and Business Administration. The Centre staff consists of full-time professors and researchers, in addition to research assistants and secretarial staff. It has an extensive international network. Visits by domestic and foreign graduate students and researchers are encouraged. The Centre for Fisheries Economics undertakes research projects for both

the private and public sectors. Traditionally, most research projects have been funded by the Norwegian Research Councils and Government ministries, including the Ministry of Fisheries and the Ministry of the Environment. Research has also been carried out for the industry and international agencies.

7. Department of Biotechnology, the Norwegian Institute of Technology, University of Trondheim

Name	:	Department of Biotechnology, The Norwegian Institute of Technology, University of Trondheim.
Address	:	N - 7034 Trondheim
Head	:	Pr. Viggo Mohr
Tel.	:	07.59.33.20 **Fax**: 07.59.12.83

Type of institute
University Institute

Research activities
Biotechnology, biopolymers, microbiology, molecular biology, environmental chemistry and biology, fish chemistry and biochemistry: proteins, enzymes, biochemical processes in fish, fish quality.

Facilities
General laboratory for microbiology, molecular biology and biochemistry.

Staff · Scientists: 22 Others: 28

Research budget (ECU/year): Total 3.026.100
(including 181.600 for fishery-related activities.

Description
The Department of Biotechnology is part of the Faculty for Chemistry and Chemical Technology at the Norwegian Institute of Technology. 25 students graduate each year.

8. Department of Fisheries and Marine Biology (IFM)

Name	:	Department of Fisheries and Marine Biology (IFM).
Address	:	High Technology Center N - 5020 Bergen
Head	:	Pr. Dag Møller (1993-95)
Tel.	:	05.54.44.00 **Fax**: 05.54.44.50

Type of institute
Department connected to the Faculty of Mathematics and Natural Sciences, University of Bergen.
Education and Research.

Research activities
Fisheries biology and resource research, including population dynamics, assessment, recruitment, population genetics and behaviour. Environmental problems and their impact on resources.
Aquaculture:
1) Salmonids - controlled reproduction, smoltification physiology, toxicological effects, fish health.
2) Marine fish - egg and larval ontogeny, reproduction, visual systems, predatory behaviour, differences between wild and cultured, fish health, toxicological effects.
3) Environmental effects of aquaculture.
4) Scallop biology.

Facilities
Access to two research vessels for field investigation in open oceanic and coastal waters. Wet laboratories (fresh and salt water with temperature control) for experiments (rearing systems, climate controlled rooms etc., units for disease research). Dry laboratories for data processing and genetic analyses. Computer hardware and software.

Staff: Scientists: 88 Others: 5

Research budget (ECU/year): 641.500

Description
The main activities are fisheries sciences, marine biology and ecology. The department cooperates extensively with the Institute of Marine Research, Bergen, and with other departments at the University of Bergen on fisheries biology and aquaculture. Studies on resources for developing countries are carried out in cooperation with the Institute of Marine Research, where IFM is responsible for the educational part.

9. Division of General Physiology, Department of Biology, University of Oslo

Name : Division of General Physiology,
Department of Biology,
University of Oslo
Address : P.O.Box 1051, Blindern
N - 0316 Oslo
Tel. : 02.85.46.00 **Fax**: 02.84.46.64

Type of institute
University institute; education and research

Research activities
Basic research on fish physiology

Facilities
Good facilities for keeping saltwater and freshwater animals

Staff: Scientists: 10 Others: 3

Description
The Division of General Physiology offers courses in comparative and general physiology at undergraduate, graduate, and Ph.D. levels. Research projects for studies on chemical senses in fish, fish migration, hearing and lateral line senses in fish. Projects related to different types of stress in fish. Projects for studies on membrane transport associated with volume regulation in fish cells.

10. Institute of Marine Research

Name	:	Institute of Marine Research
Address	:	P. O. Box 1870 Nordnes N - 5024 Bergen
Head	:	Roald Vaage(Managing Director)
Tel.	:	05.23.85.00 **Fax**: 05.23.85.31

Type of institute
Public research Institute operated by the Norwegian Government through the Ministry of Fisheries.

Research activities
Study and monitor marine life and environment, and interactions in the ocean and coastal waters.
Acquire new, up-to-date understanding of the marine resources important for the fishing and aquaculture industries
Develop technological and biological bases on which to build rational, forward-looking fishing and aquaculture industries.
Provide the authorities and the fishing industry with advice on managing the marine environment and the marine resources.
Publish the results of research, for the benefit of the fishing and aquaculture industries and others.

Facilities
Three oceangoing and two fjord-operating vessels.
In addition to the main facilities in Bergen, the following facilities are associated with the Institute:
Flødevigen Marine Research Station, N-4817 His,
tel. 04.11.05.80, fax. 04.11.05.15
Austevoll Aquaculture Research Station, N-5392 Storebø,
tel. 04.38.03.42, fax. 05.38.03.98.
Matre Aquaculture Research Station, N-5198 Matredal,
tel. 05.36.60.40, fax. 05.36.61.43
Fish stock and environmental parameters computer database.

Staff Scientists: 130 Others: 310

Research budget (ECU/year): 28.082.300 (1991)

Description
The Institute of Marine Research is a national centre for research on life and the environment in the ocean and coastal waters.

11. Institute of Nutrition, Directorate of Fisheries

Name : Institute of Nutrition, Directorate of Fisheries.
Address : P. O. Box 1900, Nordnes
N - 5024 Bergen
Head : Kåre Julshamn
Tel. : 05.23.80.00 **Fax**: 05.23.80.95

Type of institute
Research Institute

Research activities
The Institute aims to be the leading national authority on: 1) Fish nutrition, 2) Nutritional quality of seafood, 3) Nutritional quality of marine resource as feed for farmed fish, and 4) Chemical analyses of marine organisms.

Facilities
Modern instruments for analyses of: Protein/amino acids/amines; lipids/fatty acids/carbohydrates; fat-soluble vitamins; water soluble vitamins and minerals/trace elements.

Staff: Scientists: 17 Others: 26

Research budget (ECU/year): 1.755.100

Description
The Institute of Nutrition is attached to the Directorate of Fisheries. The Institute's aims are: 1) to obtain knowledge on the requirement and metabolism of nutrients for salmonids and marine species, 2) to improve utilization of marine feed resources, and reduce negative environmental impact of feeds and feeding, 3) to contribute to an increased consumption of seafood, and 4) to increase the store of information on nutrition.

12. Institute of Zoology - University of Trondheim

Name : University of Trondheim, Institute of Zoology.
Address : University of Trondheim, College of Arts and Sciences
N - 7055 Dragvoll
Head : Ole Kritian Berg
Tel. : 47(0)7.59.62.97 **Fax**: 47(0)7.59.13.09

Type of institute
University institute with basic research and teaching in physiology, ecology and ethiology.

Research activities
Fish ecology
Fish physiology
Developmental biology
Aquaculture

Facilities
Brattøra Research Station, section of aquaculture
Tel. 47.7.59.63.10 **Fax**: 47.7.59.63.11

Staff: Scientists: 9 Others: 5

Research budget (ECU/year): 115.000

Description
The main research in fishery and aquaculture science is located at the Brattøra Research Station, and the activities include projects in the following fields:
- Basic salmonid ecology for sea ranching (migration, survival, energy budgets, growth).
- Physiology, immunology and ethology related to biological rhythms (especially smoltification and sexual maturation).

- Physiology and morphology related to developmental biology (especially functional morphology, egg quality, larval rearing and digestive mechanisms).
  Institute of Zoology also offers advanced studies in aquaculture and fish ecology, physiology and ethology at the cand.scient. and dr.scient.level.

13. Laboratory for Biotechnology, University of Bergen

Name : Laboratory for Biotechnology, University of Bergen
Address : Universitetet i Bergen, HIB
N - 5020 Bergen
Head : Dietmar Kemp
Tel. : 05.54.45.22 **Fax**: 05.31.49.52

Type of institute
University Institute

Research activities
Molecular genetics of marine micro-organisms (virus, bacteria and microalgae). Molecular genetics of fish.

Facilities
Laboratories for microbiology, cell culturing and gene technology.

Staff: Scientists: 4 Others: 4

Research budget (ECU/year): 36.300

Description
Basic research in molecular biology with emphasis on genetic technology. Current projects related to R&D in fish biology are:
a. Development of new antibiotic substances for fish pathonegic bacteria.
b. Improvement of microalgae with potential use as feed in aquaculture.
c. Characterization of genes which control growth and development in salmonids.

14. Nordland College and Nordland Research Institute

Name : Nordland College and Nordland Research Institute
Address : Mørkvedtråkket 30
N - 8002 Bodø

Head : Magne Haakstad
(Nordland College, Department of Fisheries and Aquaculture).
: Rolv Kristiansen
(Nordland Research Institute, Division of Aquaculture).
Tel. : 081.17.200 **Fax**: 081.17.349

Type of institute
College Institute with integrated research activities both for the public domain and research on a contract basis. Activities are integrated between Nordland College and Nordland Research Institute.

Research activities
Environmental effects on and of aquaculture activities. Industrial pollution affecting fjords. Coastal ecology in northern waters, and dynamics of phytoplankton blooms. Monitoring of harmful algae. Optimization of salmon aquaculture. Smoltification and salt tolerance of salmonids. New species in aquaculture; halibut and sea urchin. Managing of small and middle-sized companies, international strategy.

Facilities
General laboratories. aquaculture station

Staff: Scientists: 15 Others: 8

Research budget (ECU/year): 363.100

Description
Nordland College, Department of Fisheries and Aquaculture, and Nordland Research Institute, Division of Aquaculture, are parts of a regional college. The department offers courses for 1-3 years study in the subjects Fishery Administration, Fishery Technology, Aquaculture and Export of Fish Products. The institutions cooperate with several Norwegian universities and colleges in studies at doctorate level.

15. Norwegian College of Veterinary Medicine

Name : Norwegian College of Veterinary Medicine
Address : P.O.Box 8146 Dep.
N - 0033 Oslo
Street address: Ullevaalsv. 72, Oslo
Head : Kjell Gjaevenes(Adm. Director)
Tel. : 47-22964500

Type of institute
Institution for higher education and research in veterinary medecine.

Research activities
Aquatic biology, fish physiology, genetics, pharmacology, microbiology, pathology and food hygiene.

Staff: Scientists: 15 Others: 10

Research budget (ECU/year): 3.026.100

Description
The only institution for education of veterinarians in Norway, and the main Norwegian centre for veterinary medical research on fish.

16. Norwegian Institute for Nature Research (NINA)

Name : Norwegian Institute for Nature Research (NINA)
Address : Tungasletta 2
N - 7005 Trondheim
Head : Karl Baadsvik (Director General)
Tel. : 07.58.05.00 **Fax**: 07.91.54.33

Type of institute
Research and Consulting firm

Research activities
Related to fish and fisheries:
- Population biology
- Effects of radioactivity in nature
- Effects of human impacts on aquatic ecosystems
- Effects of acidification/heavy metals on aquatic organisms
- Development of ocean ranching for anadromous fishes
- Sustainable harvest of anadromous, catadromous and freshwater fish
- Behaviour ecology

Facilities
Data bases, library, general laboratories for biological investigations. Two research stations for producing salmonids for sea ranching. River laboratory for spawning studies.

Staff: Scientists: 78 Others: 83

Research budget (ECU/year): 12.104.400

Description

NINA works in the field of applied ecology and contributes to society's sustainable management of nature by serving authorities and corporations. NINA carries out research, monitoring and environmental impact analysis, in particular within population and community ecology. Central problems are pollution, environmental encroachment, nature conservation and optimal resource use (fish and game). Priority research areas are ecotoxicology, climate change, conservation of genetic resources, large herbivore ecology, landscape ecology, coastal ecology, fish optimization and outdoor recreation.

17. Norwegian Institute of Fisheries and Aquaculture

Name	:	Norwegian Institute of Fisheries and Aquaculture
Address	:	P.O.Box 2511 N - 9002 Tromsø
Head	:	Ola R. Valvåg (Managing Director)
Tel.	:	083.29000 **Fax**: 083.29100

Type of institute

Contract research institute for the Norwegian fisheries and aquaculture industries.

Research activities

Product- and process-oriented, marine biotechnology, information technology, economy, market and management, and research connected to fish stock management.

Facilities

- Laboratories for chemical, biochemical and microbiological analysis
- Unit for sensoric evaluation of food products
- Fisheries technology laboratory for pilot scale production and testing
- Aquaculture station with both fresh water and oceanic water, including a full scale fish farm, and equipment for the study of pathogenic microorganisms.

Staff: Scientists: 57 others : 56

Research budget (ECU/year): 6.778.500

Description

The Norwegian Institute of Fisheries and Aquaculture is a national institute for fisheries and aquaculture. Its main activities include industrial processing of marine raw materials for high quality food products; utilization of byproducts from the processing industry as raw material for aqua feeds or biochemicals; biotechnological methods in food and feed processing; characterization of high value biochemicals in raw materials; fish immunology and development of vaccines; aquatic biology and ecology; studies and production of planktonic organisms as feed organisms for fish larvae; stock assessment of commercially important marine species in Northern Norway and the Barents Sea; marketing research; management research and fisheries economics; information technology in fisheries.

18. Norwegian Marine Technology Research Institute MARINTEK

Name : Norwegian Marine Technology Research Institute MARINTEK
Address : Otto Nielsens veg 10, P.O. Box 4125 Valentinlyst
N - 7002 Trondheim
Head : Arnold Hansen (Managing director)
Tel. : 07.59.55.00 **Fax**: 07.59.57.76

Research activities
Fisheries:
- New fishing vessel concepts (catamarans etc.)
- Machinery and equipment on fishing vessels
- Safety and working conditions on fishing vessels
- Management and operation of fishing vessels
- Information technology for the fishing fleet
- Hydrodynamics of fishing gear

Aquaculture:
- Open and closed - floating and submerged - cage systems
- Rigid and flexible floating structures
- Mooring analysis
- Net design
- Logistics and operation

MARINTEK is also deeply involved in technology development for merchant ships, high speed vessels, offshore platforms, marine machinery such as diesel engines and technical operation of the same.

Facilities
Exclusive computer resources from CRAY computer to local PC networks.
Laboratories as follows:
- Ocean basin (L*B*D) 80*50*10m, including waves in two directions, current, wind and water purification
- Three towing tanks, 178*19*5.5, 30*2.5*1.0 and 82*10.5*10m
- Cavitation tunnels

No. 1, test section diameter: 0.36 m
No. 2, test section diameter: 1,20 m
- Machinery and process laboratory with LNG supply for gas engine research
- Corrosion and paint laboratory (located in Sandefjord)
- Cold climate laboratory
- Operation laboratory

Staff: Scientists: 105 (22 % doctorate level) Others: 100
Fisheries and aquaculture : 18 scientists

Research budget (ECU/year): 15.735.800

Description
MARINTEK's stated objective is to supply high quality marine technology R&D services to leading national and international companies and organizations in the field of Fisheries and Aquaculture Marine Machinery, Technical Operation, Offshore Floating Structures and Marine Vehicles.
The main currently fisheries research topics are:
- catamaran fishing vessel concepts
- technology to keep fish catches alive
- information systems onboard fishing vessels
- energy saving and reduction of pollution from fishing operations.

The major challenge in aquaculture technology is to develop systems with physical conditions for raising fish of high quality and new species of fish. To obtain high productivity, the structures of the fish farms should be designed for a correct level of safety and for efficient operation.

19. Research Institute Norconserv

Name	:	Research Institute Norconserv
Address	:	P.O. Box 327 N - Stavanger
Head	:	Harald Pedersen (Managing director) Per Henrik Prante (Director of R&D)
Tel.	:	04.52.90.44 **Fax**: 04.53.64.44

Type of institute
Contract research institute for the Norwegian fish products industry

Research activities
Raw material quality assessment and processing, processing technology and design, packaging materials and processes, microbiology and end product quality including marketing.

Facilities
Pilot scale laboratories for the fish products process industry; pilot scale packaging lines; laboratory for sensoric studies; chemical; biochemical and microbiological laboratory facilities.

Staff: Scientists: 25 Others: 25

Research budget (ECU/year): 1.936.700

Description
The R&D activities of Norconserv are mainly concerned with the canning industry and related food processing industry. The areas emphasized are connected with development of long term stability and packaging of fish products.

20. SINTEF Applied Chemistry

Name	:	SINTEF Applied Chemistry
Address	:	N - 7034 Trondheim
Head	:	Dr. Karl A. Almås (Division Director)
Tel.	:	07.59.28.73 **Fax**: 07.59.69.95

Type of institute
The SINTEF Group is the largest independent research organization in Scandinavia that provides the expertise required to keep business and industry updated with scientific and technological information. About 77% of the total income is generated by contract work for public and private sector.

Research activities
Chemical engineering (process simulation and evaluation, unit operations etc.).
Biotechnology (fermentation, fish preservation, marine biotechnology).
Aquaculture (fish feed and feed technology, start feed for marine fish larvae).
Environmental technology (oil pollution in marine environment).
Sensor and analytical chemistry.

Facilities
Laboratory facilities for chemical engineering.
Biotechnology pilot plant for fermentation and downstream processing.
Facilities for hatching and start feeding of marine fish species.
Laboratory for fish quality evaluation.
Extruding plant.
Arctic test-laboratory in Spitsbergen, Svalbard for marine oil spills
General and analytical laboratories.

Staff: Scientists: 105 Others: 55

Research budget (ECU/year): 10.894.000

21. SINTEF Group

Name : SINTEF Group
Address : N - 7034 Trondheim
Head : Idar Schei (Head of aquaculture research programme)
Tel. : 07.59.24.39 **Fax**: 07.59.23.76

Type of institute
Independent research institution

Research activities
Aquaculture technology:
Design and operation of seabased and landbased fish farms. Water treatment. Quality monitoring. Development of instruments. Cooling and freezing. Handling of fish.
Aquaculture biology:
Live feed production. Larvae rearing. Nutrition and food processing. Stress and effects on fish meat quality.

Facilities
Ocean basin. Laboratories for experiments with physical models, cooling and freezing, extraction of dry feed pellets. Aquaculture laboratory, small scale. Test- and demonstration station, fresh and salt water.

Staff: Scientists: 45 Others: 10

Research budget (ECU/year): 4.236.500

22. State Veterinary Laboratory for North Norway - Harstad

Name : State Veterinary Laboratory for North Norway, Harstad.
Address : P.O. Box 652
N - 9401 Harstad
Head : H.S. Norberg(Manager)

Type of institute
Regional Diagnostic Laboratory

Research activities
Research work on local fish disease problems.

Facilities
General laboratory facilities for diagnostic work.

Staff: Scientists: 6 Others: 9

23. State Veterinary Laboratory - Bergen

Name : State Veterinary Laboratory, Bergen
Address : P.O. Box 40
N - 5032 Minde
Head : Gunn Knutsen(Manager)

Type of institute
Regional Diagnostic Laboratory

Research activities
Work on infectious pancreas necrosis, pancreas disease, bacterial kidney disease and local fish disease problems, nectotizing hepatitis in Atlantic Salmon, disease surveillance in feral fish.

Facilitites
General laboratory facilities for diagnostic work

Staff: Scientists: 8 Others: 7

Description
State Veterinary Laboratory in Bergen is the regional Veterinary Laboratory for the countries Hordaland, Sogn and Fjordane, the laboratory has the same working strategy as mentioned for the regional laboratory in Sandnes.

24. State Veterinary Laboratory - Sandnes

Name : State Veterinary Laboratory, Sandnes
Address : P.O. Box 295
N - 4300 Sandnes
Head : I. Solberg(Manager)

Type of institute
Regional Diagnostic Laboratory

Research activities
Research work on local fish disease problems

Facilities
General laboratory facilities for diagnostic services

Staff: Scientists: 4 Others: 10

Description
State Veterinary Laboratory in Sandnes is the regional veterinary diagnostic laboratory aiming at a rapid and correct diagnosis work and disease control for fish in the countries Rogaland, East and West Agder.

25. State Veterinary Laboratory - Trondheim

Name : State Veterinary Laboratory, Trondheim.
Address : Tungasletta 2
N - 7005 Trondheim
Head : M. Fjølstad(Manager)

Type of institute
Regional Diagnostic Laboratory

Research activities
Research work on local fish disease problems

Facilities
General laboratory facilities for diagnostic work

Staff Scientists: 4 Others: 6

Description
State Veterinary Laboratory for Trondheim is the regional veterinary laboratory for the countries South and North Trøndelag. The laboratory has the same working strategy as given for the laboratory in Sandnes.

26. The Norwegian College of Fisheries, University of Tromsø

Name : The Norvegian College of Fisheries, University of Tromsø
Address : Dramsveien 201 B, P.O.Box 3083 Guleng
N - 9001 Tromsø
Head : Bjørn Hersoug

Type of institute
University faculty connected to the University of Tromsø

Research activities
Research integrated with the activity of the departments of fishery organization and fishery law studies; fishery economics and technology; resource and aquatic biology; marine chemistry.

Facilities
A new building for the whole faculty will be ready in 1993-94. General laboratory facilities, oceangoing vessels, access to University library and computer resources.

Staff: Scientists: 50 Others: 50

Research budget (ECU/year): 3.631.300

Description
The Norwegian College of Fisheries carries out research in all aspects of fisheries and aquaculture, and educates and graduates about 60 candidates on the B.Sc. level every year.

27. The Norwegian Herring Oil and Meal Industry Research Institute

Name	:	The Norwegian Herring Oil and Meal Industry Research Institute
Address	:	Kjerreidviken 16 N - 5033 Fyllingdalen
Head	:	Nils Urdal(Managing director)

Type of institute
Cooperative industrial research, mainly fish-meal industry.

Research activities
Three main programmes:
- Development of process technology
- Further development of existing products and markets
- New products

Facilities
Laboratories for chemical, biochemical, nutritional and microbiological analysis. Facilities for feeding experiments with domestic animals and farmed fish. Pilot scale equipment for processing of fish meal and fish oil, refining of fats and oils and processing of fish feed (extrusion, pelletizing, and granulation).

Staff: Scientists: 15 Others: 40

Research budget (ECU/year): 3.389.200

Description
The primary task of the Institute is applied research and development, working in the fields of chemistry, biochemistry, microbiology, process technology and nutrition. The Institute performs contract research for member and non-member companies. Analytical and advisory services are important parts of the total activity. The strategy and fields of competence are based on the needs of the Norwegian fish meal industry and the vessels catching fish for industrial purposes.

28. The Norwegian Institute of Technology - Division of Refrigeration Engineering

Name : The Norwegian Institute of Technology, Division of Refrigeration Engineering.
Address : N - 7034 Trondheim
Head : Prof. Per-Erling Frivik (Head of Division)
M.Sc. Nils Kristian Nakstad (Head of Food Eng. Section)
Tel. : 47(0)7.59.39.00 **Fax**: 7(0)7.59.39.26

Type of institute
Research Institute with main areas:
Refrigeration and heat pump engineering,
Low temperature process engineering,
Multiphase flow technology and food engineering.

Research activities
Fish processing
Cooling, freezing, thawing, storage and transport/distribution of food products.
Energy saving in food procesing plants.
Insulation and insulation materials.
Low temperature dewatering systems with emphasis on heat pump drying and freeze concentration technology.

Facilities
Cooling and freezing storage rooms at different temperature levels.
Freezing tunnel with 2-stage ammmonia plant.
Heat pump/fluidized bed drier.
Heat pump shelf drier.
Freeze concentration plant.
Test facilities for insulation materials, measuring thermal conductivities.

Staff: Scientists: 12 Others: 3

Research budget (ECU/year): 968.400

29. Trondheim Biological Station

Name : Trondheim Biological Station
Address : Bynesveien 46
N - 7018 Trondheim
Head : Jon-Arne Sneli
Tel. : 07.59.15.80 **Fax**: 07.59.15.97

Type of institute
Educational and research institute connected to University of Trondheim, The Museum.

Research activities
Photobiological studies on marine phytoplankton, deep-sea invertebrate taxonomy, oceanography, zoophysiology and population genetics.

Facilities
Research vessel (64 ft.), sea water facilities, wet and dry laboratories, standard marine chemistry and optical instrumentation.

Staff: Scientists: 9 Others: 22

Research budget (ECU/year): 242.100(excluding wages)

Description
The competence of the present staff covers the classical discipline of marine biology as well as special fields of expertise for which the economic funding comes mainly from Norwegian and EC research councils.

30. VESO

Name : Veterinarmedisinsk Oppdragssenter AS (VESO)
(National Centre for Veterinary Contract Research and
Commercial Services)
Head : Hans Petter Bugge(Managing Director)
Departments : VESO Vaccines
VESO VetResearch
VESO Vikan AkvaVet
Addresses : Head Office Oslo / Trondheim Office / VESO Vikan Akva Vet
Ullevålsvn 68 / Pir-Senteret / Esvalandet, Namsos
P.O.Box 8109 Dep / N-7005 Trondheim / N-7810 Alhusstrand
N-0032 Oslo
Tel. : 47.2296.4500 / 47.7351.5400 / 47.7428.4331
Fax : 47.2256.6254 / 47.7351.7084 / 47.7428.4366

Type of institute
Contract research and consulting firm.
Veterinary Vaccines distributor.

Research activities
Contract research in veterinary medecine in general, and with special emphasis on fish diseases. VESO's main objective is to provide services to the scientific community in order to enhance the utilization of veterinary competence in the community at large, and on a commercial basis.

Facilities
Competence in contractual, economic and employment questions concerning contract research. Insight in, and contracts to all veterinary institutions in Norway. Specialized laboratory for contract research on infectious fish diseases (VESO Vikan AkvaVet). Veterinary vaccines storage and distribution facilities.

Staff: Scientists: 12 Others: 13

Research budget (ECU/year): 5.000.000

Description
VESO comprises three separate units:
- *VESO Vaccines* is responsible for the distribution of all veterinary vaccines to the Norwegian market. VESO Vaccines, however, is not responsible for the marketing of these products, but acts as an impartial link between producers and buyers.
- *VESO VetResearch* conducts and negotiates contract research projects in the veterinary field. Its clients include both public and private institutions. Project implementors are most often drawn from the departments of the College of Veterinary Medecine or of the Central Veterinary Laboratory.
- *VESO Vikan AkvaVet* is a highly specialized research station for contract research in the field of infectious fish diseases, e.g. development of new vaccines and drugs for fish.

31. Veterinarinstituttet (Central Veterinary Laboratory)

Name : Central Veterinary Laboratory
Address : P.O.Box 8156 Dep.
N - 0033 Oslo
Head : B. Nass (Managing Director)

Type of institute
Diagnostics, control and research on diseases of animals, including fish.

Research activities
Multi-disciplinary approach to fish disease problems including mode of infection, pathology, bacteriology, virology, parasitology, immunology, vaccinology, epidemiology and clinical chemistry.

Facilities
Extensive libraries together with Norwegian College of Veterinary Medicine. Data bases, specialized software, good laboratory facilities for diagnostic work and research activities for diseases in aquatic organisms.

Staff: Scientists: 60 Others: 80

Research budget (ECU/year): 1.000.000

Description
The Central Veterinary Laboratory is the main veterinary reference laboratory for diagnosis and disease control for fish diseases in Norway. In addition the laboratory has an advisory role for the veterinary authorities regarding the administration of official regulations concerning fish health.
Currently the laboratory runs projects of maternal immunity in fish, virulence factors in Aeromonas salmonicida, taxonomical studies on fish parasites, infectious pancreas necroses (IPN) and pancreas disease (PD), epidemiological studies on smolt mortality, furunculosis, infectious salmon anemia (ISA) and cold water vibriosis, environmental toxicity studies, vaccination studies including efficary studies, side effects etc., clinical chemistry, studies of specific diseases such as ISA and bacterial kidney disease (BKD).

32. Vetmest - Centre of Veterinary Medicine - Tromsø

Name	:	VETMEST, Veterinærmedisinsk senter i Tromsø, Centre of Veterinary Medicine, Tromsø.
Address	:	Stakkevollveien 23 N - 9005 Tromsø
Head	:	Hans Kolbein Dahle (Director-Professor-Dr.med.vet.)

Type of institute
Research centre and coordinator of post degree training for veterinarians.

Research activities
Fish health and fish diseases related to fish farming and aquaculture
Diagnostic tests Pathogenicity of Aeromonas salmonicida
Immunologic mechanisms in fish skin related to foreign matter
Toxic effects of bacteria on fish heart
Ichtiotoxic components in marine algae
Microparasites in farmed salmon and wolffish
Stress and fish welfare and health

Facilities
Laboratories equipped for microbiology, immunology, pathology and toxicology.

Staff: Scientists: 5 Others: 3

Research budget (ECU/year): 190.000

Description
The goals for the VETMEST are to advance the level of veterinary knowledge especially in aquaculture, reindeer husbandry and wild life. The centre will carry out research that brings nationally and internationally valuable information in these fields. It will give diagnostic and scientific service to managers of fish farms, reindeer husbandry and wild life management and will participate in national and international networks for scientific cooperation.
The centre operates under the Ministry of Agriculture and reports to the Director general of the Veterinary Services.

33. Zoological Institute - University of Bergen

Name : Zoological Institute, University of Bergen
Address : Allegt. 41
N - 5007 Bergen
Head : Helge Leivestad (Institute Manager)
Tel. : 05.21.22.43 **Fax**: 05.32.91.11

Type of institute
University Department, Research and Education

Research activities
Physiological, parasitological, ecological and anatomical studies of fish and fish larvae. Effects of pollutants (acid rain, heavy metals, petrolium hydrocarbons) on fish and fish larvae. Energetics and nutrition of marine fish larvae. Amino acid dynamics in fish larvae and marine plankton.

Facilities
Laboratory equipment for physiological studies of fish and fish larvae. Thermostated cold rooms and aquaria facilities for experimental research in seawater and freshwater.

Staff: Scientists: 27 Others: 20

Description
The Zoological Institute has specialized staff for studies of environmental physiology, ecology, parasitology and ultrastructure of fish and fish larvae. The institute has a wide research network of contacts in Europe, USA and Canada, and many graduate students and doctorates takes degrees in fish biology.

34. Zoological Museum - University of Oslo

Name : Zoological Museum, University of Oslo
Address : Sars Gate, 1
N - 0562 Oslo
Head : M.E. Christiansen (Director)
Tel. : 22.68.69.60/22.85.50.50/22.85.16.00 **Fax**: 22.57.14.37

Type of institute
Zoological university museum

Research activities
General freshwater fish biology.
Impact on freshwater fish (esp. salmon and sea trout) of regulated rivers.
Parasitology of fishes.

Facilities
General laboratories for fish biology studies.
Experimental aquaria.
Molecular biology laboratory.
Large fish collections.

Staff: Scientists: 4 Others: 3

Research budget (ECU/year): 242.100

Description
Zoological Museum is a university institute with general research responsibilities including education of graduate students, maintenance of zoological collections and exhibitions for the public. Fisheries projects mostly regarding salmon, *i.e.* management and improvement of stocks, parasitology of salmon and stock impact of hydroelectric regulation of rivers.

State of Research

PORTUGAL

The Portuguese Coast is rather long and the geographic position of the Azores and Madeira Islands makes the Portuguese Economic Exclusive Zone the largest in Europe.

Fish is an important element in the national diet and Portugal ranks first in fish consumption "per capita".

The interest in marine sciences is very traditional, being more consistent since the last years of the XIX century.

At present, it is the Portuguese Marine Research (IPIMAR), former INIP, an institution depending on the Ministry of the Sea, that conducts most of the fisheries research and technology, embracing five main areas: marine biology and management of living resources, protection of the marine environment, culture of marine fish, shellfish and crustaceans, marine technology and upgrading of fishery products.

It is the INIP/IPIMAR that plays the important role of advising in fishery management, at both levels of the national administration and EC coordination programmes, and carries out research to provide basis for the establishment of TAC's and quotas.

Also, the Instituto Hidrografico (IH) of the Portuguese Navy is involved in several studies that contribute to the fisheries research, such as ocean circulation analyses, sedimentological surveys, and construction of specialised charts to the fisherman in joint collaboration with IPIMAR/INIP.

Fisheries research is also carried out in the Universities, mainly in the Biology and Veterinary Faculties, by some scattered departments. Most of the specific marine training is, nevertheless, only possible at INIP/IPIMAR.

The Portuguese Fisheries School provides suitable professional preparation to fishermen.

One of the main constraints in fisheries research comes from the lack of specialized personnel which has not been renewed at a desirable rate during the past years owing to restrictions in staff recruitment. On the other hand, funding almost totaly depends on Central Administration, with little participation of fishery industries or other private sources.

Aquaculture is developing, having to compete with other activities such as tourism, for the access to suitable sites which are, in many cases, located in protected areas and natural reserves with a severely restricted use of soil policy.

Off-shore aquaculture is, for the moment, almost not representative.

List of Institutes from PORTUGAL

1. IPIMAR - Instituto Português de Investigação Maritima
2. IPIMAR - Centro Regional de Investigação Maritima - Norte
3. IPIMAR - Centro Regional de Investigação Maritima - Sul
4. Instituto Hidrográfico - Lisboa
5. Instituto de Investgação Cientifica Tropical
6. Instituto Nacional de Meteorologia e Geofisica
7. Lab. Nacional de Investigação Veterinária
8. LNEC - Laboratório Nacional de Engenharia Civil
9. LNETTI - Laboratório de Engenharia e Tecnologia Industrial
10. Região Autónoma da Madeira
11. Secção Autónoma de Engenharia Naval - Instituto Superior Técnico
12. Universidade Algarve - Dept. Aquatic Resources
13. Universidade de Aveiro
14. Universidade do Porto
15. Universidade dos Açores
16. Univ. Lisboa - Dept. Zoologia e Antropologia - Aquacultura
17. Univ. Lisboa - Dept. Zoologia e Antropologia - Biologia Maritima
18. Univ. Lisboa - Dept. Zoologia e Antropologia - Impacto Ambiental
19. Univ. Lisboa - Dept. Zoologica e Antropologia - Ambiente e Evolução
20. Univ. Lisboa - Dept. de Fisica
21. Univ. Nova de Lisboa
22. Univ. Porto - Instituto de Ciências Biomédicas - Abel Salazar
23. Univ. Tras-Os-Montes-Dept. Zootécnia

1. IPIMAR - Instituto Português de Investigação Maritima

Name : Instituto Português de Investigação Maritima
Av. Brasilia
P - 1400 Lisboa
Head : Carlo Sousa Reis
Tel. : 351.01.3016361 **Fax**: 351.01.30155948 **Telex**: 15857 INIP P

Type of institute
Public Research Centre

Research activities
Fisheries management
Stock assessment
Multi-disciplinary approach to marine resources management
Environmental impact on resources
Marine technology
Upgrading of fishery products
Aquaculture

Facilities
Research vessels
Acoustic equipments
Specialized laboratories
Rearing tanks
Data bases, extensive libraries
Specialised software

Staff: Scientists: 87 Others: 233

Research budget (ECU/year): 7.000.000

Description
Research projects:
Life-cycles, growth and population dynamics of marine commercial species, including benthos and plankton studies, in order to correct resources management.
Development and improvement of more selective fishing gear and catching methods. Biological, physical and chemical oceanography. Detection and evaluation of biological and chemical pollution in water, sediments and organisms. Physiology of reproduction, genetics, feeding and pathology of commercially valuable farmed fishes, crustacia and mollusca.
Aquaculture/environmental interaction. Upgrading of underutilized species. Preparation of new fish products. Quality assurance of fish and fish products.

2. IPIMAR - Centro Regional de Investigação Maritima - Norte

Name : Instituto Português de Investigação Maritima
Canal das Pirâmides
P - 3800 Aveiro
Head : Manuel Sobral
Tel. : 351.034.28908 **Fax**: 351.034.381981

Type of institute
Regional Research Centre

Research activities
Fisheries management
Stock assessment
Multi-disciplinary approach to fisheries management
Environmental problems and their impact on resources
Oceanology
Aquaculture

Facilities
Research vessel
Analytical laboratories

Staff: Scientists: 6 Others: 17

Description
Stock assessment, recruitment, mortality of the most important commercially exploited fish stocks. Interaction between aquaculture and environment. Biotoxines survey programme for seashells. Monitoring of chemical, biological and physical parameters of the Portuguese North Atlantic Coast.

3. IPIMAR - Centro Regional de Investigação Maritima - Sul

Name : Instituto Regional de Investigação
Av. 5 de Outubro
P - 8700 Olhão
Head : Carlos Costa Monteiro

Tel. : 351.089.713375 **Fax**: 351.089.712514

Type of institute
Regional Research Centre

Research activities
Fisheries management
Stock assessment
Environmental impact on resources
Growth, nutrition, food composition, live food production for aquaculture.
Aquaculture/environment interaction.
Aquaculture structures, systems, techniques.

Facilities
Research vessel
Rearing tanks
Hatchery

Staff: Scientists: 7 Others: 22

Description
Interaction between aquaculture and environment. Shellfish culture management in the Ria Formnosa. Rearing of larvae and spat and growth of juveniles (*Crassostrea angulata, Rudtapes decussatus, Penaeus keraturus, Sparus aurata, Dicentrarchus labrax*). Technical assistance to the farmers. Artificial reefs programme (restocking practices and mariculture). Stock assessment of commercially exploited fish stocks in the Algarve. Biotoxines survey programme for seashells. Monitoring of chemical, biological and physical parameters of the Algarve Coast.

4. Instituto Hidrográfico - Lisboa

Name : Instituto Hidrografico
Rua das Trinas 49
P - 1296 Lisboa Codex
Head : J.A. Sarmento Gouveia
Tel. : 01.3955119/24 **Fax**: 1.3960515 **Telex**: 65990 HIDROG P

Type of institute
Research centre
Government Department/Regional or local authority
National

Research activities
Fisheries management
Environmental problems and their impact on resources

Facilities
Research vessel
Data bases, extensive libraries
Specialized software

Staff: Scientists: 50 Others: 120

Research budget (ECU/year): 3.120.800

Description
The fundamental aim of the Portuguese Hydrographic Institute is to guarantee the activities related with sciences and sea technologies and their application in military areas, in order to contribute to the country development of marine science and environment defence.

5. Instituto de Investgação Cientifica Tropical

Name : Instituto de Investigação Cientifica Tropical
R. da Junqueira, 86
P - 1300 Lisboa
Tel. : 01.364.50.71 **Fax**: 01.364.14.60

6. Instituto Nacional de Meteorologia e Geofisica

Name : Instituto Nacional de Meteorologia e Geofisica
Rua C-Aeroporto
P - 1700 Lisboa
Tel. : 01.848.39.61 **Fax**: 01.80.23.70

7. Lab. Nacional de Investigação Veterinária

Name : Laboratorio Nacional de Investigação Veterinária
Estrada de Benfica 701
P - 1500 Lisboa
Head : Dr. Matos Aguas
Tel. : 001.7162075

Type of institute
Research centre
Fish/Crustacean/mollusc farming
Government department/reg. author.

Research activities
Aquaculture
Disease, immunology, stress
Aquaculture/environment interaction upgrading of fishery products
Consumer response and demand studies

Facilities
Rearing tank
Controlled environmental chamber
Field site, electron microscope, specialized analytical apparatus

Staff: Scientists: 5 Others: 2

Description
Objectives:
- To develop appropriate methods in order to establish the nosological situation and the epizootology of the aquatic animals.
- To contribute to a better and efficient protection of the Public Health.
Research Projects:
- "Application of the ELISA Method to the Diagnosis of the Infectious Haematopoietic Necrosis (I.H.N) of Salmonids, Immunological and Biochemistry Characterization of the Polipeptides of the Virus" - "Epidemyological Study of Parasites and Parasitoses in Brackish water in the center of Portugal" - *Anguillicola crassus* in the European Eel in Portugal. - Epidemiology and Control" - "Ecopathological Study of the Infectious and Parasitary Diseases in Atlantic Eel (*Anguilla anguilla*) and Rainbow Trout (*Oncorhynchus mykiss*) in Hatcheries of the EEC Mediterranean Countries (Spain, Italy, Greece and Portugal).

8. LNEC - Laboratório Nacional de Engenharia Civil

Name : LNEC - Laboratório de Engenharia Civil
Av. do Brasil, 101
P - 1799 Lisboa
Tel. : 01.848.21.31 **Fax**: 01.849.76.60

9. LNETTI - Laboratório de Engenharia e Tecnologia Industrial

Name : LNETTI - Laboratório de Engenharia e Tecnologia Industrial
Azinhaga dos Lameiros-Estrada do Paço do Luminar-Edificio A-R/C
P - 1699 Lisboa Codex
Tel. : 01.716.51.41 **Fax**: 01.716.09.01

10. Região Autónoma da Madeira

Name : Região Autónoma da Madeira
Direcçao Regional das Pescas
Secretaria Regional da Economia
P - 9 000 Funchal Madeira
Head : D.M.B. Caldeira de Sena Carvalho
Tel. : 091.33140 **Fax**: 091.29691 **Telex**: 72183 DRPESC P

Type of institute
Research centre
Government Department / Regional or local authority
Regional

Research activities
Fisheries management
- Stock assessment
- Multi-disciplinary approach to fisheries management
- Environmental problems and their impact on resources

Fishing techniques
- Selective gears
- Static gear
- Other improvements in gears and/or fishing vessels

Facilities
Research vessel
Laboratories of fishing biology and oceanographics.
Construction of experimental fishing gears.

Staff: Scientists: 7 Others: 6

Research budget (ECU/year): 167.200

Description

Research projects:

Fisheries studies on swordfish, mackerel, horse-mackerel, shrimps, cephalopods in EEZ of Madeira. Artificial reefs construction. Studies on eggs and larvae of pelagic fishes. Environmental impact of sand extraction.

11. Secção Autónoma de Engenharia Naval - Instituto Superior Técnico

Name : Secçao Autónoma de Engenharia Naval
Instituto Superior Técnico
Universidade Técnica de Lisboa
Avenida Rovisco Pais
P - 1096 Lisboa Codex
Head : Prof. C. Guedes Soares
Tel. : 351.1.841.7000 ext 1468 **Fax**: 351.1.847.4015

Type of institute
University Department/Higher Education Institute

Research activities
Fishing techniques
Improvements in gears and/or fishing vessels
Aquaculture
Aquaculture structures, systems, techniques

Facilities
Specialised computer hardware
Specialised software

Staff: Scientists: 15 Others: 4

12. Univ. Algarve - Dept. Aquatic Resources

Name : Universidade do Algarve
Unidade de Ciências e Tecnologias dos Recursos Aquáticos (UCTRA)
Campus de Gambelas
Apartado 322
P - 8000 Faro Codex
Head : Josefina Cocelo/M.J. Bebiano
Tel. : 89.817761/817166 **Fax**: 89.818560 **Telex**: 56168 IP FARO

Type of institute
University Department/Higher Education Institute

Research activities
Fisheries management
 Stock assessment
 Multi-species biological models
 Environmental problems and their impact on resources
Aquaculture

Facilities
Specialised software

Research budget (ECU/year): 41.900

Description
The Unidade de Ciêncas e Tecnologias dos Recursos Aquáticos (Department of Sciences and Technologies of Aquatic Resources is actively involved in fisheries and aquaculture research. The department has 15 faculty members with a Ph.D., a Licenciatura(B.S.) degree program in Marine Biology and Fisheries with 60 new students a year, a M.S. degree program with 17 students, as well as a number of teaching assistants and graduate students working towards their Ph.D. in the areas of fisheries or aquaculture. Current European Community (FAR, AIR, STRIDE, ENVIREG) and Nationally funded research in fisheries and aquaculture includes:
-Bioeconomic studies of crustacean fisheries in the Algarve
-Comparative studies of Norway lobster (biology and fisheries)
-Baseline fish biology, artisanal fisheries, and community studies (SW coast of Portugal)
-Squid dynamics, recruitment and fisheries in the NE Atlantic
-Sexual cycles and growth of pargo (*Pargus pargus*)
-The nutritional and pheromonal importance of lipids in gilthead sea bream
-The use of fishpond outflow for growth of bivalves
-Environmental and hormonal control of growth in gilthead sea bream
-New species for aquaculture
-Optimization of water quality in fish ponds
-Bioenergetics and modelling

13. Universidade de Aveiro

Name : Universidade de Aveiro
Campus Universitário-Santiago
P - 3800 Aveiro
Tel. : 034.25085 **Fax**: 034.28600

14. Universidade do Porto

Name : Instituto de Zoologia Maritima "Dr. Augusto Nobre"
Faculdade de Ciências-Universidade do Porto
P - 4000 Porto
Head : J. Machado Cruz
Tel. : 02.31.02.90 **Fax**: 02.200.47.77

15. Universidade dos Açores

Name : Departamento de Oceanografico e Pescas
Cais de Santa Cruz
P - 9900 Horta-Açores
Head : J. Gil Pereira
Tel. : 092.229.88 **Fax**: 092.226.59

Type of institute
Research centre
University Department

Research activities
Fisheries management
Stock assessment
Physical and chemical oceanography
Environmental problems and their impact on resources

Facilities
Research vessel
Specialized laboratories

Staff: Scientists: 15 Others: 30

Description
Tuna fisheries management.
Demersal fisheries management and assessment.
Physical and chemical oceanography in order to identify the distribution of different species.
Environmental studies in the littoral areas.

16. Univ. Lisboa - Dept. Zoologia e Antropologia - Aquacultura

Name : Faculdade de Ciências de Lisboa
Secção de Aquacultura
Edificio C3-Campo Grande
P - 1700 Lisboa
Head : Carlos Almaça
Tel. : 01.757.31.41 **Fax**: 01.759.77.16

Type of institute
Research centre/University Department

17. Univ. Lisboa - Dept. Zoologia e Antropologia - Biologia Maritima

Name : Faculdade de Ciências de Lisboa
Secção de Biologia Maritima
Laboratório de Biologia Maritima
Forte de N. Sra. da Guia-Estrada do Guia
P - 2750 Cascais
Head : L. Caldas Saldanha
Tel. : 01.486.92.11 **Fax**: 01.486.92.11

Type of institute
Research centre
University Department

Research activities
Fisheries management
- Stock assessment
- Multi-disciplinary approach to fisheries management
- Environmental problems and their impact on resources

Facilities
Specialized laboratories

Staff: Scientists: 8 Others: 4

18. Univ. Lisboa - Dept. Zoologia e Antropologia - Impacto Ambiental

Name : Faculdade de Ciências de Lisboa
Secção de Impacto Ambiental
Edificio C2-Campo Grande
P - 1700 Lisboa
Head : M. José Costa
Tel. : 01.757.31.41 **Fax**: 01.759.77.16

Type of institute
Research centre
University Department

Research activities
Environmental problems and their impact on resources
Multi-disciplinary approach to fisheries management
Stock assessment

Facilities
Specialized laboratories
Electrofishing apparatus
Radiotracking apparatus

Staff: Scientists: 5 Others: 4

Description
Evaluate the environmental impact of human activities on natural communities. Investigate the structure and functioning of estuarine communities. Reabilitation of stocks of migrating fishes in Portuguese waters.

19. Univ. Lisboa - Dept. Zoologica e Antropologia - Ambiente e Evolução

Name : Faculdade de Ciências de Lisboa
Secção de Ambiente e Evolução
Edificio C3-Campo Grande
P - 1700 Lisboa
Head : Carlos Almaça
Tel. : 01.757.31.41 **Fax**: 01.759.77.16

Type of institute
Research centre/University Department

20. Univ. Lisboa - Dept. de Fisica

Name : Faculdade de Ciências
Secção de Oceanografia
Rua da Escola Politécnica
P - 1200 Lisboa
Head : Armando Fiuza
Tel. : 01.60.79.15

Type of institute
Research centre/University Department

21. Univ. Nova de Lisboa

Name : Faculdade de Ciências e Tecnologia
Quinta da Torre
P - 2825 Monte da Caparica
Tel. : 01.295.44.64 **Fax**: 01.295.44.61

22. Univ. Porto - Instituto de Ciências Biomédicas Abel Salazar

Name : Instituto Ciências Biomédicas "Abel-Salazar"
Dept. Produção Aquática
Largo Prof. Abel Salazar 2
P - 4000 Porto
Head : Luis Cieuve Monteiro
Tel. : 002.310359 **Fax**: 2.698736

Type of institute
University Department/Higher Education Institute
National

Research activities

Fisheries management
  Multi-disciplinary approach to fisheries management
Aquaculture
  Early stages of reared species
  Aquaculture structures, systems, techniques

Facilities

Rearing tank

Staff: Scientists: 1 Others: 2

Description

Research projects:
Growth studies on glass eels, closed circulation systems with integrated hydrophonics. Photographic documentation of the small scale fishery of Aguda (15 km. South of Porto).

23. Univ. Tras-Os-Montes - Dept. Zootécnia

Name : Universidade de Tras-Os-Montes
Dept. de Zootécnia
Divisão de Piscicultura
Quinta dos Prados
P - 5000 Vila Real
Head : Joao Oscar Branco/M.V. Cortes
Tel. : 059.25031 **Fax**: 059.74480 **Telex**: 24436 P

Type of institute

Research centre
University Department/Higher Education Institute

Research activities

Fisheries management
  Stock assessment
  Multi-species biological models
  Environmental problems and their impact on resources
Aquaculture
  Physiology
  Growth, nutrition, feed composition, live food production for aquaculture

Facilities
Research vessel
Rearing tank
Specialised software
Field site, electron microscope, specialized analytical apparatus
Laboratory for chemical water analyses
Electrofishing gears
Unit for processing water stable pellets for fish feeding
Fish feces collection device for the determination of food digestibility

Staff: Scientists: 4 Others: 8

Research budget (ECU/year): 30.000

State of Research

SPAIN

The Ministry of Agriculture, Fishery and Food (MAPA) is responsable for all the matters related to fisheries and marine farming through the Secretariat General of Marine Fisheries (SGPM) to whom is attached the Spanish Institute of Oceanography (IEO), which is the body charged with advising on research matters. The main lines of research for the IEO are: Fisheries Management, Aquaculture, Marine Environment, Oceanography and Marine Geology. The IEO headquarters is in Madrid, and along the coast there are seven Oceanographic Centers in: Santander, Coruña, Vigo, Fuengirola (Malaga), S. Pedro del Pinatar (Murcia) and in the Balearics and Canary Islands.

Within the Higher Council of Scientific Research (CSIC), which is attached to the Ministry of Education and Science, are the Marine Science Institutes with centres in Barcelona, Cadiz and Vigo, engaged in fisheries ecological and aquacultural activities. The Aquacultural Institute in Torre la Sal (Castellon) and the Centro Experimental del Frio also carry out research. All these centres tend to carry out academic as well as applied work.

Research is also carried out in laboratories depending on regional governments. These laboratories generally carry out academic as well as applied research, giving advice in response to the needs of their region's specific problems in very special circumstances. Some Universities do research work mainly on aquaculture and a few of them also in fisheries management.

There is some coordination between these bodies, but it should be further improved. This problem has been clearly defined and should be solutionned in a near future, specially thanks to the Interdepartmental Committee for Science and Technology (CICYT) and the National Programs.

The main constraints come from the lack of personnel and the lack of time for the scientists to recycle themselves on new research tools.

Specific marine training only exists at three Universities in Las Palmas (Canary Is.), Cadiz and in Vigo. Most of the Research Centers have to provide training for the new staff, which is the only efficient way to become a specialist. Moreover, there is a lack of specialists in certain branches, thus a great effort must be made to organize training in those fields.

List of Institutes from SPAIN

1. ANFHAPES - Asoc. Nac. de Fabr. de Harinas y Aceites de Pescado
2. Asociación Española de Acuicultura
3. AZTI/SIO - Servicio de Investigación Oceanográfica
4. Cabildo Insular de Gran Canaria - Instituto Canario de Ciencias Marinas
5. CEDEX - Centro de Estudios y Experimentación de Obras Públicas
6. Centro de Control da Calidade do Medio Mariño
7. Centro de Cultivos Mariños
8. CEA - Centro de Experimentacion en Acuicultura
9. Centro de Experimentación Pesquera - Direccion Regional de Pesca - Principado de Asturias
10. CIMA - Centro de Investigacions Mariñas
11. CIS - Centro de Investigaciones Submarinas
12. Conselleria Agricultura I Pesca - Estación de Acuicultura - Mallorca
13. CSIC - Inst. Acuicultura - Torre de la Sal
14. CSIC - Inst. Ciencias Marinas - Andalucía
15. CSIC - ICM - Inst. de Ciencias del Mar - Barcelona
16. CSIC - Inst. del Frío
17. CSIC - Inst. Investigaciones Marinas - Vigo
18. Grupo de Investigación en Acuicultura
19. IEO - Centro Oceanográfico de Baleares
20. IEO - Centro Oceanográfico de Canarias
21. IEO - Centro Oceanográfico de la Coruña
22. IEO - Centro Oceanográfico de Málaga
23. IEO - Centro Oceanográfico de Murcia
24. IEO - Centro Oceanográfico de Pontevedra (Vigo)
25. IEO - Centro Oceanográfico de Santander
26. IEO - Instituto Español de Oceanografía
27. Institut de Ecologia Litoral
28. IGAFA - Instituto Galego de Formacion en Acuicultura
29. Ministerio Agricultura, Pesca y Alimentación - Secr. Gen. Pesca Marítima
30. PEMARES - Plan de Explotación Marisquera y Cultivos Marinos - Sevilla
31. Principado Asturias - Cons. Agricultura y Pesca - Lab. Sanidad Animal
32. Unidad Helmintología
33. Univ. Alicante - Fac. Ciencias - Matemáticas y Estadística
34. Univ. Alicante - Fac. Ciencias - Dept. Ciencias Ambientales y Recursos Naturales
35. Univ. Baleares - Lab. Biología Marina
36. Univ. Barcelona Autónoma E.S.A.B.
37. Univ. Barcelona Autónoma - Food Technology Section - Fac. Veterinaria
38. Univ. Barcelona Autónoma - Dept. Biol. Cel. Fisiologia
39. Univ. Barcelona Autónoma - Facultad de Veterinaria
40. Univ. Barcelona - Dept. Bioguímica y Fisiologia
41. Univ. Barcelona - Fac. Biología
42. Univ. College Girona - Lab. Ictiología

43. Univ. de La Laguna - Dept. de Biología Animal
44. Univ. Granada - Dept. de Biología Animal y Ecología
45. Univ. La Coruña - Dept. de Biología Celular y Molecular
46. Univ. Madrid Complutense - Fac. Biología - Dept. Biología Animal I
47. Univ. Madrid - Complutense - Patología Animal I
48. Univ. Madrid - Complutense - Producción Animal
49. Univ. Madrid - Complutense - Progr. Acuicultura
50. Univ. Murcia - Biología Celular
51. Univ. Murcia - Fisiología y Farmacología
52. Univ. Oviedo - Dept. Biología
53. Univ. Santiago - Fac. Biología - Dept. Arca Biología Celular
54. Univ. Santiago - Fac. Biología - Dept. Fisiología
55. Univ. Santiago - Fac. Biología - Dept. Microbiología y Parasitología
56. Univ. Santiago - Fac. Biología - Lab. Tecnología Productos Pesqueros
57. Univ. Santiago - Fac. Farmacia - Parasitología
58. Univ. Santiago - Fac. Farmacia - Dept. Microbiología
59. Univ. Santiago - Fac. Farmacia - Dept. Bioquímica
60. Univ. Santiago - Fac. Farmacia - Dept. Fisiología
61. Univ. Santiago - Fac. Química - Dept. Química Orgánica
62. Univ. Santiago - La Coruña - Biología Fundamental - Area Genetica
63. Univ. Valencia - Dept. Ciencia Animal

1. ANFHAPES - Asoc. Nac. des Fabr. de Harinas y Aceites de Pescado

Name : Association Nationale des fabricants de farine et d'huile de poisson (ANFHAPES)
Address : CEA Bermúdez 14, 2E
E - 28003 Madrid
Head : J. H. Sainz-Rozas
Tel. : 91.2545340 **Fax**: 2545340 **Telex**: 49786 ANFH E

Type of institute
Fishing
Processing Industry/agro-food/diets producer

Research activities
Upgrading of fishery products
Novel processing methods for by-products, discards, and process wastes

Staff: Scientists: 10 Others: 30

2. Asociación Española de Acuicultura

Name : Asociación Española de Acuicultura
Address : Isabel la Católica 12-4D
E - 28013 Madrid
Head : Carmen Bautista Parejo
Tel. : 5423715

Research activities
Fisheries management
Multi-disciplinary approach to fisheries management

Research budget (ECU/year): 99.700

Description
Organization of national and international symposiums in several fields of aquaculture (economics, marketing production and laws).

3. AZTI/SIO - Servicio de Investigación Oceanográfica

Name : AZTI/SIO Instituto de Investigación y Tecnología para la Oceanografía, Pesca y Alimentación
Address : Txatxarramendi Irla
E - 48395 Sukarrieta-Bizkaia
Tel. : (4)687.00.51/75 **Fax**: (4)687.07.00
Address : Avda. de Satrústegui, 8
Villa Ondaretxe
E - 20008 San Sebastían (Guipúzcoa)
Tel. : (43)21.41.24 **Fax**: (43)21.21.62
Head : Jose Manuel Odriozola - General Director
Rogelio Pozo(Technical Director)

Type of institute
Government Agency
Research and Development Centre
Engineering/consulting firm
3 departments (Fisheries, Oceanography and Marine Environment, Food Technology)

Research activities
Fisheries management
  Stock assessment
  Monitoring of fisheries
  Artisanal fisheries
Fishing techniques
  Selective gears
  Other improvements in gears and/or fishing vessels
Oceanography and Marine Environment
  Monitoring
  Environmental problems and their impact in the resources
  Marine water quality
  Modelling
Aquaculture
  Aquaculture structure, systems, and technics
Upgrading food and fishery products
  Improvement of techniques for handling, storing, processing, packaging of fish
  Methods for grading and quality assessment of fish
  novel processing methods for by-products, discards and process wastes

Facilities
Research vessel
Data bases, extensive libraries
Specialised computer software and hardware
Field site, specialized analytical apparatus
Biochemistry, microbiology and sensory laboratories

Staff: · Scientists: 14 Others: 35

Research budget (ECU/year): 2.537.000

Description
Major research projects:
Fisheries Dept.:
a) Fisheries Management: Stock assessment and Biology on Demersal and pelagic fisheries in EC waters: Bay of Biscay. - International Cooperation: ICES,ICCAT and NAFO. Biology and Population Dynamics of Hake and related species. - biology and Population Dynamics of Albacore and Blue-fin Tuna in North Atlantic. Biology and Population Dynamics of Mackerel and Horse Mackerel in North Atlantic. - DEPM estimation for Anchovy Biomass - Engraulis encrasicholus in the Bay of Biscay. - Spawning of other species of commercial interest. - biology and Population Dynamics of Anchovy. - Study of the Artisanal fisheries in the Basque Country. - Cod fisheries in the Basque Country.
Oceanography and Marine Environment:
b) Temporal and Spatial Variation of the hydrography in the neritic area of the Basque Coast. - Technical Advice on Aquaculture. - Assessment on Alga Agarofita Gelidium in the Basque Coast. - Pluridisciplinar study in the shelf break off Cap Ferret (Bay of Biscay). Studies on environmental impacts of pipelines and outfalls. - Environmental surveillance of ports constructions.
Food Technology:
c) Research Projects in the area of fish are:
Maduration process of salt anchovy. - Improvement of Fish handling and discards of fish. - Safety and quality of fish products. - New products development base on fish.

4. Cabildo Insular de Gran Canaria - Instituto Canario de Ciencias Marinas

Name	:	Cabildo Insular de Gran Canaria Instituto Canario de Ciencias Marinas (ICCM)
Address	:	Apartado de Correos 56 E - 35200 Telde (Las Palmas)
Head	:	Dr. Octavio Llinas(Director of the Institute)
Tel.	:	34.28.693900/696904/693912 **Fax**: 34.28.693908

Tel. : 34.28.693900/696904/693912 **Fax**: 34.28.693908

Type of institute
Public Institute

Research activities
Fisheries
Small scale fisheries in the Canary Islands.
Fishing gears selectivity (gillnets and traps).
Technology of small scale fishing gears.
Fishing prospections in deep-sea waters around the Canary Islands.
Biology of commercial fishes and crustaceans.
Zoogeography of Canarian fishes and crustaceans.
Aquaculture:
Reproduction, broodstock management and eggs and larvae quality of commercially relevant marine fish species.
Nutritional requirements of broodstock, larval and post-larval stages of cultured marine fish species.
Pathology of cultured marine fish species, including histopathology, bacteriology and stress.
Hatchery technology.
Management and technology of floating cage systems.

Facilities
Data base about unloadings in several fishing ports of the Canary Islands.
Biology of some commercial species in the Canary Islands.
Research vessel (39/60m in length, 267 gross register tons and 1,070 h.p.).
Basic equipment for fishing research.
Basic equipment for fish and crustacean biology.
Fully equipped experimental hatchery and nursery, including tanks sytems, microscopes, binoculars, profile projector and live food culture facilities.
Feed preparation room, equiped with mixer, pelletizer, direr, balances, cold vacuum drier and miscellaneous equipment.
Four $64m^3$ in-shore and six $500m^3$ off-shore floating cages.
Laboratory of Fish Nutrition, including GLC, HPTLC, Iatroscan MK5, HPLC, Tecator Soxhlet crude Lipid Extractor, Tecator digestor and desillator (Kheldahl method), muffle furnace, Tecator fibre detector, balances, ultracentrifuge, oxygen bomb calorimeter and vacuum drier. Histopathology laboratory, including tissue processor, tissue embedder, microtome, sterioscopic zoom microscope and photomicroscope.

Staff: Fisheries: Scientists 8 Technicians 3
Aquaculture: Scientists 4 Technicians 2

Description
Fisheries: Biometric and meristic parameters, growth and reproducion of commercial fishes and crustaceans. Research on the small scale fisheries in the Canary Islands.

Aquaculture: Marine Aquaculture. Reproduction, nutrition, pathology and culture technology of marine fish species.

5. CEDEX - Centro de Estudios y Experimentación de Obras Públicas

Name : CEDEX - Centro de Estudios y Experimentación de Obras Públicas
Address : Alfonso XII, 3
E - 28014 Madrid
Head : F. Martínez Martínez
Tel. : 34.1.3357500 **Fax**: 34.1.5280354 **Telex**: 45022 E

Type of institute
Research centre
Government Department (Ministry of Public works and Transport)

Research activities
Fisheries management
Environmental problems and their impact on resources

Facilities
Data bases, extensive libraries
Specialised computer hardware
Specialised software
Field site, electron microscope, specialized analytical apparatus
Simulation techniques
Full scale models

Staff: Scientists: 320 Others: 530

Research budget (ECU/year): 35 000.000

Description
Port, coastal and oceanographic engineering;
Hydrographic studies; Surfaces and roadbeds;
Building structures and materials;
Use and protection of the environment;
History and archeology of the public works;
Training and documentation.

6. Centro de Control da Calidade do Medio Mariño

Name : Centro de Control da Calidade do Medio Mariño
Depending from: Conselleria de Pesca, Marisqueo e
Acuicultura - Xunta de Galicia
Address : Peirao de Vilaxoan s/n - Vilaxoan
E - 366 Vilagarcia de Arousa (Pontevedra)
Head : Joaquin Marino Cadarso
Tel. : 986.512320/512322 **Fax**: 986.512300

Type of institute
Center for the control of toxic plankton and red tides

Research activities
Related with its function, as primary productivity and its relation with oceanographical conditions, origin and evolution of red tides and plankton blooms, etc.

Facilities
A very well equipped laboratory for plankton studies
A well provided laboratory for nutrient analysis in sea water
Equipment for sea water sampling

Staff: Scientists: 2 Others: 7

7. Centro de Cultivos Mariños

Name : Centro de Cultivos Mariños
Depending from: Conselleria de Pesca Marisqueo e
Acuicultura - Xunta de Galicia
Address : Muelle de Porcillan s/n
E - 27700 Ribadeo (Lugo)
Head : Dorotea Martinez Patiño
Tel. : 982.110100/130391 **Fax**: 982.130492

Type of institute
Marine Aquaculture Research Institute

Research activities
Bivalve larval development
Bivalve hatchery technology development
Bivalve farming

Facilities
Two big greenhouses with a big number of fiberglass tanks for plankton production under natural light
A research bivalve hatchery
A well equipped laboratory for bivalve biology and research

Staff: Scientists: 2 Others: 110

8. CEA - Centro de Experimentación en Acuicultura

Name	:	Centro de Experimentacion en Acuicultura Depending from: Conselleria de Pesca, Marisqueo e Acuicultura - Xunta de Galicia
Address	:	Punta de Couso - Aguiño E - 15965 Santa Uxia de Ribeira (A Coruña) Postal address: Apartado oficial s/n E - 15965 Santa Uxia de Ribeira (A Coruña)
Tel.	:	981.841600/841750 **Fax**: 981.841516

Type of institute
Marine Aquaculture Research Institute

Research activities
Rotifer research and enrichment - finfish nutrition and larval development - turbot aquaculture - microalgae mass production - bivalve larval development, aquaculture and farming.

Facilities
Well equipped laboratories for finfish and bivalve larval aquaculture research and development
A big area of different sized fiberglass tanks for any kind of aquaculture mass production research
Facilities for temperature control of sea water supply

Staff: Scientists: 4 Others: 12

9. Centro de Experimentación Pesquera - Direccion Regional de Pesca - Principado de Asturias

Name : Consejería de Agricultura y Pesca
Centro de Experimentación Pesquera
Address : Avda. Príncipe de Asturias, s/n
E - 33212 Gijón (Asturias)
Head : A. Vizcaíno Fernández
Tel. : 985.314652 **Fax**: 985.312899

Type of institute
Research centre

Research activities
Fisheries management
Stock assessment
Fishing techniques
Aquaculture
Upgrading of fishery products
Consumer response and demand studies

Facilities
Rearing tank
Controlled environmental chamber
Data bases, extensive libraries
Specialised computer hardware
Specialised software
Field site, electron microscope, specialized analytical apparatus
Small vessel (8 m 0.L., 120 HP) for support.

Staff: Scientists: 11 Others: 5

Description
Research projects:
Studies on red sea bream (Pagellus bogaraveo) in Asturias. Biometry, reproduction, feeding and fisheries (2 years, 6 scientists involved). Studies on the fishery of elver in Asturias. Biology, fishing grounds, fishing techniques and influence of the environment (3 years, 2 scientists involved) - Biology of Gelidium sesquipedale, cartography, monitoring of algae fields; studies on beaching. (3 years, 7 scientists involved) - Deep-water resources appraisal: Fauna, new commercial species and fishing grounds. Marketing of new species (3 years, 7 scientists involved).
Fishery statistics in ports from Asturias (routinary, 1 scientist).
- Advice to the Administration on environmental impact, technical measures, fishing planification, etc.. Deep-water resources: bathimetry (1 year, 4 scientists). Deep-water resources: design and adaptation of selective years (traps and lines) (3 years, 3 scientists)-

Study on the condition index (CI) in oyster and clam as spawning index (3 years, 2 scientists) Experimental hatchery and phytoplankton camera for production of seeds of oyster and clam (routinary, 2 scientists) - Quality control of water for mollusc farming (routinary, 1 sc.) - Oyster and clam farming at different tide levels. Preparation of substratum. (3 year, 2 scientists)

10. CIMA - Centro de Investigacions Mariñas

Name : Centro de Investigaciones Mariñas (CIMA)
Address : Piedras de Coron s/n
E - 36620 Vilanova de Arousa (Pontevedra)
Postal address: P.O.BOX 208
E - 36600 Vilagarcia de Arousa (Pontevedra)
Tel. : 986.500155/500161 **Fax**: 986.506788

Type of institute
Research Institute related with fisheries, aquaculture and estuarine ecology.

Research activities
Pathology and biology of molluscs - aquaculture of molluscs and finfishes (biology, nutrition and larval development) - mussel farming and biology - bivalve biology and aquaculture (clams, cockles oysters and scallops) - exploitation of bivalve natural beds - population dynamics of bivalve - estuarine oceanography and primary production.

Facilities
Well equipped laboratories of finfish, mussel biology, bivalve molluscs biology and larval aquaculture, bivalve pathology and population dynamics, oceanography and primary productivity and phytoplankton.
A small research boat (12 m. length) for estuarine sampling
Library

Staff: Scientists: 20 Others: 19 Boat crew: 3

11. CIS - Centro de Investigaciones Submarinas

Name : Centro de Investigaciones Submarinas (CIS)
Address : Rep. de El Salvador, 23 - 8A
E - 15701 Santiago de Compostela
Head : Juan C. Soleto Solana/ C. Durán Neira
Tel. : 981.59.40.78 **Fax**: 59.81.72

Type of institute
Research centre
Engineering/consulting firm

Research activities
Fisheries management
Stock assessment
Multi-species biological models
Multi-disciplinary approach to fisheries management
Environmental problems and their impact on resources
Aquaculture
Growth, nutrition, feed composition, live food production for aquaculture
Aquaculture structures, systems, techniques

Facilities
Research vessel
Data bases, extensive libraries
Specialised software
Field site, electron microscope, specialized analytical apparatus
Boat (8 units)
Photographic material
Mobile laboratory

Staff: Scientists: 4 Others: 5

Description
Research projects:
Physical and biological surveys in NW Spanish coast.
Cartography of resources (oysters and algae).
Aquaculture studies: mussel, artificial reefs.

12. Conselleria Agricultura I Pesca - Estación de Acuicultura - Mallorca

Name : Estación de Acuicultura
Address : Camino del Faro
E - 07071 Puerto Andratx/Mallorca
Head : Sr. D. Sebastián Pou

Type of institute
Research centre
Government Department / Regional or local authority

Research activities

Aquaculture

- Early stages of reared species
- Disease, immunology, stress
- Aquaculture/environment interaction
- Aquaculture structures, systems, techniques

Facilities

Research vessel
Rearing tank

Staff: Scientists: 4 Others: 3

13. CSIC - Inst. Acuicultura - Torre de la Sal

Name : CSIC-Instituto de Acuicultura de Torre de la Sal
Address : Ribera de Cabanes
E - 12595 Torre de la Sal (Castellón)
Head : Dra. Pilar Alvarez Pellitero
Tel. : 964.310325 **Fax**: 64.310350

Type of institute

Research centre
National

Research activities

Aquaculture

- Physiology
- Reproduction
- Genetics and conservation of gametes
- Early stages of reared species
- Growth, nutrition, feed composition, live food production for aquaculture
- Disease, immunology, stress
- Aquaculture structures, systems, techniques

Facilities

Rearing tank
Controlled environmental chamber
Specialised computer hardware
Specialised software
Field site, electron microscope, specialized analytical apparatus

Staff: Scientists: 8 Others: 27

Description
There are five research groups:
PATHOLOGY RESEARCH GROUP: The group carries out several projects dealing with parasitic and infectious diseases.
FISH PHYSIOLOGY RESEARCH GROUP: The main lines of work are reproduction and growth in sea bass and sea bream.
AUXILIARY SPECIES RESEARCH GROUP: This group carries out research studies on different biological aspects of ARTEMIA and use in larviculture
CRUSTACEAN RESEARCH GROUP: The main lines of work are nutrition and reproduction in peneids
SHELLFISH RESEARCH GROUP: The group works in several projects dealing with different aspects of shellfish biology and culture.

14. CSIC - Inst. Ciencias Marinas - Andalucía

Name : Consejo Superior det Investigaciones Científicas
Instituto de Ciencias Marinas de Andalucía
Address : 11510 Puerto Real
Apartado Oficial
E - Cádiz
Head : D. E. Pascual Vázquez
Tel. : 34.56832612 **Fax**: 34.56834701

Type of institute
Research centre

Research activities
Aquaculture
- Physiology
- Reproduction
- Early stages of reared species
- Disease, immunology, stress
- Aquaculture/environment interaction
- Aquaculture structures, systems, techniques
- Pollution, effects in marine organisms
- Litoral ecosystems

Facilities
Seawater supply, several tanks, 6-meters boat, seawater autoanalyser

Staff: Scientists: 12 Others: 20

Research budget (ECU/year): 191.700

Description

Research projects:

Ecological aspects of aquaculture, assessment of factors affecting larval development of seabream in culture, physiology of microalgae under culture.

- subtethal effects of heavy metals and detergents on marine organisms.
- sole reproduction physiology

15. CSIC - ICM - Inst. de Ciencias del Mar - Barcelona

Name : Instituto de Ciencias del Mar
CSIC - ICM
Address : Paseo Nacional s/n
E - 08003 Barcelona
Tel. : 34.3.3194328 **Fax**: 34.3.3199842 **Telex**: 59367 E

Type of institute

Research centre

Research activities

Fisheries management
  Stock assessment
  Multi-species biological models
  Multi-disciplinary approach to fisheries management

Facilities

Research vessel
Data bases, extensive libraries
Specialised software
Field site, electron microscope, specialized analytical apparatus

Staff: Scientists: 27 Others: 4

Research budget (ECU/year): 1.657.900

Description

Research projects:

Demersal fisheries in ICSEAF area, Spanish Mediteranean coast, Pacific Ocean Mexico), and Beagle's channel.

16. CSIC - Inst. del Frío

Name : Instituto del Frío-CSIC.
Refrigeración y congelación de Carnes y pescados
Address : Ciudad Universitaria
E - 28040 Madrid
Head : Antonio Moral Rama
Tel. : 91.2445600 **Fax**: 4113077 **Telex**: 42182 CSTC E

Type of institute
Research centre

Research activities
Upgrading of fishery products
Improvement of techniques for handling, storing, processing, packaging fish
Methods for grading and quality assessment of fish
Novel processing methods for by-products, discards, and process wastes

Facilities
Fish technology pilot; Biochemistry Laboratory; Microbiology laboratory and rheology laboratory.

Staff: Scientists: 10 Others: 4

Research budget (ECU/year): 115.000

Description
Our main research programs are:
- Relation between protein functionality of muscle and structure of myofibrilar protein.
- Importance of muscular collagen on texture of the muscle.
- Improvement of gel characteristics made with fish muscle.
The group also works on projects financed by the industry dealing with different topics.
The staff also participates as professor in different courses and conferences.

17. CSIC - Inst. Investigaciones Marinas - Vigo

Name : Instituto de Investigaciones Marinas
CSIC-Consejo Superior de Investigaciones Científicas
Address : Eduardo Cabello 6
E - 36208 Vigo
Head : U. Labarta
Tel. : 86.231930 **Fax**: 86.292762

Type of institute
Public research centre

Research activities
Aquaculture
- Experimental physiology
- Early stages of reared species
- Growth, nutrition, diet composition, live food production for aquaculture
- Larvae enzimology
- Disease, immunology, stress
- Deseases: viral, bacterial, parasitic, immunological
- Metabolic and nutritional phatology

Oceanography
- Chemical oceanography of coastal and deep sea waters.
- Biological and biogeochemical oceanography.
- Antarctic research

Upgrading of fishery products
- Improvement of techniques for handling, storing, processing, packaging fish. Novel processing methods for by-products, discards and process wastes. Modelling and engineering studies on food processing.

Fisheries management
- Stock assessment
- Multi-species biological models. Multi-disciplinary approach to fisheries management.

Biology of shellfish, cephalopods and fish.

Facilities
Sea water aquaria. Research vessel. Data bases, extensive libraries. Specialized sofware and hardware. Specialized analytical apparatus. Technicon. Laboratories dedicated to biotechnology, nutrition, virology, micro-biology, parasitology, oceanography, and fish processing. Gas and liquid chromatography equipement. Mass spectroscopy. Scintillation counter. Ultracentrifuge. Infrastructure for environmental and oceanographical data acquisition.

Staff: Scientists: 44 Others: 46

Research budget (ECU/year): 1.785.000

Description
The Institute is a multidisciplinary research center. The main research lines can be described as: Aquaculture, Fisheries and Biology, Oceanography and Food Technology. The research in the field of Aquaculture is focused on the most important local cultures such as mullusc culture (mussels, oyster and clams) and turbot culture. Several parameters are being studied to optimize the production in both cases such as: the dynamics of mussel spat settlement and adult stock reproduction, the effect of food quality and composition on mollusc and fish culture are being studied to preserve and increase the production (virus, bacteria and parasites).
In Fisheries and Biology there are several ongoing projects on stock assessment and management of fisheries in the NAFO fishing grounds and on different aspects of fishery and biology of cephalopods in european waters. The department of oceanography is currently involved in the description and characterization of the dynamics and modelling of the exchange between the ocean and the rias and the mechanisms of development of red tides and algal blooms.
Food technology is engaged on study of different ways of recycling the waste of the mussel processing chain. The identification of canned fish and the influence of different parameters on the quality of processed chain. The identification of canned fish and the influence of different parameters on the quality of processed fish are other aspects that are currently the objective of this research area.
These different research interest are being financed by different CEE programs (MAST, FAR, AIR) and by the Spanish National Agency CICYT (Comisión Interministerial Ciencia y Tecnología). There are also several contracts with private industries and producers organizations to develop projects with immediate applied interest.

18. Grupo de Investigación en Acuicultura

Name : Grupo de Investigación en Acuicultura
Address : Facultad de Ciencias del Mar
Campus Universitario de Tafira
E - 35017 Las Palmas de Gran Canaria
Head : Dr. María Soledad Izquierdo López
(Marine Aquaculture Professor of Las Palmas University)
Tel. : 34.28.693900/452908 **Fax**: 34.28.693908

Type of institute
Research Centre

Research activities
Reproduction, broodstock management and eggs and larvae quality of commercially relevant marine fish species.
Nutritional requirements of broodstock, larval and post-larval stages of cultured marine fish species.
Pathology of cultured marine fish species, including Histopathology, Bacteriology and Stress.
Hatchery technology.
Management and technology of floating cage systems.

Facilities
Fully equiped experimental hatchery and nursery, including tank systems, microscopes, binoculars, profile projector, live food culture facilities and related miscellaneous equipment.
Feed preparation room, equiped with mixer, pelletizer, drier, balances, cold vacuum drier and miscellaneous equipment.
Four 64m3 in-shore and six 500m3 off-shore floating cages.
Laboratory of Fish Nutrition, including GLC, HPLTC, Iatroscan MK5, HPLC, Tecator sochlet Crude Lipid Extractor, Tecator digestor and destillator (Kjedldahl method), oven, muffle furnace, Tecator fibre detector, balances, ultracentrifuge, oxygen bomb calorimeter and vacuum drier.
Histopathology laboratory, including tissue processor, tissue embedder, microtome, sterioscopic zoom microscope, photomicroscope and miscellaneous.

Staff: Facultad de Ciencias del Mar:
Scientists: 5 Technicians 1
Instituto Canario de Ciencias Marinas:
Scientists: 4 Technicians 2

Research budget (ECU/year): 25.170.500

Description
Marine Aquaculture, Reproduction, Nutrition, Pathology and Culture Technology

19. IEO - Centro Oceanográfico de Baleares

Name	:	Centro Oceanográfico de Baleares
Address	:	Muelle de Pelaires, s/n-Ap 291 E - 07080 Palma de Mallorca
Head	:	P. Oliver
Tel.	:	34.71.401561 **Fax**: 34.71.404945 **Telex**: 69493

Type of institute
National Research Centre depending of the I.E.O.(MAPA)

Research activities
Fisheries management
Stock assessment
Environmental problems and their impact on resources
Fishing techniques
Electronics/acoustics

Facilities
Research vessel
Specialised computer hardware
Specialised software
Field site, electron microscope, specialized analytical apparatus
Ecosounder and ecointegrators for biomass evolution.
Research vessel: B/O ODON DE BUEN, 23mO.L., 62t GRT, 400 HP / B/O AROLA 7m. / B/O ACARTIA 6m.
1 IBM 6150, + 4 workstations, 9 PC.
AIX 2.1. operating system, SOL database, UNIPLEX, GraPHIGS.
Several microscopes. Data bases (long term series, librairie).
Acoustic equipments. Analytical apparatus.

Staff: Scientists: 9 Others: 17

Description
Research projects:
Fisheries management:
Pelagic fisheries in Balearic Islands, Cataluna and Valencia.
Internat. Coop.: GFCM
Fisheries in the Mediterranean sea. Internat. Coop.: CIESM
Biology of exploited fishes. Internat. Coop.: DGXIV CEE.
Fisheries in Western Mediterranean sea.
Marine environment:
Study of the pelagic environment in bays of Mallorca.
Tides in the Balearic Islands.
Study on the currents in the Western Mediterranean.
Micropaleontology and Atlanto-Mediterranean Glacioeustatisme.

20. IEO - Centro Oceanográfico de Canarias

Name	:	Centro Oceanográfico de Canarias
Address	:	Carretera de San Andrés s/n
		E - 38120 Santa Cruz de Tenerife
Head	:	Ma Angeles Rodríguez Fernández
Tel.	:	3422.549439 **Fax**: 3422.549554 **Telex**: 92573

Type of institute
Research centre depending to the I.E.O. (MAPA)

Research activities
Fisheries management
Stock assessment
Environmental problems and their impact on resources
Aquaculture
Reproduction
Early stages of reared species
Growth, nutrition, feed composition, live food production for aquaculture

Facilities
Rearing tank
Specialised computer hardware
Specialised software
Electronic mail
Several microscopes
Data bases (long term) series
Library
Field site, electron microscope, specialized analytical apparatus
1 micro VAX II + 19 Workstations VMS Oper.system, 9 PC.

Staff: Scientists: 13 Others: 18

Description
Research projects:
Fisheries management:
- study on biology and Canary fisheries.
- study on the sardine of west Africa
- study on the cephalopoda of West Africa. Internat. coop.: CECAF
- study on the Sparidae of CECAF division 34.1.3. Intern. Coop.: CECAF Tropical Tuna fishes. Int. Coop.: ICCAT Tuna fishes of the Canary Islands. Int. coop.: ICCAT Fish. of South East Atlantic-South West Indian Ocean. In. coop. ICSEAF Antartic fisheries. Intern. Coop.: CCRVMP/CCAMLR Marine environment: Ecological status of the marine environment in Canary Islands Research and monitoring of pollution in ICES division IXa South.
Oceanography of coastal waters in Canary Islands
Aquaculture: fish farming in the Canary shoreline

21. IEO - Centro Oceanográfico de la Coruña

Name : Centro Oceanográfico de La Coruña
Address : Muelle de Animas-Apartado de Correos 130
E - 15080 La Coruña
Head : N. González García-Estrada
Tel. : 3481.211451 **Fax**: 3481 229077 **Telex**: 86070

Type of institute
Research centre depending to the the I.E.O. (MAPA)

Research activities
Fisheries management
Stock assessment
Environmental problems and their impact on resources
Aquaculture
Reproduction
Early stages of reared species
Growth, nutrition, feed composition, live food production for aquaculture

Facilities
Research vessel
Rearing tank
Specialised computer hardware
Specialised software
Field site, electron microscope, specialized analytical apparatus
Two research vessels: B/O F. DE P. NAVARRO, 30.5 m O.L, 275 t GRT, 750 HP . / B/O LURA, 14.3 m O.L., 34t GRT, 160 HP. Rearing tanks. / 1 IBM 6150, + 4 workstations, 14 PC / AIX 2.1 Oper. System., SQL database, UNIPLEX, GraPHIGS.

Staff: Scientists: 14 Others: 28

Description
Research projects:
FISHERIES MANAGEMENTS: Pelagic fisheries in EC waters. International cooperation: ICES. Demersal fisheries in EC waters. Intern. Coop.: ICES. Marine mammals. Int. Coop.: ICES. Swordfish in the Atlantic and Mediterranean Sea. Int. Coop.: ICCAT, SEFC(NOAA). Albacore fisheries in North Atlantic. Int. Coop. ICCAT
MARINE ENVIRONMENT: Long-term study of benthic communities at ría de La Coruño (COST 647).
Research and monitoring of red tides.
Factors involved in spawning and larval of sardine (SARP)
AQUACULTURE: Optimal conditions for hatching of (Macropipus puber).
Role of feeding in larval development and growth of bivalve seeds.

Pathology of molluscs.

22. IEO - Centro Oceanográfico de Málaga

Name : Centro Oceanográfico de Málaga
Address : Puerto Pesquero s/n. Apartado 285
E - 29640 Fuengirola (Málaga)
Head : Juan A. Camiñas Hernández
Tel. : 3452.478148 **Fax**:3452.463808 **Telex**: IEOA E 79280

Type of institute
Research centre depending of the I.E.O. (MAPA)

Research activities
Fisheries management
Stock assessment
Environmental problems and their impact on resources
Fishes, crustacean and mollusc biology
Physical oceanography
Marine geology
Marine ecology

Facilities
Specialised computer hardware 1 IBM 6150 + 8 workstations, 14 PC
Specialised software
Field site, microscope, specialized analytical apparatus

Staff: Scientists: 19 Others: 10

Description
Research projects:
a) Fisheries Management: stock assessment on demersal fisheries, anchovy and other small pelagic fishes in ICES Division IX. International cooperation: ICES. Assessment of stock of sardine and anchovy. Acoustic detection and evaluation of pelagic.
Trawl fisheries. Selectivity of trawl. Artisanal fishery. Int. Coop.:
GCFM, CIESM. Sardine and anchovy recruitment project. Int. Coop.: ICES. Biology and stock assessment of hake, shrimp, and other crustaceans in CECAF area. Int. Coop. CECAF/FAO: Swordfish and bluefin tuna in Atlantic and Mediterranean sea. Int. Coop.:

ICCAT,CFPM.

b) MARINE ENVIRONMENT: Environmental factors in Alboran Sea affecting productivity. The coastal circulations. The marine environmental parameters and its relation with pelagic fishes. Int. Coop.: IOC, PRIMO Project, ICES. Pollution at ría de Huelva and Spanish South Atlantic area. Int. Coop.: ICES, Oslo and Paris convention. The east Spanish continental slope and margin at the Mediterranean Sea recent evolution. Int. Coop.: IUGS (UNESCO). Benthos fauna and Marine Geology in Antarctica. Int. Coop.: CCMLAR. New prospection systems applied to Sea-Nedd.
Pelagic production in an upwelling system: Galicia.
Environmental conditions in South-Atlantic region: Gulf of Cadiz.

23. IEO - Centro Oceanográfico de Murcia

Name : Centro Oceanográfico de Murcia
Address : c/Magallanes s/n.-Murcia
E - 30740 San Pedro del Pinatar (Murcia)
Head : L. Arévalo
Tel. : 3468.180500 **Fax**: 3468.184441 **Telex**: 67745 IEOM

Type of institute
Research centre depending to the I.E.O. (MAPA)

Research activities
Fisheries management
Environmental problems and their impact on resources
Aquaculture

Facilities
Research vessel
Rearing tank
Specialised computer hardware
Specialised software
Field site, electron microscope, specialized analytical apparatus
1 IBM 6150, 11 PC. Software: AIX 2 1 Op.Syst., SQL data base, GmPHIGS.

Staff: Scientists: 9 Others: 22

Description
One of the 7 coastal centres of the IEO. Highly specialised in aquaculture.

Research projects:
Fisheries management:
Mediterranean fisheries at Murcia and Valencia. Marine environment:
Monitoring and spreading of pollution in the Mediterranea Sea.
Aquaculture:
Reproduction and larval farming of sea bream at Mar Menor. Pectinidae farming.

24. IEO - Centro Oceanográfico de Pontevedra (Vigo)

Name	:	Centro Oceanográfico de Pontevedra (Vigo)
Address	:	Cabo Estay - Canido - Apdo. 1552 E - 36280 Vigo
Head	:	Beatriz Reguera
Tel.	:	3486.49211 **Fax**: 3486 492351 **Telex**: 83444

Type of institute
Research centre depending of the I.E.O. (MAPA)

Research activities
Fisheries management
  Stock assessment
  Environmental problems and their impact on resources
Fishing techniques
  Electronics/acoustics
Aquaculture

Facilities
Research vessel
Rearing tank
Specialised computer hardware
Specialised software
Field site, electron microscope, specialized analytical apparatus
Research vessel: B/O J.M. NAVAZ,16 mOL, 30tGRT, 430HP.
1 IBM 6150,4 workstation, 26 PC
Software:AIX 2.1 Op.System, SQL data base, UNIPLEX, GmPHIGS.

Staff: Scientists: 25 Others: 25

Description
Research projects:
Fisheries:
ICES fisheries-Galicia Norte. International cooperation: ICES
Demersal fisheries-EC waters. International cooperation: ICES

Spawning and larval biology of sardine (SARP). Int. cooper.: CBI (USA)
Marine mammals. International cooperation: ICES Swordfish in the Atlantic and Medit.Sea: int. coop. ICCAT,SEFC (NOAA) Albacore in North Atlantic. Int. Cooperation: ICCAT. Fisheries in Southwest Atlantic. Int. coop.: FIFD (Malvinas) ICES Fisheries in the Antarctic. Intern. Cooperation: CCRVMA/CCAMLR Marine environment: Red tides at Vigo Pollution in Cantabrian and Galician Coast. Coastal flow at Galicia in upwelling conditions.
Aquaculture: Turbot farming

25. IEO - Centro Oceanográfico de Santander

Name : Centro Oceanográfico de Santander
Address : Apartado de Correos 240-Promontorio San Martín
E - 39080 Santander
Head : O. Cendrero Uceda
Tel. : 34.42.274043 **Fax**: 34.42.275072 **Telex**: 35998 IEOS E

Type of institute
Research centre depending on the I.E.O (MAPA)

Research activities
Fisheries management
Stock assessment
Multi-species biological models
Oceanographical factors and their impact on resources
Aquaculture

Facilities
Rearing tank
Specialised computer hardware
Specialised software
Research vessels (2 units)
Field site, specialized analytical apparatus
Three vehicles, of which two four-wheel drive
1 IBM and 6 PC
research vessel: B/O JOSE RIOJA, 16 m O.L., 32 t GRT, 430 HP
B/O A. GONZALEZ DE LINARES, 12m O.L., 15 GRT, 60 HP
1 IBM 6150 + 4 workstations, 13 PC, AIX 2.1 Operating system, SQL database, UNIPLEX, GraPHIGS.

Staff: Scientists: 16 Others: 32

Description
Research projects:
Fisheries Management:
Assessment of fishing resources in EC waters: Cantabrian coast.
International cooperation: ICES.
Demersal fisheries in EC waters. International Cooper.: ICES.
Blue-fin tuna fisheries in Western Atlantic and Mediterranean Sea
Int. Coop.: ICCAT Albacore fisheries in North Atlantic. International coop.: ICCAT. Tuna fisheries in Indian Ocean. International cooperation: IPTP. (Malvinas), ICES. North-Atlantic fisheries (NAFO - Norway - Svaldbard). International cooperation: NAFO, ICES.
Aquaculture:
Farming of turbot and red sea-bream.
Farming of agraphytic algae: (Gelidium and Gracilaria).
Study of algae beaching.

26. IEO - Instituto Español de Oceanografía

Name : IEO-Instituto Español de Oceanografía
Headquarters: Madrid
Address : Avda del Brasil 31
E - 28020 Madrid
Head : R.Robles
Tel. : 91.34.1.597.4443 **Fax**: 91.34.1.597.4770 **Telex**: 44460 E

Type of institute
Public Research Centre
Directorate of the seven Coastal Oceanographic Centres
Appointed to the Ministry of Agricultural, Fisheries and Food (MAPA)

Research activities
Fisheries management
  Stock assessment
  Multi-species biological models
  Environmental problems and their impact on resources
Aquaculture
  Project management
Marine environment
  Physical oceanography
  Marine ecology
  Marine geology
  Pollution
  Harmful algal 6100m

Facilities

7 Oceanographic coastal laboratories

Centro Oceanografico de Baleares
Centro Oceanografico de Canarias
Centro Oceanografico de la Coruña
Centro Oceanografico de Malaga (Fuengirola)
Centro Oceanografico de Murcia
Centro Oceanografico de Pontevedra (Vigo)
Centro Oceanografico de Santander

Extensive library (more than 10.000 books, about 2000 scientific journals, 650 of which are living)
Data Center
9 Research Vessels (see below their distribution)
Computer hardware: 1 IBM 6150 + 7 working stations, 37 Pc.
Specialised software
AIX 2.1 Op. System, SQL database, UNIPLEX, GraPHIGS, etc.

Staff: Scientists: 129 Others: 268

Research budget (ECU/year): 25.170.500

Description

Headquarters of the IEO; Administration and coordination of research in other centers. Coordination of national programmes on marine items. Representative of the state at international level.

Research projects:

Fisheries management:
Spaning and larval biology of sardine(SARP) Intern. Coop: ICES
Pelagic fisheries EC waters. Int.coop.: ICES
Tropical tuna fisheries. Int. Coop.:ICCAT.
Population dynamics.
Marine environment:
Oceanography-fish relationship at Galicia: Microbiological aspects
Transformation of pure Mediterranean water into Atlantic
Mediterranean water at Gibraltar.
Fishing charts - Antarctic Area - Gibraltar Strait

27. Institut de Ecologia Litoral

Name	:	Institut d' Ecologia Litoral
Address	:	Ctra. Benimagrell 5
		E - 03560 El Campello (Alicante)
Head	:	Alfonso A. Ramos Esplá

Tel. : 34.6.565.76.90 **Fax**: 34.6.594.19.22

Type of institute
Research centre

Research activities
Upgrading of fishery products
Consumer response and demand studies
Oceanographical parameters and phitoplanktonic populations
Benthic bionomical cartography
Management of the littoral marine living resources, artificial reefs and marine protected areas (fishes and decapods crustaceans)
Small-scale fisheries around protected areas

Facilities
Data bases, extensive libraries
SCUBA diving equipment
Laboratory

Staff: Scientists: 4 Others: 3

Research budget (ECU/year): 74.000

Description
Research centre depending of the Alicante University (scientific stand) and the littoral Municipalities of the Alicante Province (economic stand).
FIELD OF RESEARCH: Marine living resources of the littoral sector between O and 5O m depth.
RESEARCH AREA: Valencia region (Alicante, Castellón and Valencia provinces, East and Southeastern Iberian Peninsula).
RESEARCH PROGRAMS (in fisheries) and main objectives:
-Faunistic groups; fishes and decapod crustaceans (adults, juveniles and larvae).
-Seagrass ecosystems. Role of the seagrass ecosystems (mainly (Posidonia oceanica) meadows) as nursery grounds.
-Management of the living marine littoral resources in protected areas (marine reserves, ex Marine Reserve of Tabarca) and artificial reefs (attraction/concentration and antitrawling reefs)

-Small-scale fisheries inside and around protected areas.
-Design artificial reefs and scientific pursuit.

28. IGAFA - Instituto Galego de Formacion en Acuicultura

Name	:	Instituto Galego de Formacion en Acuicultura (IGAFA) Depending from: Conselleria de Pesca, Marisqueo e Acuicultura - Xunta de Galicia
Address	:	Punta Niño do Corvo - Illa de Arousa E - 36626 Vilanova de Arousa (Pontevedra)
Head	:	Miguel Cuña Casabellas
Tel.	:	986.527101/527102 **Fax**: 986.527161

Type of institute
Professional Teaching of Marine Aquaculture

Research activities
Associated with the teaching, there are some activities of practical studies on marine aquaculture.

Facilities
A big area of fiberglass tanks to work on marine species aquaculture
Floating cage
Mollusc floating raft

Staff: Scientists: 8 Others: 16

29. Ministerio Agricultura, Pesca y Alimentación - Secr. Gen. Pesca Marítima

Name	:	Ministerio de Agricultura, Pesca y Alimentación Secretaría General de Pesca Marítima
Address	:	C/Ortega y Gasset, 57, 5a planta E -28006 Madrid
Head	:	Eladio Santaella
Tel.	:	91.402.50.00 **Fax**: 91.402 02 12 **Telex**: 47457 SGPM/E

Type of institute
Government Department

Research activities
Fisheries management
Environmental problems and their impact on resources

Aquaculture
  Aquaculture structures, systems, techniques
  Upgrading of fishery products
  Consumer response and demand studies

Facilities
Data bases, extensive libraries, reports

Staff: Scientists: 7 Others: 5

Description
Eladio Santaella is now General Subdirector. This Subdirection, within the Public Administration, has only a limited link with investigation, which is moreover an indirect one, in the following aspects:
- Redaction of programs of pluriannual orientation in aquaculture and artificial reefs.
- National plans in aquaculture.
- Studies of water quality for the hatching of molluscs.
- Studies of medioenvironmental problems affecting aquaculture and fishery (red tides, wastes, gravel extration, E.I.A. in hatching).
- Creation and management marine reserves in external waters.

30. PEMARES - Plan de Explotación Marisquera y Cultivos Marinos - Sevilla

Name : Plan de Explotación Marisquera
y de Cultivos Marinos de la Región Suratlántica (PEMARES)
Address : Avda. S. Francisco Javier, 24
Ed. Sevilla 1. 8a planta
E - 41071 Sevilla
Head : Dr. C. Manzano
Tel. : 95.4657511 **Fax**: 95.4657511

Type of institute
Government Department / Regional or local authority

Research activities
Fisheries management
  Stock assessment of shellfish resources
  Multi-disciplinary approach to shellfisheries management
  Environmental problems (pollution, diseases, parasites) and their impact on shellfish resources
Aquaculture

Facilities
Research vessel
Rearing tank
Controlled environmental chamber
Data bases, extensive libraries
Field site, specialized analytical apparatus

Staff: Scientists: 36 Others: 39

Research budget (ECU/year): 3.074.500

Description
Research projects:
Aquaculture: spat production on bivalves, juvenils production of prawns, seabream and seabass. Growth, nutrition and pathology on prawns, bivalves and commercial fishes.
Environmental studies (physico-chemical analysis of waters in Spanish South-Atlantic coast)
Economical analysis in aquaculture products.

31. Principado Asturias - Cons. Agricultura y Pesca - Lab. Sanidad Animal

Name : Consejería de Agricultura y Pesca del Principado de Asturias
Laboratorio de Sanidad Animal
Address : C/ Jove del Medio
E - 33299 Gijón-Asturias

Research activities
Fisheries management
  Multi-disciplinary approach to fisheries management
Fishing techniques
Aquaculture

32. Unidad Helmintología

Name : Unidad Helmintología
Address : Dr. Moliner, 50
E - Burjasot-Valencia
Head : Mr. E. Carbonell
Tel. : 96.3864606 **Fax**: 3864372

Type of institute
Research centre
University Department/Higher Education Institute

Research activities
Fisheries management
  Environmental problems and their impact on resources
Aquaculture
  Disease, immunology, stress

Staff: Scientists: 4

33. Univ. Alicante - Fac. Ciencias - Matemáticas y Estadística

Name	:	Universidad de Alicante / Facultad de Ciencias Dpto. Matematicas y Estadística
Address	:	Apartado 99 E - 03071 Alicante
Head	:	Eduardo Ferrandis Ballester
Tel.	:	96.5661150 **Fax**: 96.5668867 **Telex**: 66616

Type of institute
University Department/Higher Education Institute

Research activities
Fisheries management
  Stock assessment
  Multi-disciplinary approach to fisheries management
  Marine ecosystems analysis

Facilities
Research vessel
Data bases, extensive libraries
Specialised software

Staff: Scientists: 14

Description
1) Stochastic models of fish population dynamics.
Elaboration of Gaussian Processes which provide an objective estimation of fishing mortality rates and stocks.
Research developed in the context of the Statistics Committee and the Methods Group of the ICES.

2) Description and simulation of multispecies exploited marine ecosystems submitted to dynamic environmental and human interactions.
Research developed in the context of an international cooperation with the Peruvian Main Institute.

34. Univ. Alicante - Fac. Ciencias - Dept. Ciencias Ambientales y Recursos Naturales

Name : Universidad de Alicante / Facultad de Ciencias
Dept. Ciencias Ambientales y Recursos Naturales
Address : Apartado 99
E - 03080 Alicante
Head : Dr. Alfonso Ramos Esplá
Tel. : 34.6.5903668 **Fax**: 34.6.5903464 **Telex**: 66616

Type of institute
University Department/Higher Education Institute

Research activities
Marine Proteced Areas
Artificial Reefs
Seagrass Ecosystems
Ichthyology and Ichtyoplankton
Fisheries Management
Small Scale Fisheries

Facilities
Research vessel
Diving equipment
Data bases

Staff: Scientists: 4

Description
Research programs:
Impact of bottom trawling on seagrass ecosystems and the effect of its protection by antitrawling reefs on fishes and macroinvertebrates.
Marine protected areas effects on fish populations and surrounding fisheries.
Role of seagrass ecosystems as nursery grounds.
Taxonomy on faunistic marine groups.
Biology of demersal fishes.
Education: Higher education on marine ecology and fisheries.

35. Univ. Baleares - Lab. Biología Marina

Name : Universidad de les Illes Balears - Facultad de Ciencias
Laboratorio de Biología Marina
Address : E - 07071 Palma de Mallorca
Head : Isabel Moreno
Tel. : 971.207111(340) **Fax**: 971.200741

Type of institute
University Department/Higher Education Institute

Research activities
Plankton of the coastal waters. Artificial Reefs in Balearic coastal shelf. Benthic coastal communities,

Staff: Scientists: 2 Others: 3

Research budget (ECU/year): 8.000

Description
The major objective is the education of the University students in this subject, carrying on classes, field work and laboratory practicals.
The research projects carried on, up till now, have been specially on benthic coastal communities and plankton of coastal waters. At the moment, the laboratory is monitoring the artificial reefs established on the coastal shelf, S. of Majorca, E. of Ibiza and N. and S. of Formentera.
There is a small group of biologists, integrated in the Dept. of Environmental Biology, that work together, usually in local projects, sponsored by the local authorities and, at present, by a foundation from Madrid.

36. Univ. Barcelona Autónoma E.S.A.B.

Name : Escuela Superior d'Agricultura de Barcelona E.S.A.B.
Dept. de Produccions Agràries
Address : c/Comte d'Urgell 187
E - 08036 Barcelona
Head : Rosa Flos
Tel. : 34.34304207 **Fax**: 34.34192601

Type of institute
University Department/Higher Education Institute

Research activities
Aquaculture
Growth, nutrition, feed composition, live food production for aquaculture
Disease, immunology, stress

Facilities
Rearing tank
Small tanks with closed circuit
Sea water tanks have been built to share with other Universities Departments

Staff: Scientists: 2 Others: 1

Research budget (ECU/year): 76.000

Description
Research activities:
Natural feeding and body composition of sole (Solea solea) from the Ebro Estuary. Financed by Diputacio de Barcelona, five scientists involved during 3 years.
Natural feeding and Gody Composition of Sole (Solea senegalensis) from the Ebro estuary. Financied by CIRIT, three scientists involved for 1 year.
Flesh quality in sea bream related to food composition in farmed fish. Body composition in wild seabream relative to nutritional requirements. Four scientists involved in several other studies on stress of fishes.

37. Univ. Barcelona Autónoma - Food Technology Section - Fac. Veterinaria

Name : Univ. Barcelona - Facultad Veterinaria
Food Technology Section
Address : E - 08028 Barcelona
Head : B. Guamis
Tel. : 3.5811446 **Fax**: 3.5812006

Type of institute
University Department/Higher Education Institute

Research activities
Upgrading of fishery products
Methods for grading and quality assessment of fish
Novel processing methods for by-products, discards, and process wastes.

Facilities
Data bases, extensive libraries
Specialised software
Field site, electron microscope, specialized analytical apparatus

Staff: Scientists: 3 Others: 1

Description
This section has hardly started research in this field.

38. Univ. Barcelona Autónoma - Dept. Biol. Cel. Fisiologia

Name : Universidad Autónoma de Barcelona.
Facultad de Ciencias
Dept. Biol. Cel. i Fisiologia
Address : E - 08193 Bellaterra (Barcelona)
Head : D. Lluis Tort
Tel. : 3.5811914 **Fax**: 3.5812003 **Telex**: 52040 E

Type of institute
University Department/Higher Education Institute
National

Research activities
Aquaculture
Fish physiology
Immunology and stress in fish
Cardiorespiratory physiology

Facilities
Specialized analytical apparatus
Univ. close circuit laboratory
Open circulation seawater laboratory
HPLC, automated system for acquisition and analysis of biological signals in fish.

Staff: Scientists: 4 Others: 2

Research budget (ECU/year): 23.400

Description

Research projects:

- Effects of heavy metals on fish physiology.
- Stress procedures and their influence on hormone, metabolic and immune response in (Sparus aurata).
- Cardiorespiratory physiology of fishes.

39. Univ. Barcelona Autónoma - Facultad de Veterinaria

Name : Universidad Autónoma de Barcelona
Facultad de Veterinaria, Lab. de Biologia
Address : Bellaterra
E - 08193 Barcelona
Head : Silvia Crespo
Tel. : 3.5811746 **Fax**: 5812006

Type of institute

University Department

Research activities

Fisheries management
  Environmental problems and their impact on resources
Aquaculture
  Disease, immunology, stress

Facilities

Field site, electron microscope, specialized analytical apparatus

Staff: Scientists: 4

Description

Financial support is obtained for specific research projects.
Research projects in the last three years:

- Ecotoxicological, histopathological and genetoxic effects of textile effluents from wool processing. Financial support from the DGICYT (Ministerio de Educacion y Ciencia): 37.110 ECU/3 years.
- Development of turbot larvae: histopathological aspects. Financial support from the CICYT (Comision Interministerial de Ciencia y Tecnologia): 43.750 ECU/1 year
- Epitheliocystis disease in the amberjack (Seriola dumerili). Financial support from the CIRIT (Generalitat de Catalunya): 3.907 ECU/1 year.

40. Univ. Barcelona - Dept. Bioguímica y Fisiologia

Name : University of Barcelona, Dept. of Bioguímica y Fisiologia
Unidad de Fisiologia
Address : Avda. Diagonal 645
E - Barcelona 08071
Head : J. Fernandéz - Borrás
Tel. : 3.4110358 **Fax**: 3.4112967

Type of institute
Research centre
University Department/Higher Education Institute

Research activities
Aquaculture
Physiology
Growth, nutrition, feed composition, live food production for aquaculture

Facilities
Controlled environmental chamber
Research tanks, metabolites and hormones analysis equipment
Metabolism studies equipment

Staff: Scientists: 4 Others: 3

Research budget (ECU/year): 38.300

41. Univ. Barcelona - Fac. Biología

Name : Universidad de Barcelona
Facultad de Biología. Dept. Microbiología
Address : Avda. Diagonal 645
E - 08028 Barcelona
Head : Juan Jofre/Jorge Guinea/Felipe Fernández
Tel. : 3308851 **Fax**: 3307157

Type of institute
University Department/Higher Education Institute

Research activities
Aquaculture
Disease, immunology, stress

Facilities
We have all the facilities for microbiological studies

Staff: Scientists: 5 Others: 1

Research budget (ECU/year): 102.800

42. Univ. College Girona - Lab. Ictiología

Name : Universidad de Barcelona (Autónoma)
University College of Girona (Biology)
Address : Plaza Hospital 6
E - 17071 Girona
Head : Dr. Carle Pla Zanuy
Tel. : 72.210210 **Fax**: 72.216406

Type of institute
University Department/Higher Education Institute

Research activities
Fisheries management
Stock assessment
Aquaculture
Genetics and conservation of gametes

Facilities
Protein and DNTA analysis
Specialized analytical apparatus
Power supplies for electrophonesis low, high and ultra speed centrifuges. Gel driers, incubators, transilluminators - climatic chambers ultrafreezer-electrofishing apparatus

Staff: Scientists: 6 Others: 1

Research budget (ECU/year): 17.100

Description
The laboratory of Ichthyology is coming into its third year of existence. The main research field is genetics, for use in ecology stock identification and aquaculture.
RESEARCH WORK:
Genetics:
4 scientists involved in differential population studies:
genetic structure of brown trout populations (1987-1989)
genetic differentiation between mediterranean and atlantic hake populations (1988-1989)
Ecology:
2 scientists allocated, mainly in population dynamics:
studies on feeding ecology of Triglidae, based on analysis of gut contents (1986-1988).
Fish population dynamics in Lake of Banyoles (Girona) (1989-1991)
Genetics and ecology:
status and planting program of Aphanins items in Spain(1988-1989)

43. Univ. de La Laguna - Dept. de Biología Animal

Name : Universidad de La Laguna
Dept. de Biología Animal
U.D.I. de Ciencias Marinas
Address : Avda. Astrofísico Francisco Sánchez s/n
E - 38206 La Laguna-Tenerife-Islas Canarias
Head : Dr. Ignacio and J. Lzano Soldevilla
Tel. : 34.22.60.37.55/56/57/58 **Fax** 34.22.25.33.44

Type of institute
University Department/Higher Education Institute

Research activities
Fisheries management
Stock assessment
Eggs and larvae of fishes (Ichthytioplankton)
fish ecology
Small-scale Fisheries of the Canary Islands

Facilities
Recent database about partial landings of commercial species of the Canary Islands.
Biology of some commercial species of the Canary Islands.
Artificial Reef information in the Canary Islands.
Basic equipment for fishery and fish and crustacean biology research.
Small ship (about 9m length).
Scuba diving equipment.
Others related with the Service of the University.

Staff: Scientists: 5 Others: 3

Description

Research projects:

Biometric and meristic parameters, growth, age and reproduction of commerical crustaceans and fishes. Influence of the artificial reefs on fish populations. Biology, stock status, fishing methods and concentration of some liver chemical compounds of the deep-sea sharks. General overview and research on the small-scale fisheries of the Canary Islands.

44. Univ. Granada - Dept. de Biología Animal y Ecología

Name : Universidad de Granada
Facultad de Ciencias
Dept. Biología Animal y Ecología
Address : Campus Fuente Nueva
E - 18071 Granada
Head : M. de la Higuera
Tel. : 34.58.243240 **Fax**: 34.58.243238

Type of institute

University Department/Higher Education Institute

Research activities

Aquaculture
- Physiology and Biochemistry
- Growth, nutrition, feed composition, live food production for aquaculture
- Aquaculture/environment interaction

Facilities

Controlled environmental chamber
Specialised computer hardware
Specialised software
Specialised analytical apparatus
Extensive aquarium facilities

Staff: Scientists: 15 Others: 2

Research budget (ECU/year): 45.000

Description

Research projects:

Protein turn-over and intermediary metabolism responses to diet composition. Nutritive evaluation of alternative protein sources for fish culture. Microencapsulation applied to diet supplementation and larval feeding. Nutritive evaluaton of energy-yielding nutients: energy balance studies. Feeding behaviour of fish: self-feeding activity and diet preferences determination by using a dispositive controlled by computer that also controls environmental conditions. Larval feeding.

45. Univ. La Coruña - Dept. de Biología Celular y Molecular

Name : Universidad de La Coruña
Facultad de Ciencias
Dept. de Biología Cellular y Molecular
Address : La Zapateira s/n
E - 15071 La Coruña
Head : C. Herrero and J. Abalde
Tel. : 34.981.280788

Type of institute

University Department/Higher Education Institute
Fish/crustacean/mollusc farming

Research activities

Aquaculture
- Growth, nutrition, feed composition, live food production for aquaculture
- Aquaculture/environment interaction.
- Heavy metal contamination. Bioaccumulation. Bioremediation.

Facilities

Controlled environmental chamber
Rearing tanks
Specialized computer hardware
Specialized software
Field site, electron microscope, specialized analytical apparatus
Atomic absorption spectrophotometer; RMN spectrometer; beta and gamma counters. Flow cytometer. Capillary electrophoresis system.

Staff: Scientists: 5 Others: 1

Description
Optimisation of the cultivation of marine microalgae for aquacultural systems and other biotechnological uses.
Heavy metal contamination in the sea: effect on the first trophic level and bioaccumulation in the food chain.
Bioremediation.

46. Univ. Madrid - Complutense - Fac. Biología - Dept. Biología Animal I

Name : Universidad Complutense de Madrid
Facultad de Biología - Dept. de Biología Animal I
Address : Ciudad Universitaria
E - 28040 Madrid
Head : J. Gosálbez

Type of institute
University Department/Higher Education Institute

Research activities
Fisheries management
Stock assessment

47. Univ. Madrid - Complutense - Patología Animal I

Name : Universidad Complutense
Facultad Biología
Dept. de Patología Animal I
Address : Ciudad Universitaria
E - 28040 Madrid
Head : J. G. Noguera
Tel. : 915491318 **Fax**: 915447426

Type of institute
University Department/Higher Education Institute

Research activities
Fisheries management
Stock assessment

48. Univ. Madrid - Complutense - Producción Animal

Name : Universidad Complutense de Madrid
Dept. de Producción Animal
Fac. de Veterinaria
Address : Avda de Puerta de Hiero S/N
E - 28040 Madrid
Head : J. Thos Ruhi
Tel. : 91.4491600

Type of institute
University Department/Higher Education Institute

Research activities
Aquaculture
Growth, nutrition, feed composition, live food production for aquaculture
Aquaculture structures, systems, techniques

Staff: Scientists: 2

49. Univ. Madrid - Complutense - Progr. Acuicultura

Name : Universidad Complutense de Madrid
Vicerrectorado de Investigación,
Programa de Acuicultura
Address : c/ Vallhermoso 15
E - 28015 Madrid
Head : Carmen Bautista
Tel. : 4453714

Type of institute
University Department/Higher Education Institute
Institute for professional training/Secondary school

Research activities
Fisheries management
Stock assessment
Environmental problems and their impact on resources

Aquaculture
- Disease, immunology, stress
- Aquaculture/environment interaction
- Aquaculture economics

Staff: Scientists: 16

Research budget (ECU/year): 122.700

Description
Research projects:
Organization of aquaculture stages, publication of scientific papers about pathology, economy and culture techniques, impact of teluric overflows in aquaculture.

50. Univ. Murcia - Biología Celular

Name : Universidad de Murcia
Dept. Biología Celular
Address : c/ Santo Cristo 1
E - 30071 Murcia
Head : Dr. Blanca Aguilleiro Díaz, Professor
Tel. : 968.833000 **Fax**: 34.68.835418

Type of institute
University Department/Higher Education Institute

Research activities
Aquaculture
- Physiology
- Reproduction
- Early stages of reared species
- Disease, immunology, stress

Facilities
Electron microscopy service. Cell culture service. Histology laboratories

Staff: Scientists: 7 Others: 1

Research budget (ECU/year): 23.000

Description
Research projects:
Diet composition influence on the histophysiology of the gastro-entero-pancreatic system and liver in the seabream and seabrass.
Ontogeny of the endocrine gastro-entero-pancreatic system in seabass and seabream.
Production and use of the monoclonal antibodies against immunoglobulins of seabass and seabream.
Morphofunctional characterization of different nutritional conditions of seabrass.
In vitro study of the immune phagocytic defense of seabrass and seabream against infections caused by different pathogenic bacteria.

51. Univ. Murcia - Fisiología y Farmacología

Name	:	Universidad de Murcia, Dept. Fisiología y Farmacología Faculdad de Biología
Address	:	Campus de Espinardo E - 30100 Espinardo
Head	:	S. Zamora Navarro
Tel.	:	968.833000 **Fax**: 968.835418

Type of institute
University Department/Higher Education Institute

Research activities
Aquaculture
Physiology
Growth, nutrition, feed composition, live food production for aquaculture
Aquaculture/environment interaction

Facilities
Rearing tank
Field site, electron microscope, specialized analytical apparatus

Staff: Scientists: 10

Research budget (ECU/year): 58.800

Description
Main research lines:
Substitution of fish meal by agroindustrial by-products in trout feeding (1983-1987) 4 scientists involved.
Influence of food composition on the growth of sea-bream and sea-bass (1984-1989) 7 scientists involved.
Influence of the energetical component of the diet on different aspects of metabolism of the rainbow trout (1987-1991) 3 scientists involved.
Effect of ergogenic substances on growth and strong exercise in teleost fishes (rainbow trout, sea-bram and sea bass) (1991), 10 scientists involved.

52. Univ. Oviedo - Dept. Biología

Name	:	Universidad de Oviedo. Sec. Zoología Dept. de Biología de Organismos y Sistemas
Address	:	Jesus Arias de Velasco s/n E - 33005 Oviedo
Head	:	L. Villegas
Tel.	:	34.85233200 **Fax**: 85.233911 **Telex**: 84322

Type of institute
University Department/Higher Education Institute

Research activities
Fisheries management
  Stock assessment
  Eggs and larvae of fishes (Ichthyoplankton)

Facilities
Controlled environmental chamber
Field site, electron microscope, specialized analytical apparatus
Environmental study on salmonids Asturias
Biology and catches of Trisopsterus luscus, and other commercial fishes

Staff: Scientists: 2 Others: 5

Description
Research projects:
Biometric and meristic parameters, growth, age and reproduction of commercial and littoral fishes.
Influence of environmental parameters on trout.

53. Univ. Santiago - Fac. Biología - Dept. Arca Biología Celular

Name : Universidad de Santiago de Compostela
Facultad de Biología
Dept. Arca Biología Celular
Address : E - 15706 Santiago de Compostela
Head : M. Ramón Anadón
Tel. : 81.563100 ext. 2560 **Fax**: 81.596904

Type of institute
University Department/Higher Education Institute

Research activities
Study of fish nervous system
Fish protozoology
Histology and histopathology associated to healthy and diseased marine and freshwater fish

Facilities
Aquaria
Light & Fluorescence microscopy
Electron microscopy

Staff: Scientists: 8

54. Univ. Santiago - Fac. Biología - Dept. Fisiología

Name : Universidad de Santiago
Facultad de Biología. Dept. Fisiología
Address : E - 15706 Santiago
Head : M. Aldegunde
Tel. : 981.563100

Type of institute
University Department/Higher Education Institute

Research activities
Fisheries management
  Environmental problems and their impact on resources
Aquaculture
  Physiology

Facilities
Rearing tank
Field site, electron microscope, specialized analytical apparatus

Staff: Scientists: 7

55. Univ. Santiago - Fac. Biología - Dept. Microbiología y Parasitología

Name : Universidad de Santiago de Compostela
Facultad de Biología
Dept. de Microbiología y Parasitología
Address : E - 15706 Santiago de Compostela
Head : M. Juan L. Barja/Alicia E. Toranzo
Tel. : 34.9.81.563100 ext. 2548 **Fax**: 34.9.81.596904

Type of institute
University Department

Research activities
Phenotypic and serological studies of the microflora associated to healthy and diseased marine and freshwater fish
Biochemical and molecular characterization of bacterial and viral pathogens for fish and shellfish
Ecological studies on Marine Microbiology. Impact of the pollution in the Aquaculture.
Prevention and control of diseases in Aquaculture systems.

Facilities
Equipment for Bacterial & Virological analysis. Visible-UV spectrophotometer, ELISA reader, FPLC. Equipement for Molecular Biology. Ultracentrifuge and high speed centrifuges. Fluorescence microscopy. Electron Microscopy, Aquaria.

Staff: Scientists: 12 Others 2

Research budget (ECU/year) : 78.000

56. Univ. Santiago - Fac. Biología - Lab. Tecnología Productos Pesqueros

Name : Universidad de Santiago de Compostela
Facultad de Biología
Tecnología de Productos Pesqueros
Address : Campus Universitario s/n
E 15706 - Santiago de Compostela
Head : Ma de la Cruz Pascual López
Tel. : 81.563100

Type of institute
University Department/Higher Education Institute

Research activities
Aquaculture
Reproduction and growth of Bivalves
Aquaculure/environment interactions
Upgrading of fishery products
Improvement of techniques for handling, storing, processing and packaging of fishery products.
Quality assessment of raw and processed fishery products
Provision of information and advice to industry

Staff: Scientists: 13

Research budget (ECU/year): 50.000

Description
Research projects (1988-1990)
Influence of biodeposition of mussels in Galician rias (1988-1989).
Study of biodeposition of mussels (1988).
Anchorying and derived products (1989).
Upgrading of by-products from canning industries (1989).
Use of by products from canning industries as food for trout farming (1989).
Reproduction of Japanese clam in Galicia (1989).
Biogenic amines in fish. Microbiology, technology and toxicology (1990-1992).
Optimisation of fish sileage processes by enzymatic ad microbiological means (1989).
MAIN CONSTRAINTS:
Budgetary. Lacking of essential apparatus as HPLC, Atomic Absortion Spectrophotometer, etc...

57. Univ. Santiago - Fac. Farmacia - Parasitología

Name : Universidad de Santiago
Facultad de Farmacia
Lab. de Parasitología
Dept. de Microbiología y Parasitología
Address : E 15706 - Santiago de Compostela
Head : F.M. Ubeira/M. Sanmartín Durán/J.F. Casal
Tel. : 81.563100 **Fax**: 81.562569

Type of institute
University Department/Higher Education Institute

Research activities
Fisheries management
Environmental problems and their impact on resources
Aquaculture
Parasitic diseases, immunology, stress

Facilities
Controlled environmental chamber

Staff: Scientists: 8 Others: 4

Research budget (ECU/year): 30.000

Description
In the field of marine biology and aquaculture, our parasitology laboratory is conducting 3 principle research programs:
a) Experimental chemotherapy for protozoans and helminths. Different chemotherapeuthic agents are studied for their antiparasitic activity against protozoans and helminths, isolated from Galician fish farms, to provide adequate treatment for each particular case.
b) Production of monoclonal antobodies for diagnosis in marine aquaculture. The research is centered on commercially-important marine species such as the turbot, gilhead, seabass etc.
c) Analysis of the frequency and distribution of fish parasites in the Galician rias and their incidence in the farming of marine species.

58. Univ. Santiago - Fac. Farmacia - Dept. Microbiología

Name	:	Universidad de Santiago Facultad de Farmacia Dept. de Microbiología
Address	:	E 15706 - Santiago de Compostela
Head	:	J. Fabregas

Type of institute
University Department/Higher Education Institute

Research activities
Fishing techniques
  Electronics/acoustics
Aquaculture
  Reproduction
  Early stages of reared species
  Growth, nutrition, feed composition, live food production for culture
  Aquaculture structures, systems, techniques
Upgrading of fishery products
  Novel processing methods for by-products, discards, and process wastes
Economics

Facilities
Rearing tank
Specialised software
Field site, electron microscope, specialized analytical apparatus

Staff: Scientists and Others: 70

Description
1- Jaime Fabregas. Massive production of marine microalgae with different source of N. 1988-1989. Budget: 26.000 ECU.
2- Jaime Fabregas. Design of nutrients for marine microalgae using ammoniacal sources. 1989-1990. Budget: 9.000 ECU.
3- Jaime Fabregas. Computer controlled oyster larval cultivation. 1989-1990. Budget: 20.000 ECU.
4- Jaime Fabregas and Teresa Galhano. Intensive culture of larvae and spat af clams with artificial and controlled blooms of marine microalgae. 1990-1992. Budget: 190.000 ECU.
Other research projects are undertaken in the university.

59. Univ. Santiago - Fac. Farmacia - Dept. Bioquímica

Name : Universitad de Santiago
Facultad de Farmacia
Dept. de Bioquímica
Address : E 15706 - Santiago de Compostela
Head : J.L. Sánchez
Tel. : 81.563100 **Fax**:: 81.588522

Type of institute
University Department/Higher Education Institute

Research activities
Aquaculture
Early stages of reared species

Staff: Scientists: 6 Others: 2

Research budget (ECU/year): 22.800

60. Univ. Santiago - Fac. Farmacia - Dept. Fisiología

Name : Universidad de Santiago de Compostela
Facultad de Farmacia
Dept. de Fisiología
Address : E - 15706 Santiago de Compostela
Head : J. Espinosa
Tel. : 981.563100 **Fax**: 81.594912

Type of institute
University Department/Higher Education Institute

Research activities
Aquaculture
Physiology
Reproduction
Growth, nutrition, feed composition, live food production for aquaculture

Facilities
Field site, electron microscope, specialized analytical apparatus.

All instrumental facilities for endocrinology studies.

Staff: Scientists: 4 Others: 1

Research budget (ECU/year): 11.600

Description
The laboratory of mollusc histophysiology has a research group engaged in the study of molluscan endocrinology, with two main objectives:

1) endocrinology regulation of reproduction, development and growth particularly in (Mytilus genus). It aims at improving mollusc farming.
2) to produce postgraduate and postdoctoral people trained in that field.

61. Univ. Santiago - Fac. Química - Dept. Química Orgánica

Name : Universidad de Santiago de Compostela
Facultad de Química
Dept. de Química Orgánica
Address : Avda. de las Ciencias
E 15706 - Santiago
Head : R. Riguera
Tel. : 981.591085 **Fax**: 3481562569

Type of institute
University Department/Higher Education Institute

Facilities
Specialized equipment for organic chemistry

Staff: Scientists: 5 Others: 3

Research budget (ECU/year): 38.200

Description
Research is done on marine natural products from molluscs, echinoderma, seaweeds and sponges.

62. Univ. Santiago - La Coruña - Biología Fundamental - Area Genetica

Name : Dept. de Biología Fundamental
Area Genética
Colegio Universitario de La Coruña
Address : Universitad de Santiago de Compostela
E - 15071 La Coruña
Head : Dra. Josefina Méndez Felpeto
Tel. : 981280788

Type of institute
University Department/Higher Education Institute
Fish/Crustacean/mollusc farming

Research activities
Aquaculture
Genetics and conservation of gametes

Facilities
Controlled environmental chamber

Staff: Scientists: 7 Others: 1

63. Univ. Valencia - Dept. Ciencia Animal

Name : Universidad Politécnica Valencia
Dept. de Ciencia Animal
Unidad de Alimentación
Escuela de Ingenieros Agrónomos
Address : Camino de Vera, 14
E - 46071 Valencia
Head : J.F. Carmona/Dr. Julio Fernández Carmona
Tel. : 34.6.3877432/34 **Fax:** 34.6.3877439 **Telex**: 62808

Type of institute
University Department/Higher Education Institute

Research activities
Aquaculture
Growth, nutrition, feed composition, extruded feedstuffs.
Aquaculture structures, systems, techniques

Staff: Scientists: 4 Others: 6

State of Research

SWEDEN

Responsibilities

There are several levels of responsibility within the field of fisheries and aquaculture.

Main responsibilities:
The Swedish Council for Forestry and Agricultural Research is the main body responsible for overall basic research and development programmes. It has a general responsibility to give support to basic research.

The national Board of Fisheries is the main body responsible for overall applied fishery research and development. In addition to this the Board carries out such activities at its own laboratories and fisheries research stations. It has three research offices performing investigations in cases involving applications for structures in waters which might be harmful to fisheries.

The Swedish Council for Forestry and Agricultural Research and the National Board of Fisheries cooperate within the fishery research field with regard to commercial fishery, recreational fishery, aquaculture and fishery management. The cooperation is related to biological, technical as well as economic research and development matters.

Subordinated responsibilities:
The University of Agriculture (SLU) has a responsibility for research related to aquaculture of fish for consumption.

The Swedish Environmental Protection Agency (SNV) supports environmentally-related research, which may be of relevance to fish and fisheries.

The Swedish Natural Science Research Council (NRF) and the Swedish Council for Planning and Coordination of Research support basic research in general.

The Swedish National Board for Industrial and Technical Development (NUTEK) supports technical research and development.

The Swedish Transport Research Board and the Swedish Work Environment support technical research related to commercial fishery.

Resources
From the viewpoint of economy there is a total of 5.693.000-6.832.000 ECU available for fishery research each year. One major part or 4.554.000-5.124.000 ECU is directed towards the commercial and the recreational fishery. The rest of the total sum is related to basic research at the universities etc.

List of Institutes from SWEDEN

1. Baltic Sea Research Station - National Board of Fisheries
2. Department of Aquaculture - Swedish University of Agricultural Sciences
3. Fisheries Research Station - National Board of Fisheries - Älvkarleby
4. Fisheries Research Station - National Board of Fisheries - Kälarne
5. Institute of Coastal Research - National Board of Fisheries
6. Institute of Freshwater Research - National Board of Fisheries
7. Institute of Marine Research - National Board of Fisheries
8. Institutes for Research on Fish as Food
9. Institutes for Research on Fish Diseases
10. Institutes for Technical Research
11. Salmon Research Institute
12. University Institutes for Ecological Research
13. University Institutes for Genetic Research
14. University Institutes for Physiological Research

1. Baltic Sea Research Station - National Board of Fisheries

Name : Fiskeriverket Östersjölaboratoriet, Karlskrona
Address : Utövägen 5,
S - 371 37 Karlskrona
Head : Mr Gunnar Sellerberg
Tel. : 46.45514230 **Fax**: 46.45581523

Type of institute
Research Institute

Research activities
The institute is a laboratory subsidiary of the Institute of Marine Research in Lysekil with a main task to investigate the cod stock in the Baltic. The staff amounts to 2 scientists and 1 engineer.

Research budget (ECU/year): 200.600

2. Department of Aquaculture - Swedish University of Agricultural Sciences

Name : Vattenbruksinstitutionen Sveriges Lantbruksuniversitet, Umeå
Address : S - 901 83 Umeå
Head : Prof. Lars-Ove Eriksson
Tel. : 46 90127676 **Fax**: 46.90123729

Type of institute
Research centre

Research activities
Fish biology
Selectivity programmes
Technical development

Staff: Scientists: 10

Research budget (ECU/year): 724.400

Description
The main research objects are fish biology linked to aquaculture, selectivity programmes and technical development of rearing systems. In addition to this there is carried out basic

research such as theories on the life history and migrations of salmonids, reproduction strategies, intraspecies interactions and socio-ecological studies, salt water adaptation, the sexual maturity of salmonids, breeding experiments on char and rainbow trout, quality criteria and optimization.
Research on aquaculture is also carried out at several university departments, but in relation with marine fish species such research is only conducted at the University of Göteborg in cooperation with Kristineberg Marine Biological Station, Fiskbäckskil.
At Tjärnö Marine Biological Laboratory, belonging to the University of Göteborg, there is project on oyster aquaculture.
Research on the environmental effects of aquaculture is carried out at the Universities of Lund, Göteborg and Umeå and at the Swedish Environmental Research Institute.

3. Fisheries Research Station - National Board of Fisheries - Älvkarleby

Name	:	Fiskeriverket Försöksstationen Älvkarleby
Address	:	Brobacken
		S - 810 70 Älvkarleby
Head	:	Mr Nils Steffner
Tel.	:	46.2672670 **Fax**: 46.2682170

Type of institute
Rearing station

Research activities
Experiments

Facilities
Fluvarium

Staff: Scientists: 1 Others: 8

Research budget (ECU/year): 111.500

Description
The station at the river Dalälven is a rearing station for sea trout with a new, large fluvarium for experiments, and the station is subsidiary to the Institute of Freshwater Research.

4. Fisheries Research Station - National Board of Fisheries - Kälarne

Name : Fiskeriverket Försöksstationen Kälarne
Address : S - 840 64 Kälarne
Head : Dr. Jan Henriksson
Tel. : 46.69640405 **Fax**: 46.69640029

Type of institute
Rearing station

Research activities
Preservation
Genetics

Staff: Scientists: 2 Others: 8

Research budget (ECU/year): 1.003.100

Description
The station is a rearing station for breeding of char and rainbow trout and for preservation and restauration of different stocks of salmonids. One main study object is the effect on the genetic quality of long time breeding. The station is subsidiary to the Institute of Freshwater Research.

5. Institute of Coastal Research - National Board of Fisheries

Name : Fiskeriverket Kustlaboratoriet, Öregrund
Address : P.O.Box 584
S - 740 71 Öregrund
Head : Dr. Erik Neuman
Tel. : 46.17331305 **Fax**: 46.17330949

Type of institute
Research and monitoring institute

Research activities
Fisheries management
Stock assessment
Monitoring of environmental impacts

Facilities
Research vessel
Rearing tanks

Staff: Scientists: 6 Others: 13

Description
The main research objects are: assessment of coastal species, environmental impacts caused by the forest industry and the nuclear power stations and the effects of climate changes.

6. Institute of Freshwater Research - National Board of Fisheries

Name : Fiskeriverket Sötvattenslaboratoriet, Drottningholm
Address : S - 171 11 Drottningholm
Head : Dr. Stellan Hamrin
Tel. : 46.87590040 **Fax**: 46.87590338

Type of institute
Research centre

Research activities
Methods
Acidification
Stock assessments
Monitoring

Facilities
Research vessels
Aquarium

Staff: Scientists: 15 Others: 10

Research budget (ECU/year): 891.600

Description
The main research fields are the development of methods for the preservation, conservation and enhancement of fish species, the study of acidification processes and liming operations, stock assessments in the great lakes and monitoring of fish stocks. The institute has a branch institute in Örebro, Trädgårdsgatan 13, S - 702 12 Örebro. There are 2 scientists at this institute Tel. and fax 46.19103525. Head: Dr. Per Nyberg.

7. Institute of Marine Research - National Board of Fisheries

Name : Fiskeriverket Havsfiskelaboratoriet, Lysekil
Address : P.O.Box 4
S - 453 21 Lysekil
Head : Dr. Bernt Ingemar Dybern
Tel. : 46.52314180 **Fax**: 46.52313977 **Telex**: 27108 NATFISH S

Type of institute
Research centre

Research activities
Fisheries management
 Stock assessment
 Multi-species biological models
Environmental impacts
Fish diseases
Fishing techniques
Aquaculture

Facilities
Research vessels
Rearing tank

Staff: Scientists: 10 Others: 45 (crews included)

Research budget (ECU/year): 2.451.900

Description
The institute is responsible for fisheries and marine research and gives advice concerning fisheries management, stock enhancement, and environmental problems in the marine environment. The main research activities are: Fisheries biology including data collection, monitoring and assessment of fish stocks (including international cooperation with ICES, ACFM), multispecies and fish gear research. Marine ecology: biological data collection, eutrophication and pollution effects on fish stocks, fish diseases. The institute is also engaged in development assistance programmes.

8. Institutes for Research on Fish as Food

Swedish Institute for Food Research (SIK), Göteborg

The research on fish constitutes only a minor part of the activities at the Institute. The main subject in the fisheries field is the quality of fish products when stored. New products of herring are also developed. On an average there are 3-4 persons working on fish research and the annual budget amounts to 55.700 ECU.

9. Institutes for Research on Fish Diseases

National Veterinary Institute (SVA), Uppsala
The institute carries out investigations on diseases in wild and reared fish populations, especially on immunology and development of vaccines. The staff amounts to 6 persons. The institute cooperates with the Salmon Research Institute on problems related to bacteriological, immunological and pathological questions with regard to kidney inflammations, vibrosis and effects related to low pH values and aluminium, particularly of fish in aquaculture.

Departments of Microbiology at the Universities in Göteborg and Umeå
Infections of marine fishes and the molecular biology of the bacterium Vibrio.

Departement of Zoology, University of Uppsala
The effects on fish of discharges from the pulp industry.

Department of Zoology, University of Stockholm
Studies of parasites (Gyrodactylus).

Department of Physiological botany
Diseases and immunological reactions of crayfish.

10. Institutes for Technical Research

In Sweden there is no specific institute or department for technical research, but research is performed in several institutes such as *Chalmers Technical University*, Göteborg, where studies are going on concerning trawl boards and technical matters related to the distribution of fresh fish. At *SSPA Maritime Consulting AB*, Göteborg, research is carried out on fishing vessel models and trawls. *Projekt Lindholmen*, Göteborg, carries out research on ergonomics and fishery occupational injuries.

11. Salmon Research Institute

Name : Laxforskningsinstitutet, Älvkarleby
Address : S - 810 70 Älvkarleby
Head : Dr. Björn Wahlberg
Tel. : 46.2672600 **Fax**: 46.2672664

Type of institute
Research centre

Research activities
Stock assessment
Genetics

Staff: Scientists: 5 Others: 12

Research budget (ECU/year): 891.600

Description
The institute is a foundation between the National Board of Fisheries and the Waterpower Industries. The main studies concern assessments of salmon stocks and sea trout, and development of the rearing and ranching of salmon. The institute also performs research on the genetic aspects of rearing and diseases of salmon.

12. University Institutes for Ecological Research

Research related to fish ecology is conducted at practically all Swedish universities.

Department of Limnology, University of Lund
Intra- and interspecies interaction in lake ecosystems, particularly size structures. The effects of fish populations on the environment (biomanipulation).

Department of Zoology, University of Lund
Effects of predation in the marine environment.

Department of Zoology, University of Göteborg
Reproduction biology and population dynamics of salmonids. Effects of acidification.

Department of Zoology, University of Stockholm
Factors structuring populations, construction of models for simulation. The importance of eutrophication to pelagic fish stocks. Ecological studies on Baltic herring. The reproduction and growth of crayfish.

Department of Limnology, University of Uppsala
Ecological research on crayfish. The ecology of char. Environmental effects on micro-structures of otoliths.

Department of Zoology, University of Uppsala
Measures increasing the productivity of salmonid populations.

Department of Zoology, University of Umeå
The ecology and biology of anadromous fish species. Intra- and interspecies interactions in limnic ecosystems.

Kristineberg Marine Biological Station (belonging to the Royal Swedish Academy of Science)
The effects of eutrophication on fish and crayfish.

13. University Institutes for Genetic Research

Seven departments at the universities of Umeå, Uppsala, Stockholm, Lund and at the Swedish University of Agricultural Sciences in Uppsala (SLU) carry out genetic research related to fish and shellfish. The research covers the following topics:
Population genetics, breeding-consumption, breeding-fishery management, preservation of genetic resources, taxonomy/systematics, gene technology, sex differentiation.
The research is only related to anadromous species in fresh water.

14. University Institutes for Physiological Research

Five departments at the universities of Uppsala, Göteborg and Lund and at SLU perform physiological fish research related to nutrition, endocrinology, growth physiology and toxicology primarily in relation to aquaculture and environmental control. Only the department in Göteborg carries out studies on marine fish species. Physiological studies on algae are conducted at the departments in Göteborg, Umeå and Lund, e.g. on micro algae as fish food in aquaculture.

State of Research

UNITED KINGDOM

Marine science and technology in the U.K. depend on several Ministries, each with its specific competence. The greatest part of financial help comes from the two Ministries of Agriculture: the Ministry of Agriculture, Fisheries and Food (MAFF) for England and Wales, and the Department of Agriculture and Fisheries for Scotland (SOAFD). The programmes are coordinated internally between the two Ministries.

Conduct of Fisheries Research and Technology takes place in five categories:
* management of living resources, funded by MAFF, SOAFD.
* protection of the marine environment, funded by MAFF, SOAFD.
* marine fish and shellfish culture, policy support funded by MAFF and SOAFD, industrial development funded by Sea Fish Industry Authority (SFIA).
* fisheries technology and processing, funded by government and industry.
* marine biology, funded by MAFF, SOAFD.

The policy divisions of the Ministry will determine the research requirements to support their policy objectives and set the priorities for competing demands on research resources. Supporting specialist research of a more basic nature outwith the competence or facilities of the government laboratories will be funded extramurally by the Ministry's Chief Scientist Group with universities, research associations, other government departments or other appropriate associations such as the Sea Fish Industry Authority.

The executive branch of the Ministries commissions each project from contracting laboratories, either their own laboratories or from other research organisations. (Additional funding can be provided by a series of agencies like the research council, the EC...)

Most research and development related to fisheries remains in the public sector, and is directed towards EC policy. The industry, however, is bound to take a growing part in research funding for topics from which it is likely to be the major beneficiary, as it is a current policy of U.K. government not to fund research which is classified as "near-market". However, the government will fund fifty percent of the cost of the so-called pre-competitive research in order to improve the technological base of the industry.

Fisheries in the U.K. is conducted mainly by the Fisheries Laboratories at Aberdeen and Lowestoft. These laboratories conduct almost all government and industry funded research work and are responsible to the Scottish Office and the Ministry of Agriculture, Fisheries and Food. The government departments act as customers, seeking the advice and information necessary to manage the fish stocks. Informal links with the industry exist to exchange views and information. Industry takes always an active role both in funding and in the decision making process of U.K. research and development in fishery.

The U.K. programme of fisheries research is formally coordinated by the Controller of Fisheries Research and Development. Fisheries scientists in the U.K. have many contacts throughout Europe and in the rest of the world, with organisations engaged in R&D in Fisheries.

The executive branch of the Ministries have also an annual meeting with the industry representatives to keep each other informed of their respective views of research progress and priorities.

Research particularly revelant to fisheries utilisation in developing countries is funded by the Overseas Development Administration, in part at the Overseas Natural Resources Research Institute, and at other centers in the U.K.

Educational support for fisheries in the U.K. is provided in different universities which form specialists.

List of Institutes from UNITED KINGDOM

1. AFRC - Institute of Arable Crops Research
2. Aquaculture Vaccines Limited
3. Aquatic Sciences Research Division - Dept. Agriculture North Ireland
4. BP Nutrition (UK) Ltd.
5. British Antarctic Survey
6. Campden Food and Drink Research Association
7. Centre of Veterinary Investigation - Penrith
8. Dunstaffnage Marine Laboratory
9. Environmental Consultants - Marenco Ltd.
10. Fisheries Research Laboratory - Dept. Agriculture
11. IERC - International Ecotechnologie Research Center
12. Institute of Freshwater Ecology - River Lab.
13. Institute of Freshwater Ecology - Windermere Lab.
14. Institute of Hydrology
15. Institute of Oceanographic Sciences - Deacon Lab.
16. JM Associates
17. Leatherhead Food Research Association
18. MAFF - Directorate of Fisheries Research
19. MAFF - Fisheries Laboratory - Lowestoft
20. MAFF - Fisheries Laboratory - Burnham-on-Crouch
21. MAFF - Fisheries Laboratory - Conwy
22. MAFF - Fish Diseases Laboratory - Weymouth
23. MAFF - Torry Research Station
24. Marine Harvest Ltd. - Edinburgh
25. Marine Resources Assessment Group - MRAG
26. Marine Station Marenco
27. MARR - Technical Services Ltd.
28. National Museums & Galleries on Merseyside
29. National Rivers Authority - Bristol
30. National Rivers Authority - Welsh Region
31. Natural History Museum
32. Nature Conservancy Council
33. Nautilus Consultants
34. NERC - Natural Environment Research Council
35. NERC - Plymouth Marine Laboratory - PML
36. NERC - Sea Mammal Research Unit
37. Orkney Water Test Centre Ltd.
38. Poly Enterprises Plymouth Ltd.
39. Polytechnic of Central London - Applied Ecology Research Group
40. Polytechnic of Humberside - School of Food, Fisheries and Environmental Studies
41. Polytechnic of Portsmouth
42. Polytechnic of Portsmouth - Biological Sciences
43. Polytechnic of Portsmouth - Dept. of Economics

44. Polytechnic of Sheffield City
45. Polytechnic South West - Dept. Biological Sciences
46. Powergen Freshwater Biology Group
47. RRAG - Renewable Resources Assessment Group
48. RVAS - Aquatic Resource Unit
49. Scottish Agricultural College
50. Sea Fish Industry Authority - Argyll
51. Sea Fish Industry Authority - Edinburgh
52. Sea Fish Industry Authority - Hull
53. Sir Alister Hardy Foundation for Ocean Science - SAHFOS
54. SOAFD - Freshwater Laboratory
55. SOAFD - Marine Laboratory
56. The Scottish Association for Marine Science - SAMS
57. UnicoMarine
58. Unilever Research Lab.
59. Univ. Aberdeen - King's College - Fac. Arts and Social Science
60. Univ. Aberdeen - Marischal College - Biochemistry Dept.
61. Univ. Aberdeen - Zoology Dept.
62. Univ. Belfast - Queens University
63. Univ. Birmingham - School of Biological Sciences
64. Univ. Bradford - Biomedical Sciences
65. Univ. Bristol - Zoology Dept.
66. Univ. Brunel - Biology and Biochemistry Dept.
67. Univ. Buckingham - Dept. of Life Sciences
68. Univ. Cambridge - Physiological Lab. - RGMER
69. Univ. College North Wales - School of Ocean Sciences
70. Univ. Durham - Dept. of Biological Sciences
71. Univ. Edinburgh (Heriot-Watt) - Biological Sciences
72. Univ. Edinburgh (Heriot-Watt) - Civil Engineering Dept.
73. Univ. Edinburgh (Heriot-Watt) - Inst. of Offshore Engineering
74. Univ. Essex - Biology Dept.
75. Univ. Glasgow - Dept. of Naval Architecture & Ocean Engineering
76. Univ. Glasgow - Genetics Dept.
77. Univ. Glasgow - Microbiology Dept.
78. Univ. Glasgow - Zoology Dept.
79. Univ. Hull
80. Univ. Keele - Biological Sciences
81. Univ. Lancaster - Biological Sciences
82. Univ. Leeds - Dept. Biochemistry
83. Univ. Leeds - Dept. of Electronic and Electrical Engineering
84. Univ. Leicester - Biological Sciences Zoology Dept.
85. Univ. Liverpool - Environmental and Evolutionary Biology
86. Univ. Liverpool - Port Erin Marine Lab.
87. Univ. London - Queen Mary College - CRAB
88. Univ. London - Royal Holloway & Bedford New College - Biology Dept.
89. Univ. Loughborough - Dept. Chemical Engineering

90. Univ. Manchester - Agricultural Economics Dept.
91. Univ. Marine Biological Station Millport
92. Univ. Newcastle upon Tyne - Dept. Biology - Dove Marine Lab.
93. Univ. Oxford - Dept. Engineering Science
94. Univ. Oxford - Dept. Zoology
95. Univ. Reading - Animal and Microbial Sciences
96. Univ. Salford - Dept. Biological Sciences
97. Univ. Southampton - Dept. Biology
98. Univ. Southampton - Dept. Oceanography
99. Univ. St Andrews - Gatty Marine Lab. - Biological and Preclin Medicine
100. Univ. Stirling - Institute of Aquaculture
101. Univ. Strathclyde - Dept. Statistics and Modelling Science
102. Univ. Strathclyde - Ship and Marine Technology
103. Univ. Surrey - School Biological Sciences - Food Safety Research Group
104. Univ. Ulster - Freshwater Lab.
105. Univ. Wales - School of Pure and Applied Biology
106. Univ. Wales - Swansea
107. Univ. East Anglia - School of Biological Sciences
108. Virus Reference Laboratory
109. Water Research Centre - Medmenham Lab.

1. AFRC - Institute of Arable Crops Research

Name : AFRC Institute of Arable Crops Research
Aquatic Weeds Research Unit
Address : Sonning Aquatic Research Centre
Broadmoor Lane, Sonning
UK - Reading RG4 0TH
Head : P.R.F Barrett(Head of unit)
Tel. : 0734 690072

Type of institute
Research centre
National

Research activities
Fisheries management
Environmental problems and their impact on resources
Control of Aquatic Weeds by mechanical, chemical and biological means

Facilities
Flume tank
Mesocosm
Field site, electron microscope, specialized analytical apparatus

Staff: Scientists: 3 Others: 3

Research budget (ECU/year): 225.000

Description
The purpose of the Aquatic Weeds Research Unit is to develop better methods of controlling aquatic weeds.
The result can be of benefit to fisheries, land drainage, potable supplies or other users.
Current projects include:
Developing improved techniques for pesticides application.
Developing a nodel method of mechanically harvesting aquatic weeds.
Developing the use of barley straw as a means of selectivelly controlling algae.

2. Aquaculture Vaccines Limited

Name : Aquaculture Vaccines Limited
Address : 24-26 Gold Street
UK - Saffron Walden-Essex CB1O 1EJ

Head : Dr. P.D. Smith
Tel. : 799.28167 **Fax:** 799.25546 **Telex:** 265871 MONREF G

Type of institute
Veterinary Pharmaceuticals Producer

Research activities
Aquaculture
Disease, immunology, stress

Staff: Scientists: 8

Research budget (ECU/year): 21.400

Description
Over the past 3 years AVL has been involved in a number of research projects to develop new vaccines and diagnostic kits for the major diseases of farmed fish.
Major projects concern:
Development of vaccine against bacterial kidney disease (BKD).
Analysis of strains and development of vaccines against the major bacterial diseases of farmed fish in Norway.
Immune response of the Atlantic salmon to the salmon louse (Lepeotheirus) with a view to vaccine development.
Development of vaccines for oral administration.
Development of vaccines against the major bacterial diseases of peracid shrimp.
Investigation of the immune status of carrier fish.
Investigation of the immune response of fish to the causative agent of PKD with a view to vaccine develpment.

3. Aquatic Sciences Research Division - Dept. Agriculture North Ireland

Name : Aquatic Sciences Research Division
Departement of Agriculture for Northern
Ireland
Address : Agriculture and Food Science Centre
Newforge Lane
UK - Belfast BT9 5PX
Head : Dr. S. Ivan Heaney

Type of institute
Government

Research activities
Fisheries and Environment related research

Facilities
Well found laboratory with 45 m Research Vessel "Lough Foyle"

Staff: Scientists: 37 Others: 3

Research budget (ECU/year): 1.580.000

Description
The Aquatic Sciences Research Division undertakes fundamental and applied research in the aquatic environment, thus providing the scientific base by which Government can formulate policy for the efficient and cost-effective management and exploitation of aquatic resources on a sustrainable basis consistent with the protection of, and improvement to, the environment.

4. BP Nutrition (UK) Ltd.

Name : BP Nutrition (UK) Ltd.
Address : Wincham
UK - Northwich, Cheshire
Head : Dr. John Roberts
Tel. : 606.41133 **Fax:** 606.41963 **Telex:** 668994

Type of institute
Fish/Crustacean/mollusc farming
Processing Industry/ agro-food / diets producer

Research activities
Aquaculture
- Early stages of reared species
- Growth, nutrition, feed composition, live food production for aquaculture
- Disease, immunology, stress
- Aquaculture/environment interaction

Facilities
Rearing tank

Staff: Scientists: 3

5. British Antarctic Survey

Name : British Antarctic Survey
Address : High Cross-Madingley Road
UK - Cambridge CB3 OET

Type of institute
Research centre

Research activities
Fisheries management
Stock assessment
Multi-disciplinary approach to fisheries management
Fishing techniques
Electronics/acoustics

Facilities
Research vessel
Field site, electron microscope, specialized analytical apparatus

Research budget (ECU/year): 14.500

6. Campden Food and Drink Research Association

Name : Campden Food and Drink Research Association
Address : Chipping Campden
UK - Glos GL55 6LD
Head : Prof. Colin Dennis

Type of institute
Research Association

Research activities
Research, Development and services for the food, drink and allied industries.

Facilities
Laboratories for chemistry, biochemistry, microbiology, food and drink packaging and processing, process engineering, sensory analysis and microscopy.
Other facilities include mathematics and computing services, agriculture, pilot scale plant, library and information services and training.

Staff: Scientists: 150 Others: 50

Research budget (ECU/year): 7.540.000

Description
The RA has been involved with fish and fish products for many years. Research services include: thermal processing (canning) and freezing of fish; microbiological analysis of fish and shellfish products including a survey of pathogenic organisms; immunological techniques for speciation; spoilage tests and process determination; consultancies to the sardine and tuna canning industries in North Africa and Indonesia.

7. Centre of Veterinary Investigation - Penrith

Name : Veterinary investigation centre
Address : Merrytought, Calthwaithe
UK - Penrith, Cumbria CA11 9RR
Head : David Counter
Tel. : 076.885295 **Fax:** 076.885314

Type of institute
Government Department / Regional or local authority

Research activities
Aquaculture
Disease diagnosis

Facilities
Specialized analytical apparatus relevant only to disease diagnosis - bacteriology virology - histopathology.

Staff: Scientists: 0 Others: 0

Description
The laboratory is engaged in fish disease diagnosis.
Nobody is working full time on fish disease research.

8. Dunstaffnage Marine Laboratory

Name : Dunstaffnage Marine Laboratory
Address : P.O. Box 3
UK - Oban, Argyll PA34 4AD, Scotland
Head : J.B.L. Matthews

Type of institute
Research centre

Research activities
Aquaculture
Early stages of reared species
Growth, nutrition, feed composition, live food production for aquaculture
Aquaculture/environment interaction
Physical oceanography of NE Atlantic
Physical and biological studies of sea lochs and coastal waters
Predator/prey interactions of larval and juvenile fish
Marine technology

Facilities
Research vessel
Rearing tank
Controlled environmental chamber
Field site, electron microscope, specialized analytical apparatus
Extensive library
Sea Cage Site

Staff: Scientists: 39 Others: 18

Research budget (ECU/year): 2.700.000

Description
A component institute of the Natural Environment Research Council whose aim is to conduct physical and biological oceanography and marine biology in coastal and offshore waters and applied research related to management of the marine environment and the use and conservation of marine life. Current research includes work on:
a) the coupling of water column and sediment processes in sea lochs and, related to this, interactions between aquaculture and the environment,
b) interactions between larval and juvenile fish, their predators and their prey,
c) the hydrodynamics and current of semi-enclosed coastal and offshore waters,
d) the structure of oceanic communities with particular reference to the Rockall Trough.

9. Environmental Consultants - Marenco Ltd.

Name : Marenco Ltd.
Environmental Consultants
Address : 27 College Gardens
UK - Belfast BT9 6BS

Head : Dr. Howard P. Platt

Type of institute
Research Centre

Research activities
Fisheries management
Environmental problems and their impact on resources
Aquaculture
Aquaculture/environment interaction
Upgrading of fishery products
Consumer response and demand studies

Facilities
Tanks and flow through seawater
Field site, specialised analytical apparatus

Staff: Scientists: 7 Others: 1

Research budget (ECU/year): 250.000

Description
MARENCO is an environmental company offering a fully integrated aquaculture and consultancy service. With direct access to the expertise and facilities of Queen's University of Belfast this firm will offer advice, feasibility studies, environmental impact assessments and pilot plant for ventures ranging from experimental projects to high seas fish farms.
With the department of Geography it can provide impact assessment of land use change on local recreational fisheries. With Kirk McClure Morton, Consulting Engineers, it can offer complete design, feasibility studies and environmental impact assessment for high seas fish farming. With QUBIS Ltd. it can offer a market assessment service and financial advice for feasibility studies for new ventures.

10. Fisheries Research Laboratory - Dept. Agriculture

Name : Fisheries Research Laboratory
Dept. of Agriculture for Northern Ireland
Address : 38 Castleroe Road
UK - Coleraine, Co Londonderry BT51 3RL
(North. Ireland)
Head : G.J.A. Kennedy
Tel. : 0265.44521 **Fax:** 0265.43301

Type of institute
Research centre
Fishing

Research activities
Fisheries management
Stock assessment
Multi-species biological models
Multi-disciplinary approach to fisheries management
Environmental problems and their impact on resources
Fishing techniques
Electronics/acoustics
Selective gears
Other improvements in gears and/or fishing vessels
Aquaculture
Reproduction
Genetics and conservation of gametes

Facilities
Research vessel
Rearing tank
Specialised software
Field site, electron microscope, specialized analytical apparatus

Staff: Scientists: 4 Others: 10

Description
Research activities concern 2 areas:
- FRESHWATER FISHERIES, whitch mainly deals with studies about salmon in the River Bush (Northern Ireland)+coarse fish stocks in Lough Erne (N. Ireland).
- MARINE FISHERIES, which treats growth and reproduction in Irish sea, and analyse data concerning sea fish with computer means.

Facilities:
- RV Lough Foyle 43 m research vessel.
- Salmonid rearing and culture facilities at River Bush Salmon Station and Movanagher fish farm.
- Specialised software for marine fisheries data logging and population analysis.
- Electrophoresis/biochemistry laboratory, water quality/radiobiology laboratories available at Fisheries Research Laboratory.

11. IERC - International Ecotechnology Research Center

Name : International Ecotechnology Research Center-
IERC
Cranfield Institute of Technology
Address : UK - Bedford MK43 OAL
Head : Prof. P.M. Allen

Type of institute
Research centre
University Department/Higher Education Institute

Research activities
Computer Based Decision Support Systems; National, Regional and City Planning; National Resources Management

Facilities
Specialised software

Staff: Scientists: 7 Others: 2

Research budget (ECU/year): 500.000

Description
Specialised software.
The software which has been developed is a multi-fleet, multi-species spatial, dynamic computer model of a fishery. It can handle at present eight fleets, eight fish species and 35 spatial zones, although these can easily be extented. It is a decision support system which enables the effects of different management policies to be explored, and evaluated. Quotas, license limits, enterprise allocation, closed areas, mesh sizes, are among the options that the software user can examine.
In addition, the model takes into account the market for the different fish, and the effects of price fixing, or subsidy can also be studied. In short, we have developed a framework which links together dynamically and spatially the ecosystem, the fishery, and the market and clearly, issues connected with pollution, spills and even employment can all be taken into account.

12. Institute of Freshwater Ecology - River Lab.

Name : Institute of Freshwater Ecology
The River Laboratory
Address : East Stoke
UK - Wareham/Dorset BH20 6BB
Head : A.D. Berrie
Tel. : 0929.462314 **Fax:** 0929.462180 **Telex:** 94070672 WARE G

Type of institute
Research centre

Research activities
Fisheries management
Stock assessment
Environmental problems and their impact on resources

Facilities
Flume tank
Mesocosm
Field site, electron microscope, specialized analytical apparatus

Staff: Scientists: 15 Others: 11

Research budget (ECU/year): 276.900

Description
Major research projects:
Baseline survey of coarse fish populations in the lower River Avon, Hampshire. Ecology of juvenile cyprinid fish in the Great Ouse. Factors influencing salmonid stocks in southern chalk streams. The ecological characteristics of coarse fish in rivers. Effects of fish farms on water quality and natural communities of rivers. The impact of water abstraction on river fish and the development of fisheries in proposed storage lakes. Salmon counting in the River Itchen.

13. Institute of Freshwater Ecology - Windermere Lab.

Name : Institute of freshwater Ecology
Windermere Laboratory
Address : Far Sawrey
UK - Ambleside, Cumbria LA22 OLP
Head : Prof. J. Gwynfryn Jones
Tel. : 05394.42468 FAX.: 05394.46914

Type of institute
Research centre

Research activities
Research into all aspects of freshwater science
Fish population dynamics and management
Physico-chemical processes in catchments
Management of lakes and reservoirs
Assessment and prediction of change in freshwaters

Facilities
Laboratory facilities: Microbiological and Chemical Laboratories; Electron Microscopy; Fish hatchery and other experimental facilities; Library and information services; Field sites and data bases.

Staff: Scientists: 39 Others: 28

Research budget (ECU/year): 5.000.000

Description
Undertakes contracts for customers, including Government departments, EEC, water industry, industrial concerns etc. Activities include research into all areas of freshwater science, consultancy work, environmental impact assessment, design and construction of instruments and sampling equipment.

14. Institute of Hydrology

Name	:	Institute of Hydrology
Address	:	Maclean building
		UK - Wallingford, Oxfordshire OX10 8BB
Head	:	Dr. Paul Whitehead
Tel.	:	0491.38800 **Fax:** 0491.32256 **Telex:** 849365

Type of institute
Research centre

Research activities
Fisheries management
- Environmental problems and their impact on resources
- Assessment of critical loads of sulphur to ecosystems
- Collaborative acidification research
- Impacts of climate change on surface water quality
- Modeling persistent organics

Facilities
Data bases, extensive libraries
Specialised software

Staff: Scientists: 7

15. Institute of Oceanographic Sciences - Deacon Lab.

Name	:	Institute of Oceanographic Sciences Deacon Laboratory
Address	:	Brook road Wormley - Godalming UK - Surrey GU8 5UB
Head	:	Dr. Colin Summerhayes
Tel.	:	042.8794141 **Fax:** 042.8793066 **Telex:** 858833

Type of institute
Research centre

Research activities
Deep sea oceanography

Facilities
Historical discovery collections. Wave tank. Pressure testing, instrument calibration. Information Service.

Staff: Scientists: 130 Others: 60

Research budget (ECU/year): 1.211.000

Description
Major projects: World Ocean Circulation Experiment. Fine Resolution Antarctic Model. AUTOSUB - Autonomous Robotic Vehicle Project. Plus research into all aspects of deep sea oceanography.
The Institute of Oceanographic Sciences Deacon Laboratory (IOSDL) at Wormley, near Godalming, Surrey, is a component body of NERC and is the centre for deep sea oceanography in the UK. Its scientists carry out physical, chemical, geological, geophysical and biological research in the deep ocean. Much of the sophisticated equipment designed to work at depths of up to 10.000 metres is designed and built by IOSDL engineers. The Institute is currently engaged on several exciting projects including the study of the interactions between the oceans and atmosphere, running a supercomputer model of the Southern Ocean and designing a robotic vehicle for deep sea surveys. The laboratory also houses the National Oceanographic Library.

16. JM Associates

Name : JM Associates
Address : Gwel-an-Carn, Heather Lane, Canonstown, Hayle
UK - Cornwall TR27 6NQ
Head : J.W Munday
Tel. : 0736 741026

Type of institute
Private organisation

Research activities
Population surveys for both white fish and shell fish.
Assessment of and means of control of pollution both into and out of fish farms.
Investigations into gill net fishing techniques including methods of reduction of bi-catches of birds and cetaceans.

Facilities
Fully equipped 11 metre fishing/research vessel

Staff: Scientists: 2 Others: 5

Description
This is a small private organisation specialising in research into inshore fisheries.

17. Leatherhead Food Research Association

Name : Leatherhead Food Research Association
Address : Randalls Road
Leatherhead
UK - Surrey KT22 7RY
Head : Dr. Mark Kierstan(Director)

Type of institute
Industrial Research Association

Research activities
Research on food technology, food science and related disciplines.

Facilities
Pilot plant facilities, processing equipment etc.

Staff: Scientists: 140 Others: 71

Description
Research Projects carried out in areas of Fisheries and Aquaculture are:
1. Rapid quality determination: Standardised sensory methodology and size measurements of raw fish.
2. Modified-atmospheres: A new approach to improving the safety and quality of fresh fish and fish products.

18. MAFF - Directorate of Fisheries Research

Name	:	Ministry of Agriculture, Fisheries and Food Directorate of fisheries Research
Address	:	Pakefield road UK - Lowestoft, Suffolk NR33 OHT
Head	:	Dr. D.J. Garrod
Tel.	:	502.562244 **Fax:** 502.513865 **Telex:** 97470 G

Type of institute
Research centre of Government Department

Research activities
Fisheries management
  Stock assessment
  Multi-species biological models
  Multi-disciplinary approach to fisheries management
Marine biology of commercial species
  Physiology, behaviour and genetics
  Reproduction, recruitment, ecology
  Plankton and productivity
Oceanography
  Physical and chemical monitoring
Modelling
Nutrient conversion in estuaries
Aquaculture
  Reproduction and juvenile stages of reared species
  Genetics
  Enhancement of stocks
  Fish and shellfish disease monitoring and research
  Hygiene standards for live bivalve molluscs
Environmental protection-monitoring and research
  Impact of radioactive and non-radioactive pollutants on aquatic resources
  Contaminant behaviour and pathways in the aquatic

environment
Deep ocean disposal of radioactive wastes
Biological impact of wastes dumped at sea
Development of sensitive assays for pollutants
Effect of human activities; gravel dredging, oil exploration, coastal construction works, etc
Impact of fishing on benthic communities
Modelling waste dispersion in coastal seas

Facilities
Two ocean-going research vessels, 72m and 53m
Flume tank, rearing tank and sea-water aquaria
Mesocosm
Specialised computer hardware and software
Specialised libraries and extensive databases
Sophisticated chemical and radioactivity analytical facilities
Electronic engineering laboratory with large acoustic test tank

Staff: Scientists: 285 Others: 158

Research budget (ECU/year): 28.000.000

Description
The primary aims of the Directorate are to provide scientific information and advice as a basis for effective management and sustainable exploitation of the stocks of fish and shellfish in and around the British Isles and to protect the marine environment. The programme of work includes the monitoring of fish and shellfish stocks by fisheries surveys, biological sampling of landings and the collection of fisheries statistics. This is complemented by research into the physiology, behaviour and genetics of marine organisms, the ecology of adults, their eggs and larvae and the assessment, modelling and prediction of multi-species population fluctuations. Monitoring and research on the diseases of aquatic organisms is necessary for compliance with EC directives, in support of the aquaculture industry and for the protection of natural stocks. Research on artificial culture systems for fish and shellfish is particularly directed to the effective management of UK coastal resources and includes the evaluation of ranching and other methods of stock enhancement. More than half of the effort of the Directorate covers monitoring and research on the distribution and effects of radionuclides, hydrocarbons, metals, pesticides, nutrients and other pollutants on the aquatic environment and its biota and on potential pathways to man.
The Directorate's programme is supported by research in physical and chemical oceanography, advanced computing facilities and innovative electronic engineering group. All laboratories are able to accept students for post-graduate research degrees.

19. MAFF - Fisheries Laboratory - Lowestoft

Details as above (MAFF-Directorate of fisheries research)

Description
The main fishery assessment and radioactivity monitoring and research programmes are directed from this site together with those on the biology of aquatic organisms and oceanography. The central scientific services of research vessels, computing, electronic engineering, library, publications and administration are also based at Lowestoft. The laboratory has close links with the University of East Anglia at Norwich and can accept students for research degrees in biological and environmental subjects.

Staff: Scientists: 191 Others: 133

20. MAFF - Fisheries Laboratory - Burnham-on-Crouch

Name : MAFF, Fisheries Laboratory
Address : Remembrance Avenue, Burnham-on-Crouch
UK - Essex CMO 8HA
Head : Dr. J. E. Portmann
Tel. : 621.782658 **Fax:** 621.784989 **Telex:** 995543

Description
This section concentrates on the protection of the aquatic environment from the use and deliberate or accidental disposal of non-radioactive contaminants and from the effects of other activities such as offshore oil and marine gravel extraction. The laboratory has advanced analytical capabilities for measuring a wide range of contaminants in the 8000 or more samples processed each year. Well established experimental facilities are used for testing the effects of contaminants on various organisms both in sea and freshwater.

Staff: Scientists: 36 Others: 11

21. MAFF - Fisheries Laboratory - Conwy

Name : MAFF - Fisheries Laboratory
Address : Benarth Road, Conwy
UK - Gwynedd LL32 8UB
Head : Dr. S.J. Lockwood
Tel. : 492 593883 **Fax:** 492 592123

Description
The research at this laboratory is focussed on the rational management of the coastal zone including the artificial cultivation of marine fish and shellfish. It is concerned with assessing the effect on fisheries of coastal construction projects including coast protection and flood defence, discharges, conservation activities etc. and with the influence of fisheries activities on the marine ecosystem. Enhancement of coastal zone productivity with cultured fish, molluscs and lobsters is the subject of long-running research programmes. The laboratory has extensive sea water aquaria, holding ponds and small mesocosms. It has close links with the University of Wales at Bangor.

Staff: Scientists: 17 Others: 7

22. MAFF - Fish Diseases Laboratory - Weymouth

Name : MAFF - Fish Diseases Laboratory
Address : 14 Albany Road, Weymouth
UK - Dorset DT4 9TH
Head : Dr. B. J. Hill
Tel. : 305 772137 **Fax:** 305 770955 **Telex:** 417104

Description
This team is currently housed in temporary accomodation while a new laboratory is under construction. It is responsible for the monitoring of and research into the diseases of cultured and native fish and shellfish in sea and fresh water. The work programme has recently expanded substantially to meet Community directives which require the regular inspection of fish farms and shellfish beds. Specialised facilities for the study of viruses and other micro-organisms, including tissue cultures, advanced histological techniques and an electron microscope are available.

Staff: Scientists: 41 Others: 7

23. MAFF - Torry Research Station

Name : MAFF - Ministry of Agriculture Fisheries & Food
Torry Research Station
Address : PO Box 31-135 Abbey road
UK - Aberdeen AB9 8DG
Head : Dr. G. Hobbs
Tel. : 0224.877071 **Fax:** 0224.874246 **Telex:** 739719 MAFTRS G

Type of institute
Research centre
Engineering/consulting firm
Institute for professional training/Secondary school

Research activities
Upgrading of fishery products
Improvement of techniques for handling, storing, processing, packaging fish
Methods for grading and quality assessment of fish
Novel processing methods for by-products, discards, and process wastes
Consumer response and demand studies

Facilities
Specialised computer hardware
Specialised software
Electron microscopes, specialized analytical apparatus
including GC/MS, many GCG, water activity "Malthus" (rapid bacteriological method)
Data bases on fish and food processing technology, extensive library.
Fully equipped production test are with chill and cold
storage, freezing smoking and other process equipment.

Staff: Scientists: 50 Others: 43

Research budget (ECU/year): 5.556.000

Description
Torry Research Station is a UK government funded research institute. Since 1929, it has worked on improvements in the handling and processing of fish. In late 1990, following a transfer to the Food Safety Directorate of the Ministry of Agriculture, Fisheries and Food, the work emphasis has changed to include all food and is now directed towards safety. In addition, analyses, R&D, training and information consultancy are offered on a commercial basis. Current projects are on - pathogenic micro-organisms; natural constituents of foods; food contaminants and residues; additives; toxicology and risk assessment; improved method of analysis, and food authenticity and adulteration.

24. Marine Harvest Ltd. - Edinburgh

Name : Marine Harvest Limited
Address : Craigcrook Castle
UK - Edinburgh EH4 3TU

Head : Dr. R. Alderson
Tel. : 031.3361777 **Fax:** 031.3361198 **Telex:** 727664 G

Type of institute
Fish/Crustacean/mollusc farming

Research activities
Aquaculture
Genetics and conservation of gametes
Growth, nutrition, feed composition, live food production for aquaculture
Disease, immunology, stress
Aquaculture/environment interaction

Facilities
Rearing tanks, cages.
Laboratory and field sites.

Staff: Scientists: 4 Others: 6

Research budget (ECU/year): 200.000

Description
Major Research Projects:
- Major programme on farmed salmon stock improvement and understanding of genetics of key factors in production performance.
- Improvement in understanding of salmon nutrition and feeding.
- Investigations into pathogenicity of A. salmonicida with the aim of vaccine production and improvement.
- Investigations into the immune defence mechanisms of salmon in order to better understand and improve the development and application of vaccines against salmon diseases.
- Phosphorus input into freshwater lochs.
- Effect of farming activities on sea bed pollution and recovery.

25. Marine Resources Assessment Group - MRAG

Name : Marine Resources Assessment Group Ltd.
Address : 27 Campden street
UK - London W8 7EP
Head : J.R. Beddington
Tel. 071.225.8934 **Fax:** 071.5895319 **Telex:** 929484 IMPCOL G

Type of institute
Consulting firm

Research activities
Fisheries management
Stock assessment
Multi-species biological models
Multi-disciplinary approach to fisheries management
Environmental problems and their impact on resources
Access negotiation and licensing

Facilities
Data bases, extensive libraries
Specialised computer hardware
Specialised software

Staff: Scientists: 20 Others: 6

Research budget (ECU/year): 552.400

Description
The Marine Resources Assessment Group Ltd. is a consulting firm (associated with the Renewable Resources Assessment Group, Imperial College and other groups) of 26 members with a multi-disciplinary capability.
MRAG Ltd. undertakes short and long term projects for a wide variety of clients worldwide in fisheries; including stock and yield assessment, licensing regimes and access negociations, environmental impact of fisheries, fisheries development policy and planning, economic assessment and evaluation etc. The firm manages the Fish Management Science Programme of the UK Overseas Development Administration which includes funding for practically oriented research and development systems for developing countries.

26. Marine Station Marenco

Name : Marenco
Marine Station
Address : Portaferry
UK - BT22 1PF Co Down - North. Ireland
Head : Mrs L. Blackiston-Houston
Tel. : 0247728909 **Fax:** 0247728902

Type of institute
Research centre

Research activities
Fisheries management
- Environmental problems and their impact on resources

Aquaculture
- Aquaculture/environment interaction
- Upgrading of fishery products
- Consumer response and demand studies

Facilities
Tanks and flow through seawater
Field site, specialised analytical apparatus

Staff: Scientists: 7 Others: 1

Description
MARENCO an environmental company based at Queen's University of Belfast Marine Station offers a fully integrated aquaculture service.
With direct access to the expertise and facilities of Queen's University of Belfast this firm will offer advice, feasibility studies, environmental impact assessments and pilot plant for ventures ranging from experimental projects to high seas fish farms. With the Department of Geography it can offer impact assessment of land use change on local recreational fisheries. With Kirk McClure Morton, Consulting Engineers, it can offer complete design, feasibility studies and environmental impact assessment for high seas fish farming. With QUBIS Ltd. it can offer a market assessment service and financial advice for feasibility studies for new ventures.

27. MARR - Technical Services Ltd.

Name	:	MARR - Technical Services Ltd.
Address	:	St Andrew's Dock UK - Hull, North Humberside HU3 4PN
Head	:	Mr. J. Hind
Tel.	:	0482.27873 **Fax:** 0482.20609 **Telex:** 592214

Type of institute
Engineering/consulting firm
Fishing
Equipment manufacturer

Research activities
Fisheries management
  Fishery protection
  Stock assessment
  Multi-species biological models
  Multi-disciplinary approach to fisheries management
  Environmental problems and their impact on resources
Fishing techniques
  Testing techniques/instruments
  Selective gears
  Electronics/acoustics (e.g. positioning of fishing vessels, detection and recognition of species, ...)

Facilities
Research vessels
Data bases, extensive libraries

Staff: Scientists: 6 Others: 3

28. National Museums & Galleries on Merseyside

Name	:	National Museums & Galleries on Merseyside Liverpool Museum
Address	:	William Brown Str. UK - Liverpool L3 8EN
Head	:	Mr. E.F. Greenwood/Dr. I.D. Wallace
Tel.	:	051.2070001 **Fax:** 051.2073759

Type of institute
National Museum

Facilities
Collection of Preserved Fish, Diseased Fish and Parasites
Aquarium with display and holding tanks.

Staff: Scientists: 2 Others: 3

29. National Rivers Authority - Bristol

Name : National Rivers Authority
Address : Rivers House
Waterside Drive, Aztec West
UK - Almondsbury Bristol BS
Head : Mr. J. Wheatly

Type of institute
Public Non-Governmental body

Research budget (ECU/year): 697.000

Description
Under Section 141 of the Water Act 1989, the NRA is given duties to maintain, improve and develop salmon, trout, freshwater and eel fisheries under its jurisdiction. Section 141 also reiterates the general duties as defined in Section 28(2) of the Salmon and Freshwater Fisheries Act (SFFA) 1975, which include establishing and maintaining Regional Fisheries Advisory Committees CRFAC's) and consulting them in the manner in which the NRA' Fisheries duties are performed. Further duties include making and maintaining fish passes securing safe passage of migratory fish, regulation by licensing of salmon, trout, freshwater and eel fisheries, the movement and introduction of fish to waters and Fisheries Protection measures. Under certain circumstances, the NRA has duties to manage sea fisheries and to notify MAFF of the occurrence of certain fish diseases and carry out fish disease work.

30. National Rivers Authority - Welsh Region

Name : National Rivers Authority, Welsh Region
H.Q. Fisheries & Conservation department
Address : Rivers House, St Mellons Business Park
St Mellons
UK - Cardiff CF3 0LT

Head : Nigel Milner/John Stoner/A Winstone
Tel. : 222.770088 **Fax:** 222.798555

Type of institute
Government Department / Regional or local authority

Research activities
Fisheries management
Stock assessment
Environmental problems and their impact on resources

Facilities
Laboratories for water quality analyses and fish examination.
Fish culture units - salmonids.

Staff: Scientists: 6 Others: 4

Description
Major research activities:
1. Stock assessment:
River Dee Stock Assessment Programme.
Microtagging of Juvenile Salmonids.
Conwy crossing monitoring.
Sea trout stocking project.
Salmon stocking Project.
Regional Juvenile Salmonid Monitoring Programme.
2. Environmental problems:
Movements of migratory salmonids in the rivers and estuaries. Restoration of water quality and fisheries in acid lakes and streams. River habitat restoration. Impact of estuary barrages on migratory fisheries.

31. Natural History Museum

Name : The Natural History Museum
Address : Cromwell Road
UK - London SW7 5BD
Head : Dr. C.R. Curds/ Dr. D.I. Gibson / Mr. N.R. Merrett
Tel. : 1.9388985 **Fax:** 1.938875

Type of institute
Research centre
Government Department

Research activities
Fisheries management
  Stock assessment
  Environmental problems and their impact on resources
Aquaculture
  Reproduction
  Disease, immunology, stress
  Early stages of reared and wild species.

Facilities
Data bases, extensive libraries
Biological reference Collections

Staff: Scientists: 14 Others: 5

Research budget (ECU/year): 1.200.000

Description
Research activities: Taxonomy, biology and parasitology of fish and shellfish. Recent projects include the biology of salmon lice, parasites of Mediterranean sparids, accidental introductions of fish parasites and larval development of crabs and prawns. Life cycles, feeding behaviour and taxonomy of crustacean parasites on marine and freshwater fish. SE Asia freshwater fish systematics and ecology. Deep-sea ichthyology and ecology. Systematics and biogeography of gadoids and cyprinids. Ecophysiological assessment of cockle population dynamics in the Somme. Biology and taxonomy of helminth parasites of commercial fish. Host specificity & co-evolution of helminth parasites of reef fish. Systematics, biology and pathogenesis of helminth parasites of wild and farmed freshwater fish. Fish helminth parasites as indicators of deep-sea demersal fish stocks. Molecular biology of the parasites of deep-sea demersal fishes. Guide to the trematode parasites of Canadian fishes.

32. Nature Conservancy Council

Name	:	Joint Nature Conservation Committee Marine Conservation Branch
Address	:	Monkstone House, City Road UK - Peterborough PE1 1JY
Head	:	Dr. R. Mitchell
Tel.	:	0733.40345 **Fax:** 0733.68834 **Telex:** 265871 CQQ190 MONRES

Type of institute
Government Agency

Research activities

Aquaculture

Aquaculture/environment interaction
Nature conservation implications of fishing operations.
Impact on habitat and non-target species

Facilities

Inshore boats and diving equipment

Staff: Scientists: 14 Others: 5

Research budget (ECU/year): 400.000

Description

Research activities:

Marine ecology and potentional environmental and conservation effects of fisheries, aquaculture and other human activities.

Some projects titles:
Genetic impact of farmed atlantic salmon on wild populations.
Benthic impacts of scallop dredging.
Marine environmental impacts of fish farming.
Environmental impacts of shellfish farming.
Environmental impacts of commercial fishing on the benthos and non-target species.
Marine Nature Conservation Review.

33. Nautilus Consultants

Name	:	Nautilus Consultants Natural resource economists and management consultants specializing in fisheries
Address	:	30/6 Elbe Street Leith UK - Edinburgh EH6 7HW
Head	:	Crick Carleton
Tel.	:	031.5550660 **Fax:** 031.5545902 **Telex:** 931213476f1G

Type of institute

Research centre
Consulting firm

Research activities

Fisheries management
- Multi-disciplinary approach to fisheries management
- Environmental problems and their impact on resources
- Market assessment and econometric modelling of markets and prices.
- Economics of Fishery Management

Aquaculture
- Aquacuture economics
- Technology appraisal and application

Upgrading of fishery products
- Improving the techniques for handling, storing, processing, packaging of fish
- Methods for grading and quality assessment of fish
- Consumer response and demand studies

Environmental studies
- Environmental assessment
- Sustainability studies
- Coastal zone management

Facilities

Data bases, libraries

Staff: Scientists/economists: 7 Others: 3

Description

Nautilus Consultants is a specialist management consultancy committed exclusively to the development of the fishery industry in terms of social viability, financial profitability and environmental sustainability. It brings together consultants of international repute, from technical, managerial and financial backgrounds, to provide a comprehensive range of fishery development services to individual and corporate clients, international development agencies and government departments.

Key fields of activity are:

Investment analysis, commercial management, policy and planning, market research, market development, aquaculture, project preparation, project management, monitoring and evaluation, industrial economics, post harvest systems, training and extension environmental assessment, and resource management studies. Nautilus Consultants has been specifically structured with a view to assisting industries and companies to identify and meet market requirements and to advising governments in the rational and harmonious development of the sector. Its work encompasses all countries around the globe, whether they be recognised as producers, processors or final markets for product. Its work spans all types of production, processing and distribution systems.

34. NERC - Natural Environment Research Council

Name : NERC/Natural Environment Research Council
Marine and Atmospheric Sciences Directorate
(MASD)
Address : Polaris House-North Star Avenue
UK - Swindon, Wilts, SN2 1EU
Head : Dr. J.D. Woods/Dr. R. Paul/Mrs N. Stevens
Tel. : 0793.411500 **Fax:** 793.411545
Telex: 444293 ENVRE G

Type of institute
Research Council

Research activities
Predator/prey relationships
Community ecology
Sea loch processes
Fish farming studies

Facilities
Research vessels (3)

Staff: Scientists: 352 Others: 163

Description
MASD is composed by several laboratories:
- DUNSTAFFNAGE MARINE LAB.
- PLYMOUTH MARINE LAB.
INSTITUTE OF OCEANOGRAPHIC SCIENCES DEACON LAB.
PROUDMAN OCEANOGRAPHIC LAB.
SEA MAMMAL RESEARCH UNIT.
NOTE: Fisheries and Aquaculture studies mainly restricted to these two laboratories.
FISHERIES AND AQUACULTURE Scientists: approx. 35
Others: approx. 20

35. NERC - Plymouth Marine Laboratory - PML

Name : Plymouth Marine Laboratory - PML
Address : Prospect Place-West Hoe
UK - Plymouth PL1 3DH
Head : B.L. Bayne
Tel. : 0752.222772 **Fax:** 0752.670637

Type of institute
Research centre

Research activities
Fisheries management
  Stock assessment
  Environmental problems and their impact on resources
Aquaculture
  Disease, immunology, stress

Facilities
Research vessel
Mesocosm
Data bases, extensive libraries
Electron microscopes.
Laser scanning microscope.
Inductively coupled plasma mass spectrometer.

Staff: Scientists: 117 Others: 45

Research budget (ECU/year): 7.800.000

Description
To understand the physical, chemical and biological interactions of marine ecosystems in order to support effective management of the marine environment. Main areas research are the fate and effect of pollutants in estuarine and shelf waters, the Biogeochemical Ocean Flux Study (BOFS), the Land Ocean Interaction Study (LOIS) and the Ocean Margin Exchange programme (OMEX) and the World Ocean Circulation Experiment (WOCE).

36. NERC - Sea Mammal Research Unit

Name	:	NERC/Sea Mammal Research Unit c/o British Antartic Survey
Address	:	High Cross-Madingley road UK - Cambridge CB3 0ET
Head	:	Dr. J Harwood
Tel.	:	0223.311354 **Fax:** 0223.62616

Type of institute
Research centre

Research activities

Fisheries management
- Stock assessment
- Multi-species biological models
- Environmental problems and their impact on resources

Aquaculture
- Aquaculture/environment interaction

Facilities

Data bases, extensive libraries
Specialised software

Staff: Scientists: 15 Others: 2

Research budget (ECU/year): 1.000.000

Description

Main research projects:

Assessment of grey and harbour seal stocks in UK waters Whale Stock Assessment. Collection of data on feeding behaviour of seals and whales and the incorporation of such data into models. Investigation of the effect of pollution, disease and parasites on seals and whales.

Investigation of the interaction between seals and marine fish farms.

37. Orkney Water Test Centre Ltd.

Name : Orkney Water Test Centre Ltd.
Address : Flotta-Stromness
UK - Orkney KW16 3NP
Head : Mr. T.J.G. Hartmann/Mrs I.M. Nicholl
Tel. : 0856.70.451 **Fax:** 0856.70.743

Type of institute

Research Centre
Engineering/Consulting firm

Research activities

Aquaculture
- Aquaculture/environmental chamber
- Bivalve/invertabrate hatchery
- Algal production unit
- External nursery area
- Aquatic toxicology suite

Staff: Scientists: 4 Others: 5

Research budget(ECU/year): 258.000

Description
Research activities:
Hatchery techniques for bivalve/invertabrate species for culture and toxicity testing
Algal production methods
Development of toxicity testing protocols with marine invertebrate species

38. Poly Enterprises Plymouth Ltd.

Name : Poly enterprises Plymouth Ltd.
Address : Charles Cross Centre-Constantine Street
UK - Plymouth, Devon, PL4 8DE
Head : Ken Wittamore(Director)

Type of institute
Commercial Company in Polytechnic South West

Research activities
Aquaculture
Fish Nutrition

Research budget (ECU/year): 80.000

Description
Industrial support for fish nutrition research i.e. Seabream, seabass.
Pigmentation of fresh - trout.
FEC - use of pulses in carp feeds.
Development of DOI carp diets.
Single call protein in tilapia and sediments.

39. Polytechnic of Central London - Applied Ecology Research Group

Name	:	Polytechnic of Central London Applied Ecology Research Group
Address	:	115 New Cavendish St. UK - London WIM 8JS
Head	:	Brian Knights
Tel.	:	071.911.5000 **Fax:** 071.911.5087

Type of institute
Research Group within H.E. Institution

Research activities
Aquaculture
Nutrition, growth and behaviour of cultured fish (especially eels)
Population biology and fisheries management (especially eels)

Facilities
Laboratory, analytical, computing and field work.

Description
Aquaculture:
Nutrition and growth of eels in warm water aquaculture
Population Biology/Management:
Studies of eel population and migration and management in R. Thames and R. Severm

40. Polytechnic of Humberside - School of Food Fisheries and Environmental Studies

Name	:	Humberside Polytechnic School of Food, Fisheries and Environmental Studies
Address	:	Nun's Corner UK - Grimsby, S. Humberside DN34 5BQ
Head	:	Mr. P.J. Warren

Type of institute
Higher Education Establishment

Research activities
Into many aspects relating to the quality and safety of foods (e.g. fish, plant products etc) as well as projects on water and effluent quality.

Facilities
2 Research Laboratories.
Food processing laboratory; Microbiology laboratory;
Packaging laboratory; Analytical laboratory;
Fish catching facilities.

Staff: Scientists: 32 Others: 17

Research budget (ECU/year): 8.000

Description
Most of the research undertaken in the School has a post-harvest technology emphasis where, for example, in the fisheries area, the relationship between the captured raw material and its expected storage life are investigated in depth. Consultancy work into fisheries work has been carried out by staff in Africa and South East Asia.

41. Polytechnic of Portsmouth

Name : Polytechnic of Portsmouth
Address : King Henry Building
UK - Portsmouth
Head : Mr. Robert L. Fletcher

Type of institute
Research centre
University Department/Higher Education Institute
Higher Education Institute

Research activities
Fisheries management
Environmental problems and their impact on resources

Facilities
Rearing tank
Controlled environmental chamber
Field site, electron microscope, specialized analytical apparatus

42. Polytechnic of Portsmouth - Biological Sciences

Name	:	Polytechnic of Portsmouth The Marine Laboratory
Address	:	Ferry Road UK - Hayling Island-Hants PO11 0DG
Head	:	Dr. MB Culley
Tel.	:	705.465714

Type of institute
Research centre
University Department/Higher Education Institute

Research activities
Aquaculture
Marine biological and ecological studies: e.g. anti-fouling studies. Monitoring also undertaken.

Facilities
Field site, electron microscope, specialized analytical apparatus
Controled access to environmental chambers
Exposure trials raft in Langstone Harbour, U.K.
Rearing tanks with running seawater.

Description
Research has tended to environmental consultancy, but physiological research and fish biology and ecology is also undertaken. Much research done in association with undergraduate studies. Several visitors for extended periods accommodated each year, but they are self funding.

43. Polytechnic of Portsmouth - Dept. of Economics

Name	:	Polytechnic of Portsmouth - Dept. of Economics Centre for Marine Resources Economics
Address	:	Locksway Road, Milton UK - Southsea, Hants (Portsmouth) PO4 8JF
Head	:	S. Cunningham
Tel.	:	0705.827681 **Fax:** 0705.844059

Type of institute
University Department/Higher Education Institute

Research activities

Fisheries management
  Multi-disciplinary approach to fisheries management
  Environmental problems and their impact on resources
Aquaculture
  Aquacuture economics
Upgrading of fishery products
  Consumer response and demand studies
  Controled zone management

Facilities

Data bases, extensive libraries
Specialised computer hardware
Specialised software
Specialised computer hardware and software
Library

Staff: Scientists: 10 Others: 1

Description

The research activities mainly deal with economic evaluation and policy studies of fisheries and aquaculture.
Major projects research, the last 3 years:
Comparative Evaluation of Fisheries Policy in Canada and the EC; Econometric model of price formation in fresh fish landings in the UK. The Determination of Fish Imports into the UK. Economic Evaluation of the Recreational and Commercial Fishery for Bass in England and Wales.
Technological Change in the UK Herring Industry. Economic Evaluation of the recreational and Commercial Fishery for Salmon in GB. Economic Investigation into the Fisheries of Lake Chad, Africa. Technological change and the exploitation of natural resources: the case of the fishery.

44. Polytechnic of Sheffield City

Name : Polytechnic of Sheffield City
Faculty of Technology
Address : Dept. of Biomedical Sciences-Pond Street
UK - Sheffield S1 1WB
Head : Dr. Nigel Wales
Tel. : 0742.720911 **Fax:** 0742.758019 **Telex:** 54680

Type of institute

University Department/Higher Education Institute

Research activities
Aquaculture
Physiology of fish especially hormonal control of water and electrolyte balance and blood pressure.
Production of corticosteroids and prostaglandins.

Staff: Scientists: 1 Others: 2

45. Polytechnic South West - Dept. Biological Sciences

Name : Polytechnic South West
Department of Biological Sciences
Address : Drake Circus
UK - Plymouth PL8 8AA
Head : Dr. L. Jervis

Type of institute
Polytechnic

Research activities
Marine Biology, Fish Biology (ecology and physiology), Fish pathology, Fish Diseases. Aquaculture. Fish Immunology, Vaccine Development

Facilities
Light and Electron Microscopy, Quantimet Image Analysis, Holding and Experimental Aquaria, Genetic Manipulation Suite, Animal House.

Staff: Scientists: 30 Others: 26

Research budget (ECU/year): 404.000

Description
- Experience in the production of monoclonal antibodies to protozoan antigens and fish immunoglobulin and purification by gel electrophoresis and FPLC.
- Recombinant DNA Technology.
- Fish Nutrition: Least cost and novel diets for commercially important fish species.
- Fish disease
- Development of novel bioassays of metal toxicity using marine invertebrates.
- Expertise in whole plant and plant cell culture, crop improvement and manipulation.

46. Powergen Freshwater Biology Group

Name : Powergen
Biology Group
Address : Ratcliffe Technology Centre, Ratcliffe-on-Soar
UK - Nottingham NG 11 OEE
Head : Dr. T.R.K. Dalziel
Tel. : 0602.830591 **Fax:** 0602.83312

Type of institute
Research centre

Research activities
Fisheries management
Environmental problems and their impact on resources

Facilities
Research vessel
Two temperature controlled laboratories.
Mobile toxicity testing laboratory.

Staff: Scientists: 3 Others: 2

Description
Research projects on acid waters and related factors on freshwater organisms; effects of flue gas desulphurisation effluents on freshwater organisms; land reclamation using pulverized fuel ash and gypsum.

47. RRAG - Renewable Resources Assessment Group

Name : RRAG - Renewable Resources Assessment Group
Imperial College of Science, Technology &
Medicine
Address : 8 Princes Gardens
UK - London SW7 1NA
Head : Prof. John Beddington
Tel. : 01.5895111 **Fax:** 01.5895319

Type of institute
Research centre
University Department/Higher Education Institute

Research activities

Fisheries management
- Stock assessment
- Multi-species biological models
- Multi-disciplinary approach to fisheries management
- Environmental problems and their impact on resources

Aquaculture
- Disease, immunology, stress

Facilities

Specialised computer hardware
Specialised software
Data bases, extensive library

Staff: Scientists: 19 Others: 4

Research budget (ECU/year): 742.700

Description

The Group is part of the Centre for Environmental Technology at Imperial College, London. It is a training, information and research Group which also operates a consultancy service. It is committed to the provision of Services for the advancement of methods for Sustainable renewable resource use, wordwide.

48. RVAS - Aquatic Resource Unit

Name	:	RVAS - Aquatic Resource Unit Coordination Office The Rectory
Address	:	Norwood Crescent UK - Dundee DD2 1PD, Tayside
Head	:	Dr. Walesby
Tel.	:	0382.66900 **Fax:** 0382.66900

Type of institute

Engineering/consulting firm
University Department/Higher Education Institute

Research activities

Fisheries management
- Environmental problems and their impact on resources
- Egg quality assessment in fin-fish and shellfish

Aquaculture
- Physiology
- Reproduction
- Fin fish and shellfish reproduction
- Aquaculture/Mariculture and Environmental Issues
- Management Information and Environmental Issues

Facilities
Rearing tank
Data bases, extensive libraries
Specialised computer hardware
Specialised software
Field site, electron microscope, specialized analytical apparatus, Image analysis, antibody production

Staff: Scientists: 4 Others: 5

Research budget (ECU/year): 60.000

Description
RVAS is a network of physiologists, biotechnologists, toxicologists and management scientists within various UK institutions. It offers a wide range of facilities and expertise in aquaculture, fisheries research and aquatic resource management. Its principal laboratory facilities are in (1) diagnostic and experimental histology/pathology (2) fish cell culture/biotechnology (3) respiratory physiology and (4) water quality management. Other key work includes the development of expert systems for aquaculture management. RVAS' special expertise is in the networking of professional biologists with management scientists and practising managers. Some of this is directed towards environmental problems of the aquaculture/fisheries industry. The group also offers university/polytechnic short courses in these management areas.

49. Scottish Agricultural College

Name : The Scottish Agricultural College Aberdeen
Address : 581 King Street
UK - Aberdeen, Scotland AB9 1UD
Head : Mr. W.G.W. Paterson
Tel. : 224.480291 **Fax:** 224.491989 **Telex:** 73538

Type of institute
Research centre
Higher Education Institute

Research activities

Aquaculture
  Aquaculture economics
  Market analysis

Staff: Scientists: 2

Description

Research projects:
- Reference Price System for Salmon
- Preparation of Model Budgets for a Range of Fin-fish and Shell-fish Enterprises.
- Study of Viability of a Proposed Pump-ashore Salmon Production Unit in Fetlar.
- Survey of Costs of Salmon Production in Scotland, 1987/88/89/90.
- Review of the French Salmon Market in 1988/89.
- Examination of Current and Future Economic Viability of Pump Ashore Units.
- Analysis of the Consumer Market for Scottish Salmon.
- Survey of costs of production of Mussels, Oysters and Scallops in Scotland 1988/1989, 1989/1990.

50. Sea Fish Industry Authority - Argyll

Name	:	Sea Fish Industry Authority Marine Farming Unit
Address	:	Ardtoe UK - Acharacle, Argyll PH36 4LD
Head	:	M J.S. Gillespie
Tel.	:	96.785666 **Fax:** 96.785343

Type of institute

Research centre
Fish/Crustacean/mollusc farming/stock enhancement

Research activities
Aquaculture
  Physiology
  Reproduction
  Early stages of reared species
  Growth, nutrition, feed composition, live food production
  for aquaculture
  Development of cultivation systems
  Provision of Information and Advice to Industry

Facilities
Rearing tank
Field site, specialized analytical apparatus

Staff: Scientists: 9 Others: 10

Research budget (ECU/year): 750.000

Description
The marine farming unit of SEAFISH was established at Ardtoe in '65 to investigate marine fish cultivation. This remote location was chosen to take advantage of natural coastal features with access to unpolluted seawater. The main projects underway in 1993 involve: Halibut cultivation.
Wrasse breeding. Lobster stock enchancement. Scallop, mussel and oyster cultivation. Flatfish seacage engineering.
Shellfish cultivation in exposed sites. Industry training and information service.
Facilities:
Finfish Broodstock Tanks. Hatchery Systems for Halibut, Wrasse, Lobsters. Onshore Rearing Tanks, Offshore cages. Suspended and Intertidal sites for Mollusc Cultivation. Field Laboratories. Work boats and Engineering Workshop.

51. Sea Fish Industry Authority - Edinburgh

Name	:	Sea Fish Industry Authority
Address	:	18 Logie Mill-Logie Green Road
		UK - Edinburgh EH7 4HG
Head	:	Peter D. Chaplin
Tel.	:	031.558.3331 **Fax:** 031.558.1442

Type of institute
Research centre

Research activities
Maintenance of high quality of fish and fishery products
National and international marketing opportunities for UK fish products
Consumer attitude and consumer demand studies
Statistics
Economics and econometric analysis
National and international policy studies

Facilities
Data bases
Specialised software

Staff: Scientists: 12 Others: 29

Research budget (ECU/year): 600.000

Description
Projects:
- study of household demand of fish
- relationship between fish quality and price
- contribution to World Groundfish model
- EEC studies concerning North Sea, Irish Sea, also broad regional studies and cost-benefit analysis of infrastructure projects;
- Costs and earnings surveys
- Increasing market for underutilised species
- Also model research of household and catering demand for fish; forecasting developments in the fish industry up to year 2000; study of effect of the Single European Market on the fish industry.

52. Sea Fish Industry Authority - Hull

Name	:	Sea Fish Industry Authority Seafish Technology
Address	:	St Andrew's Quay UK - Hull HU3 4QE
Head	:	J.E. Tumilty
Tel.	:	44.482.27837 **Fax:** 44.482.223310 **Telex:** 597261

Type of institute
Public Research Organisation

Research activities

Resources
- Multi-disciplinary approach to management of marine resources
- Environmental problems and their impact on resources

Fishing techniques
- Testing techniques/instrumentation
- Selectivity of fishing gears
- Hydrodynamics of fishing gears

Upgrading of fishery products and post harvest care
- Improvements in handling and processing of fish and fish products
- Harvesting and handling systems for molluscs
- Purification techniques for bivalve molluscs
- Transport and holding systems for live fish and shellfish

Engineering ship systems and engineering development
- Underwater recording devices for gear monitoring and research

Industry Development
- Regional fisheries studies
- Port development
- Advice to industry

Seabed Information Service
- Advice to users on all matters relating to seabed activity
- Fishing Charts

Facilities

Flume tank
Specialised computer hardware and software
Laboratory scale live holding and purification facilities for crustacea and molluscs
Fish technology laboratory with controlled environmental chamber
Electronics Laboratory
Measurement and recording equipment (for gear and vessel parameters)

Staff: Scientists: 13 Others: 37

Research budget (ECU/year): 2.440.000

Description

Major research activities:
Regional studies aimed at improving potential of fish industry by attention to improved management, development of techniques of catching, processing and distribution.
Development of testing techniques for fishing vessels and fishing gears.
Selectivity of trawls and set nets.

Studies of fishing operations, catches, mixtures of species, discards, by-catches. Development of improved otterboards and other components of fishing gear with a view to reducing energy requirements of fisheries. Improved sonar techniques to provide information on species and size mixtures as an aid to selective fishing techniques. Assisting the industry achieve new standards in quality control of fish and fish products. Maximising the value of marine resources by better post harvest care. Maintaining an up to date information service on seabed activities for fishermen and the offshore industries. Production of fishery charts.

53. Sir Alister Hardy Foundation for Ocean Science - SAHFOS

Name : Sir Alister Hardy Foundation for Ocean Science
Address : c/o Plymouth Marine Laboratory
The Laboratory, Citadel Hill
UK - Plymouth PL1 2PB
Tel. : 0752.222772 Fax: 0752.226865

Type of institute
Private limited company with charitable status

Research activities
Long term monitoring of marine ecosystems
Maintenance of the Continuous Plankton Recorder Survey
Distribution data base on planktonic organisms, including fish larvae

Facilities
Continuous Plankton Recorders
Environmental monitoring packages
60 year data base on North Atlantic Plankton

Staff: Scientists: 3 Others: 9

Research budget(ECU/year): 390,000

Description
SAHFOS was founded in 1990 to take over responsibility for the Continuous Plankton Recorder Survey. Its aims are to extend the scope of the Survey both geographically and in its application to on-going research and fisheries interests. The CPR Survey is the only extensive oceanic basin-wide survey of its type. Data for the CPR are used to interpret the fisheries data, anthropogenic effects and long-term climatic effects.

54. SOAFD - Freshwater Laboratory

Name : Scottish Office Agriculture & Fisheries Department (SOAFD) Freshwater Fisheries Laboratory
Address : Faskally, Pitlochry, UK - Pertshire PH16 5LB
Head : Dr. R.G.J. Shelton
Tel. : 0796.47.2060 **Fax:** 0796.47.3523

Type of institute
Research centre

Research activities
Fisheries management
- Stock assessment
- Environmental problems and their impact on resources
- Fish population dynamics

Aquaculture
- Physiology
- Early stages of reared species
- Aquaculture/environment interaction
- Genetics

Facilities
Data bases, extensive libraries
Specialised computer hardware
Specialised software
Field site, electron microscope, specialized analytical apparatus
Rearing facilities

Staff: Scientists: 20 Others: 13

Research budget (ECU/year): 2.270.000

Description
Main research activities are recording and monitoring national and international fisheries for salmon, sea trout, non migratory trout, eels; studying genetics and ecology of exploited stocks, studying physiology and behaviour of exploited and cultivated stocks (especially salmon); evaluating methods of salmonid stock enhancement; monitoring and studying the scottish FW environment with special reference to land use change and surface acidification.

55. SOAFD - Marine Laboratory

Name	:	Scottish Office Agriculture and Fisheries Department (SOAFD)- Marine Laboratory
Address	:	Victoria Road - PO Box 101 UK - Aberdeen AB9 8DB
Head	:	Dr. A. Hawkins
Tel.	:	224.876544 **Fax:** 224.295511 **Telex:** 73587 MARLab. G

Type of institute
Research centre

Research activities
Fisheries management
  Stock assessment
  Multi-species biological models
  Environmental problems and their impact on resources
Fishing techniques
  Selective gears
  Hydrodynamics of gears
  Other improvements in gears and/or fishing vessels
Aquaculture
  Reproduction
  Disease, immunology, stress

Facilities
Research vessel
Rearing tank
Mesocosm
Data bases, extensive libraries
Field site, electron microscope, specialized analytical apparatus

Staff: Scientists: 99 Others: 69

Research budget (ECU/year): 17.190.000

Description
The Marine Laboratory undertakes a wide range of research projects on commercially important fish and shellfish stocks, aimed at the rational exploitation of marine resources and related environmental problems. Advice is given to the UK Government and other agencies on management measures such as catch quotas and fishing gear regulations. The research programme covers four main topics: assessment of fish and shellfish stocks; fishing technology especially the selectivity of fishing gear; environmental effects of pollution and human activities such as offshore oil production; aquaculture and disease diagnosis/control in farmed fish and shellfish.

56. The Scottish Association for Marine Science - SAMS

Name : The Scottish Association for Marine Science - SAMS
Address : P.O. Box 3
UK - Oban Argyll PA34 4AD, Scotland
Head : Prof. J.B.L. Matthews
Tel. : 631.62244 **Fax:** 631.65518

Type of institute
Scientific Association

Research activities
Deap sea benthic and pelagic ecology
Fundamental studies of marine biology

Facilities
Research vessel
Rearing tank
Controlled environmental chamber
Field site, electron microscope, specialized analytical apparatus
Extensive library
Sea Cage Site

Staff: Scientists: 11 Others: 4

Research budget (ECU/year): 400.000

Description
A scientific association that aims to promote marine science through research and education, with particular reference to Scotland. In addition to its research programme it organizes scientific meetings and is developing an educational programme.

57. UnicoMarine

Name : UnicoMarine
Address : 7 Diamond Centre-Works Rd., Letchworth
UK - Herts SG6 1LW
Head : M.F. Dyer/N.C. Grist
Tel. : 0462.675559

Type of institute
Engineering/consulting firm

Research activities
Fisheries management
  Environmental problems and their impact on resources

Staff: Scientists: 8

Description
Activities:
UnicoMarine Ltd. is an independent consultancy specialising in environmental surveys - Expertise includes analysis of plankton and Benthos samples.
Facilities:
Inshore boat, diving, underwater photography, benthos and plankton collection and analysis.

58. Unilever Research Lab.

Name : Unilever Research Laboratory
Agrobusiness research
Address : Colworth House
UK - Sharnbrook, Bedford MK44 1LQ
Head : R. Mawson
Tel. : 0234.222297 **Fax:** 0234.222202 **Telex:** 82229

Type of institute
Research centre
Processing Industry/ agro-food / diets producer
Fish/crustacean/mollusc farm

Research activities
Aquaculture
  Physiology
  Reproduction
  Genetics and conservation of gametes
  Early stages of reared species
  Growth, nutrition, feed composition, live food production for aquaculture
  Disease, immunology, stress
  Aquaculture/environment interaction
Upgrading of fishery products
  Improvement of techniques for handling, storing, processing, packaging fish
  Novel processing methods for by-products, discards, and process wastes
  Consumer response and demand studies

Facilities
Rearing tank
Data bases, extensive libraries
Specialised computer hardware
Specialised software
Field site, electron microscope, specialized analytical apparatus

Staff: Scientists: 5 Others: 12

59. Univ. Aberdeen - King's College - Fac. Arts and Social Science

Name : University of Aberdeen - King's College
Faculty of Arts and Social Science
Address : Elphinstone Rd.
UK - Aberdeen AB9 2UF
Head : Dr. Don Fox/J. R. Coull
Tel. : 0224.275100 **Fax:** 224.276054 **Telex:** 73458 UNIABN G

Type of institute
University Department/Higher Education Institute

Research activities
Fisheries management
Concequences for fisheries management of extension of territorial limits to 200 miles
Aquaculture
Aquaculture structures, systems, techniques

Staff: Scientists: 1

Description
Fisheries Management around North Atlantic: North Sea, Canada, Norway; detailed study of management in Scotland.
Facilities:
University library.
Library of Torry Marine Laboratory, and liaison with its scientists.
Liaison with Scottish Fishery Inspectorate.

60. Univ. Aberdeen - Marischal College - Biochemistry Dept.

Name	:	University of Aberdeen Department of Molecular and Cell Biology
Address	:	Marischal College UK - Aberdeen AB9 1AS
Head	:	Dr. W.T. Melvin
Tel.	:	224/273110 **Fax:** 224/276054 **Telex:** 73458

Type of institute
Research centre
University Department/Higher Education Institute

Research activities
Aquaculture
Research into genetics and pathology of fish.
Study of fish immune system.
Vaccine development.

Facilities
Well equiped laboratories for biochemistry, microbiology and molecular biology.

Staff: Scientists: 100 Others: 4

Research budget (ECU/year): 2.000.000

Description
Broad range of research in biochemistry, genetics, microbiology and biotechnology.

61. Univ. Aberdeen - Zoology Dept.

Name	:	University of Aberdeen Dept. of Zoology
Address	:	Tillydrone Avenue UK - Aberdeen AB9 2TN
Head	:	P.A. Racey
Tel.	:	0224.27.28.58 **Fax:** 0224.27.23.96 **Telex:** 73458 UNIABN G

Type of institute
University Department/Higher Education Institute
Research Centre

Research activities
Biology of fish, shellfish and marine mammals; physiology - nutrition, growth, diet composition; ecology - food webs, trophic interactions, quantitative methods; fisheries - stock assessment methods especially for clupeoids and cephalopods, multispecies biological models; techniques/instrumentation; aquaculture - parasites and disease control, immunology; environment - pollution, biological effects monitoring, experimental studies.

Facilities
Sea and freshwater aquaria, field site, electron microscope, specialised analytical apparatus, extensive libraries, specialised software.

Staff: Scientists: 56 Others: 33

Research budget (ECU/year): 403.700

62. Univ. Belfast - Queens University

Name	:	The Queen's University of Belfast
Address	:	Belfast, UK-Northern Ireland BT7 1NN
Head	:	D.M. Blair/Dr. A. Ferguson
Tel.	:	44.232.661111 **Fax:** 44.232.247895 **Telex:** 74487

Type of institute
University Department/Higher Education Institute

Research activities
Fisheries management
- Stock assessment
- Environmental problems and their impact on resources

Aquaculture
- Genetics and conservation of gametes
- Aquaculture/environment interaction

Upgrading of fishery products
- Consumer response and demand studies

Facilities
Research vessel
Rearing tank
Data bases, extensive libraries
Field site, electron microscope, specialized analytical apparatus

Staff: Scientists: 8 Others: 2

Description
Research activities mainly concern fish genetics, especially isozyme, mitochondrial DNA and nuclear DNA studies. For example: Population genetic structure and homing in anadromous brown (sea) trout and Atlantic salmon. Hypervariable DNA markers for salmonid population and fitness studies. Biochemical population genetics of brown trout in northwestern Europe. Biology and conservation of the unique fish community of Lough Melvin.

63. Univ. Birmingham - School of Biological Sciences

Name : University of Birmingham
School of Biological Sciences
Address : PO Box 363
UK - Birmingham B15 2TT
Head : Dr. E.W. Taylor
Tel. : 021.4145472 **Fax:** 021.4145925 **Telex:** 338938 G

Type of institute
University Department/Higher Education Institute International

Research activities
Fisheries management
  Environmental problems and their impact on resources
Aquaculture
  Physiology of fish and crustaceans
  Responses of fish to environmental pollution, temperature, salinity and oxygen availability
  Diet manipulation in aquaculture
  Stress in fish
Upgrading of fishery products
  Improvement of techniques for handling, storing, processing, packaging fish

Facilities
Controlled environmental chamber
Field site, electron microscope, specialized analytical apparatus
Cannulation, blood sampling and analysis, respiration etc.

Staff: Scientists: 8 Others: 2

Research budget (ECU/year): 13.400

Description
Research activities - Studies on:
The physiological basis of ammonium and heavy metal toxicity to fish in varying salinities. The nervous and hormonal control of ventilation and heart rate in fish.
Respiration and water, ion and acid-base balance in marine, freshwater and semi-terrestrial crustaceans.
Nervous and hormonal control of circulation and hydromineral balance in crustaceans. The environmental cues for the hormonal control of muscle growth and the physiological changes associated with moulting. Factors limiting the survival of shellfish during their live storage handling and shipment to commercial markets. Applied study of crabs and lobsters together with commercial companies and the Sea Fish Industry Authority.
Facilities: Specialized surgical techniques.

64. Univ. Bradford - Biomedical Sciences

Name	:	University of Bradford Department of Biomedical Sciences
Address	:	W. Yorkshire BD7 1DP UK - Bradford
Head	:	Dr. R. Ash
Tel.	:	0274.733466 **Fax:** 0274.309742 **Telex:** 51309 UNIFD G

Type of institute
Research centre
University Department/Higher Education Institute

Research activities
Nutrition; digestive physiology; oral delivery systems (protein and peptide uptake); aquatic toxicology
Protein absorption in intact free-swimming fish, in relation to nutrition, immuno-modulation, desease treatment, maturation and growth
Aquatic toxicology

Facilities
Expertise in dual cannulation procedures for free swimming fish
Surgical Expertise - Dual cannulation Procedures/Absorption study
Use of isolated hepatocytes for hepato-toxicological studies

Staff: Scientists: 1 Others: 1

Research budget (ECU/year): 13.400

65. Univ. Bristol - Zoology Dept.

Name : University of Bristol
School of Biological Sciences
Address : Woodland road
UK - Bristol BS8 1UG
Head : Dr. P.J. Miller/Prof. R. Tinsley
Tel. : 0242.303764 **Fax:** 0272.257374 **Telex:** 445938

Type of institute
University Department/Higher Education Institute

Research activities
Aquaculture
- Reproduction
- Aquaculture/environment interaction
- Systematics of gobioid fishes
- Biology of small fishes
- Aquariology

Facilities
Laboratory and Aquarium facilities

Staff: Scientists: 1

Research budget (ECU/year): 6.700

66. Univ. Brunel - Biology and Biochemistry Dept.

Name : Brunel University
Dept. of Biology and Biochemistry
Address : UK - Uxbridge-Middlesex UB8 3PH
Head : Professor J.P. Sumpter
Tel. : 0895.74000 **Fax:** 0895.74348

Type of institute
University Department/Higher Education Institute

Research activities
Aquaculture
- Physiology
- Reproduction
- Growth, nutrition, feed composition, live food production

for aquaculture
Disease, immunology, stress
Aquaculture/environment interaction

Facilities
Field site, electron microscope, specialized analytical apparatus

Staff: Scientists: 4 Others: 4

Research budget (ECU/year): 300.000

Description
The research work has the following objectives:
To investigate the role of hormones in controling reproduction.
To investigate the role of hormones in controling growth.
To assess the impact of stress on reproduction and growth.
To assess the factors which control the number and quality of eggs produced.
The research has both fundamental and applied aims.
It is all on trout.

67. Univ. Buckingham - Dept. of Life Sciences

Name : University of Buckingham
Address : The Clore Laboratory for Life Sciences
UK - Buckingham MK18 1EG
Head : Prof. Len Evans
Tel. : 0280.814080 **Fax:** 0280.822245

Type of institute
University Department/Higher Education Institute

Research activities
Fisheries management
Stock assessment
Environmental problems and their impact on resources
Aquaculture
Growth, nutrition, feed composition, live food production
for aquaculture
Genetics

Facilities
Specialised software
Inflatable boat & motor, Echo sounder, Electrofishing etc.
Laboratory facilities for biological analyses

Staff: Scientists: 4 Others: 1

Description
Research projects include:
(1) Studies on wild populations of the signal crayfish (Pacifastacus leninsculus) in the River Ouse, England, and (2) The population genetic structure of brown trout (Salmo trutta) in England and Wales.

68. Univ. Cambridge - Physiological Lab. - RGMER

Name	:	University of Cambridge Physiological Laboratory Research Group in Mammalian Ecology and Reproduction
Address	:	Downing Street UK - Cambridge CB2 3EG
Head	:	Prof. P.A. Jewell

Type of institute
University

Research activities
Mammalian Ecology and Reproduction (Aquatic and restrial)

Facilities
(for aquatic mammals)
Histological laboratory, field equipment for visual and
high-frequency underwater acoustic observations

Staff: Scientists: 4 Others: 2

Research budget (ECU/year): 10.000

Description
Problems in relation with cetacean entanglement in fishing nets
Tank testing of acoustic characteristics of hard pieces
Automatic recognition and analysis of dolphin click rates
Travel strategies of whales from data recorded in the Shetland Islands
Cooperation with Polish institutes in monitoring mammals in the Polish Baltic Sea

69. Univ. College North Wales - School of Ocean Sciences

Name	:	University College of North Wales School of Ocean Sciences
Address	:	Menai Bridge-Anglesey UK - Gwynedd LL59 5EY
Head	:	Prof. E. Naylor
Tel.	:	0248.351151 **Fax:** 0248.716367 **Telex:** 61100 G

Type of institute
University Department/Higher Education Institute

Research activities
Fisheries management
- Stock assessment
- Environmental problems and their impact on resources

Aquaculture
- Physiology
- Reproduction
- Genetics and conservation of gametes
- Early stages of reared species
- Growth, nutrition, feed composition, live food production for aquaculture
- Aquaculture/environment interaction
- Aquaculture structures, systems, techniques

Facilities
Simulated tidal tanks
Mariculture research unit for temperate and tropical species
Controlled environmental chambers
Specialised computer hardware
Field site, electron microscope, specialized analytical apparatus
Research vessel of 30 m length and small boats
Sea water supply, algal culture room, larval culture facilities for marine animals, controlled cage culture unit

Staff: Scientists: 12 Others: 5

Description
Research activities:
Nutritional biology and stock assessment of Irish Sea fish, including basking shark studies. Micro growth patterns and nutritional biochemistry of cultivated marine invertebrates. Artificial food technology for molluscs, crustaceans and fish. Behavioural and environmental impact studies related to fisheries and aquaculture. Molecular studies of bacterial pathogens in aquaculture. Polyploid production and gene transfer in commercial bivalves.

70. Univ. Durham - Dept. of Biological Sciences

Name	:	University of Durham Department of Biological Sciences
Address	:	Sciences Laboratories-South Road UK - Durham DH1 3LE
Head	:	Prof. K. Bowler
Tel.	:	091.374.3340 **Fax:** 091.374.3741

Type of institute
University Department/Higher Education Institute

Research activities
Fisheries management
  Environmental problems and their impact on resources
Aquaculture
  Early stages of reared species
  Growth, nutrition, feed composition, live food production for aquaculture
  Disease, immunology, stress
  Aquaculture/environment interaction

Facilities
Laboratory for monoclonal antibody facility.

Staff: Scientists: 2 Others: 0

Research budget (ECU/year): 43.500

Description
Raising of monoclonal antibodies to crayfish plague fungus.
Several strains of fungus, obtained from Prof. K. Soderhall, Sweden and Dr. D. Alderman, MAFF Laboratories are in culture. The primary aim is to develop a diagnostic kit for identification of the fungus to ensure only fungus free crayfish are transported between bodies of water.

71. Univ. Edinburgh (Heriot-Watt) - Biological Sciences

Name : Heriot-Watt University
Dept. of Biological Sciences
Address : Riccarton
UK - Edinburgh EH14 4AS
Head : Dr. B. Austin/Dr. M.G. Poxton/Dr. J.R. Stark
Tel. : 031.451 3452 **Fax:** 031.451 3009

Type of institute
Research centre
Engineering/consulting firm
University Department/Higher Education Institute
Fish/crustacean/mollusc

Research activities
Fisheries management
- Stock assessment
- Multi-disciplinary approach to fisheries management
- Environmental problems and their impact on resources

Aquaculture
- Physiology
- Early stages of reared species
- Growth, nutrition, feed composition, live food production for aquaculture
- Disease, immunology, stress
- Aquaculture/environment interaction

Facilities
Research vessel
Rearing tank
Controlled environmental chamber
Data bases, extensive libraries
Field site, electron microscope, specialized analytical apparatus
Fish Disease Suite
Aquaria

Staff: Scientists: Others: 9

Research budget (ECU/year): 200.000

Description
Research activities:
In the field of Fisheries Management:
Abundance and distribution of ichthyoplankton around the Firth of Forth; Management of coastal fishery resources in Indonesia.
In the field of Aquaculture:
Effects of stress on farmed fish; effects of natural ultraviolet radiations on farmed fish; Design and management of recycle systems for intensive fish culture.
This dept. is specialized in:
Design and management of fish tanks for intensive fish culture. Control of nutrient accumulation in recycle systems. Development of new forms of live feeds.
Development of low cost diets for marine fish. Development of rapid identification systems for fish pathogens. Development of fish vaccines. Effects of chemotherapeutants on fish and the aquatic environment.

72. Univ. Edinburgh (Heriot-Watt) - Civil Engineering Dept.

Name	:	Heriot-Watt University Dept. of Civil Engineering
Address	:	Research Park Riccarton UK - Edinburgh EH14 4AS
Head	:	Mr. B.T. Linfoot
Tel.	:	031.4495111 **Fax:** 031.4513170

Type of institute
University Department/Higher Education Institute

Research activities
Aquaculture
Aquaculture structures, systems, techniques
Engineering aspects of sea cage for round and flat-fish farming, in exposed locations.
Pollution transport modelling in coastal and estuarine waters.
Wave forecasting in sea lochs.

Facilities
Specialised computer hardware
Specialised software
Field site, electron microscope, specialized analytical apparatus.
Flume tank with random wave generator
Wave basin with wind fans and wave paddles
High speed data logging
Large capacity loading frames for component testing

Staff: Scientists: 4 Others: 2

Research budget (ECU/year): 33.600

73. Univ. Edinburgh (Heriot-Watt) - Inst. of Offshore Engineering

Name	:	Heriot-Watt University Institute of Offshore Engineering
Address	:	Research Park Riccarton UK - Edinburgh EH14 4AS
Head	:	Prof. C.S. Johnson
Tel.	:	031.4493393 **Fax:** 031.4496254 **Telex:** 727918

Type of institute
Research centre
Engineering/consulting firm
University Department/Higher Education Institute

Research activities
Aquaculture
  Aquaculture/environment interaction
  Aquaculture structures, systems, techniques

Facilities
Controlled environmental chamber
Data bases
Extensive libraries
Specialised software

Staff: Scientists: 4 Others: 4

Research budget (ECU/year): 163.000

Description

Research activities:

Concerning Fisheries Management:

Numerous studies involving fisheries issues such as:

Assessment of fishery activity.

Examination of conflicts of use.

Critical assessment of methods used in determining loss of access.

The impacts of underwater explosives and hydrocarbon contamination upon fish.

Interaction between fishing gear and pipelines.

Concerning Aquaculture/environmental interaction:

Impact of particulate wastes from salmonid sea cages on the marine benthos.

Concerning Aquaculture structures, systems, techniques:

Investigation into the feasibility of rearing the tropical tiger shrimp, PENAEUS MONODON, in a recirculating system.

Hatchery techniques for bivalve species.

74. Univ. Essex - Biology Dept.

Name	:	University of Essex Dept. of Biology
Address	:	Wivenhoe Park UK - Colchester CO4 3SQ
Head	:	Dr. C.F. Mason
Tel.	:	0206.873322 **Telex:** 98440 G

Type of institute

University Department/Higher Education Institute

Research activities

Fisheries management

- Stock assessment
- Environmental problems and their impact on resources

Facilities

Field site, electron microscope, specialized analytical apparatus

Staff: Scientists: 1 Others: 2

Research budget (ECU/year): 40.400

75. Univ. Glasgow - Dept. of Naval Architecture & Ocean Engineering

Name : Dept. of Naval Architecture & Ocean Engineering
Address : The University of Glasgow
James Watt Building
UK - Glasgow G12 8QQ
Head : Dr. R.C. McGregor/B.F. Scott
Tel. : 041.3398855 **Fax:** 041.3304808 **Telex:** 777070 UNIGIA

Type of institute
Research centre
University Department/Higher Education Institute

Research activities
Fishing techniques
Testing techniques/instruments
Hydrodynamics of gears
Other improvements in gears and/or fishing vessels

Facilities
Research vessel
Flume tank

Staff: Scientists: 6 Others: 3

Description
The major projects concerned:
Hydrodynamics of fishing vessels.
Fishing vessel design and construction.
Performance trials of fishing SWATH.
Facilities:
Research Vessel:
The SWATH which has been designed for creel fishing would be available for trials relating to this application of SWATH technology or in support of other work were the attributes of a SWATH in seakeeping etc. needed. The rental cost has not been fixed at this time.
Flume Tank:
The Hydrodynamics Laboratory of the Department includes a well equipped towing tank 77 m long by 4.6 m wide with a water depth of up to 2.4 m. The tank is equipped with a full width carriage capable of speeds up to $6.4ms^{-1}$ and a wavemaker which is able to generate wave up to 0.2 m high. The basic rental cost (excluding staff etc) is 600 per day.

76. Univ. Glasgow - Genetics Dept.

Name : University of Glasgow
Departement of Genetics
Address : Church Street
UK - Glasgow G11 5JS
Head : B.L. Cohen
Tel. : 041.339.8855 **Fax:** 041.3305994

Type of institute
University Department/Higher Education Institute

Research activities
Aquaculture
Genetics and conservation of gametes
Genetic studies on marine invertebrates using mitochondrial DNA
RFLP polymorphisms and DNA sequencing

Facilities
Molecular genetic methodologies

Staff: Scientists: 1

77. Univ. Glasgow - Microbiology Dept.

Name : University of Glasgow
Microbiology Department
Address : UK - Glasgow G12 8QQ
Head : Dr. T.H. Birkbeck
Tel. : 041.339.8855 **Fax:** 041.330.4600 **Telex:** 777070

Type of institute
University Department/Higher Education Institute

Research activities
Fisheries management
Environmental problems and their impact on resources
Aquaculture
Early stages of reared species
Growth, nutrition, feed composition, live food production
for aquaculture
Disease, immunology, stress

Facilities
Controlled environmental chamber
Field site, electron microscope, specialized analytical apparatus

Staff: Scientists: 6 Others: 4

Research budget (ECU/year): 350.000

Description
The pathogenesis of AEROMONAS SALMONICDA infections of Atlantic salmon. (T.H. Birkbeck & C. Mackie). The virulence mechanisms of A.SALMONICIDA are being studied and experimental vaccines developed for evaluation in fish. Microbiology of hatchery-reared larval fish and shellfish (T.H. Birkbeck & P. Sinclair). Microbiological problems associated with the rearing of turbot and bivalve (oyster, clam and scallop) larvae are under investigation. Toxins involved in the pathogenic mechanisms of VIBRIO species which cause infections in bivalve larvae have been identified and are being characterized. Sodium channel blocking toxins produced by marine bacteria (T.H. Birkbeck & S. Gallacher). A survey is in progress of the incidence of bacteria from British coastal waters which produce tetrodoxin and other sodium channel blocking toxins. Regulation of production of these toxins and their possible contribution to paralytic shellfish poisoning is under investigation.

78. Univ. Glasgow - Zoology Dept.

Name : University of Glasgow
Dept. of Zoology
Address : The University
UK - Glasgow G12 8QQ
Head : P. Spencer Davies/F.W. Furness
Tel. : 041.3398955 **Fax:** 041.3304808 **Telex:** 777070

Type of institute
University Department/Higher Education Institute

Research activities
Fisheries management
- Environmental problems and their impact on resources

Aquaculture
- Physiology
- Reproduction
- Early stages of reared species
- Disease, immunology, stress
- Aquaculture/environment interaction

Upgrading of fishery products
Improvement of techniques for
handling, storing, processing, packaging fish

Facilities
Field site, electron microscope, specialized analytical apparatus
Recirculating Seawater System

Staff: Scientists: 9 Others: 10

Research budget (ECU/year): 101.400

Description
Research activities:
Interactions of seabirds with fisheries and aquaculture - biology of salmonids and nephrops (particularly physiology and behaviour)
Bioassay methods for coral reefs
Research projects:
Neuropeptides and appetite control in salmon
Primary productivity and nutrient fluxes in brackish fish ponds
Life history patterns and behaviour in salmonids
Studies on seabird fisheries interaction
Post-harvest physiology of shellfish
Behaviour of farmed salmonids

79. Univ. Hull

Name	:	University of Hull Humberside International Fisheries Institute
Address	:	Cottingham road-North Humberside UK - Hull HU6 7RX
Head	:	K. H. Haywood
Tel.	:	482.466421 **Fax:** 482.466205 **Telex:** 592592 KH MAIL G

Type of institute
Institute undergraduate
Studies, post-graduate courses, professional training and
also research and consultancy

Research activities

Fisheries management
- Multi-disciplinary approach to fisheries management
- Environmental problems and their impact on resources
- Fishing techniques
- Electronics/acoustics

Aquaculture
- Growth, nutrition, feed composition, live food production for aquaculture
- Aquaculture/environment interaction

Upgrading of fishery products
- Novel processing methods for by-products, discards, and process wastes
- Consumer response and demand studies

Facilities

Specialised software

Staff: Scientists: 10 Others: 4

80. Univ. Keele - Biological Sciences

Name : Keele University
Dept. of Biological Sciences
Address : UK - Keele Staffordshire ST5 5BG
Head : Dr. D. Hoole/Prof. C. Arme
Tel. : 0782.583028 **Fax:** 0782.630077 **Telex:** 36113 UNKLIB G

Type of institute

University Department/Higher Education Institute

Research activities

Aquaculture
- Disease, immunology, stress

Upgrading of fishery products
- Consumer response and demand studies

Facilities

Field site, electron microscope, specialized analytical apparatus

Staff: Scientists: 5 Others: 2

Research budget (ECU/year): 400.000

Description
Immunological, physiological and ecological studies on the interactions between carp, *Cyprinus* carpio and two pathogenic helminths, the blood fluke, Sanguinicola inermis and tapeworm, Bothriocephalus acheilognathi. Epidemiology and control of fish parasite and affects of pollution on immune response of fish.
Immunological and endocrinological changes in cyprinid fish infected with the metacestode of Ligula intestinalis.

81. Univ. Lancaster - Biological Sciences

Name	:	University of Lancaster Dept. of Biological Sciences
Address	:	Bailrigg UK - Lancaster LA1 4YQ
Head	:	Prof. W.T.W. Potts
Tel.	:	0524.65201 **Fax:** 0524.63806

Type of institute
University Department/Higher Education Institute

Research activities
Fisheries management
  Multi-disciplinary approach to fisheries management
Aquaculture
  Physiology
  Effects of river flow commercial netting and angling in British rivers.
  Ecology of Deep Sea Fisheries

Facilities
Controlled environmental chamber
Field site, electron microscope, specialized analytical apparatus

Staff: Scientists: 4 Others: 1

Research budget (ECU/year): 50.000

Description
1. Study of ionic composition of mitochondria rich cells and respiratory epithelial cells of trout gills in relation to water composition by low temperature x-ray emission analysis.

2. Study of sodium uptake and proton output in normal and voltage clamped gills to determine whether sodium uptake is tightly linked to proton output or whether sodium is taken up across an apical membrane which is polarized by a proton pump.
3. Analysis of commercial net and rod catches of salmon and trout in Aberdeen Dee and Lune in relation to river flow.

82. Univ. Leeds - Dept. Biochemistry

Name : The University of Leeds
Department of Biochemistry and Molecular Biology
Address : UK - Leeds LS2 9 JT
Head : Prof. S.B. Brown
Tel. : 0532.333166 **Fax:** 0532.333167

Type of institute
University Department/Higher Education Institute

Research activities
Fisheries management
Environmental problems and their impact on resources

Facilities
Rearing tank
Standard biochemical techniques

Staff: Scientists: 2 Others: 1

Research budget (ECU/year): 1.400

Description
Research activities:
Characterization of haem metabolism in normal and hyperbilirubinaemic salmonids. Pollution of Scottish rivers caused "jaundice" in salmon.
The biochemistry and physiology of this disease and the mechanism of normal haem metabolism in fish were investigated.

83. Univ. Leeds - Dept. of Electronic and Electrical Engineering

Name	:	The University of Leeds Department of Electronic and Electrical Engineering
Address	:	UK - Leeds LS2 9JT
Head	:	Prof. M.J. Howes
Tel.	:	0532.332001 **Fax:** 0532.332032

Type of institute
University Department/Higher Education Institute

Research activities
Fishing techniques
Electronics/acoustics

Facilities
Device fabrication, electronic circuit design and fabrication

Staff: Scientists: 2

Description
The first major interest of the group is the design of active transducers for the location and tracking of various species. The ultimate goal of the work is to collect data at a land based station on the physical condition and behaviour of individual fish. Work to date has concentrated on the collection, storage, and transmission of data via a non-intrusive electronic tag. This has involved the design of novel transducers and transmission systems. The principles have been established; the requirement is for the investment.
The second major interest is the detection, location and recognition of species using ultrasonic techniques. This stems from Professor M.J. Howes' early experience in this field. The main problems have been identified as of signal processing and work is continuing in this area.

84. Univ. Leicester - Biological Sciences Zoology Dept.

Name	:	University of Leicester School of Biol. Sciences/Dept. of Zoology
Address	:	Adrian Building University Road UK - Leicester LE1 7RH

Head : Prof. H.C. Macgregor
Tel. : 533.523346 **Fax:** 533.522200 **Telex:** 347250 LEICUN G

Type of institute
University Department/Higher Education Institute

Research activities
Fisheries management
- Multi-species biological models
- Multi-disciplinary approach to fisheries management
- Environmental problems and their impact on resources
- Basic research on fish ecology and behaviour

Facilities
Rearing tank
Data bases, extensive libraries
Specialised computer hardware
Field site, electron microscope, specialized analytical apparatus

Staff: Scientists: 6 Others: 3

Research budget (ECU/year): 142.500

Description
The main research activities deal with:
Biological aspects of NEPHPROPS fishery in Scotland Biology and Management of Chilean horse mackerel. Fish feeding behaviour and ecology.
Environmental consequences of river chanelisation schemes.
Facilities:
Freshwater and marine culture equipment and experimental tank rooms.
University vax-cluster, departmental IBM/Macintosh network.
Advanced light and electron microscopy units.

85. Univ. Liverpool - Environmental and Evolutionary Biology

Name : University of Liverpool
Dept. of Environmental and Evolutionary Biology
Address : PO Box 147
UK - Liverpool L69 3BX
Head : Prof. Brian Moss
Tel. : 51.7944997 **Fax:** 51.7086502 **Telex:** 627095 UNIPL G

Type of institute
University Department/Higher Education Institute
of international repute

Research activities
Fisheries management
- Stock assessment
- Environmental problems and their impact on resources

Aquaculture
- Physiology
- Disease, immunology, stress
- Aquaculture and rearing
- Pollution assessment

Facilities
Research vessel
Rearing tank
Controlled environmental chamber
Data bases, extensive libraries
Field site, electron microscope, specialized analytical apparatus

Staff: Scientists: 13

Research budget (ECU/year): 320.000

Description
A group of the Academic staff of the dept. is concerned for part or all of its research activities in the fields of freshwater biology, freshwater and Marine Fishes. A wide range of expertise is available including: African freshwaters, endocrinology, environmental pollution, heavy metals, radionuclides and eutrophication, invertebrate fauna, parasites.
Physiology, population studies, aquatic macrophytes. The marine laboratory at Port Erin, Isle of Man, includes specialists on shore fauna and marine fisheries and populations.

86. Univ. Liverpool - Port Erin Marine Lab.

Name : University of Liverpool
Port Erin Marine Laboratory
Address : Port Erin
UK - Isle of Man
Head : R.D.M. Nash/T.A. Norton
Tel. : 624.832027 **Fax:** 624.835788

Type of institute
University Department/Higher Education Institute

Research activities
Fisheries management
  Stock assessment
  Multi-species biological models
Aquaculture
  Physiology
  Genetics and conservation of gametes
  Early stages of reared species
  Growth, nutrition, feed composition, live food production for aquaculture
  Aquaculture/environment interaction

Facilities
Research vessel
Rearing tank
Controlled environmental chamber
Data bases, extensive libraries
Specialised computer hardware
Specialised software

Staff: Scientists: 14 Others: 4

Research budget (ECU/year): 308.700

Description
Major research activities:
North Irish Sea herring assessments. Scallop and Queenie assessments. Scallop enhancement programme. Models of flatfish nursery grounds. Ecology of sub-tidal macrobenthos and ecological energetics of the common whelk. The effects of disturbance on scallop grounds. Physiology of scallops under rearing conditions. A variety of projects using electrophoretic techniques to examine the genetics of various organisms. Early life history of cultured and wild marine fish. Optimal feeding conditions for salmonids. Nutrition, pigmentation and weaning of turbot. Projects on the effect of shorebased and offshore fishfarms on the marine environment.
Facilities:
The Port Erin laboratory has two research vessels: RV Roagan (24 m) and RV Sula (8 m). R V Roagan is a stern trawler, equipped to undertake fishing and standard marine biological sampling. RV Sula is equipped to undertake diving, shallow-water trawling and standard shallow-water marine biological sampling. The Port Erin Marine Laboratory has a number of long-term data bases. The major data bases relate to Irish Sea temperatures various other physical and chlorophyll data, herring statistics and scallop data.

87. Univ. London - Queen Mary College - CRAB

Name : University of London, Queen Mary
and Westfield College
Centre for Research in Aquatic Biology (CRAB)

Address : Mile End Road
UK - London E1 4NS

Head : Dr. P. Denny

Tel. : 01.9755294 **Fax:** 01.9815986

Type of institute

Research centre
University Department/Higher Education Institute

Research activities

Fisheries management
- Environmental problems and their impact on resources

Aquaculture
- Physiology
- Reproduction
- Disease, immunology, stress
- Aquaculture/environment interaction
- Aquaculture structures, systems, techniques

Facilities

Controlled environmental chamber
Specialised software
Field site, electron microscope, specialized analytical apparatus

Staff: Scientists: 18 Others: 2

Description

CRAB is a centre to promote research in the field of aquatic biology.
It offers consultancy services to private firms, conservation bodies, major industries, government agencies and international organisations.
Major research projects include environmental impact assessment on freshwater, estuarine and coastal ecosystems of developments in Europe, Africa and S.E. Asia.
Pure research projects include studies on the biology of pollutant heavy metals in fish and invertebrates, with particular respect to biomonitoring.

88. Univ. London - Royal Holloway & Bedford New College - Biology Dept.

Name : University of London -
Royal Holloway & Bedford New College
Department of Biology
Address : UK - Egham, Surrey TW20 0EX
Head : J.D. Dodge
Tel. : 0784.435551 **Fax:** 0784.437520 **Telex:** 935504

Type of institute
University Department/Higher Education Institute

Research activities
Fisheries management
Multi-disciplinary approach to fisheries management
Environmental problems and their impact on resources
Aquaculture
Fish diseases
Impact of fish on reservoir ecosystems
Hydroacoustic assessment of fish stocks in lowland rivers.
Environmental technology using freshwater fish as indicators of water quality, including pathogens and pollutants
Water quality and ecosystem management
Freshwater fish pathogens

Facilities
Portable model of Biosonics Double-Beam Acoustics System.
Electric fish and trout monitor for automatic monitoring of water quality.

Staff: Scientists: 4 Others: 4

89. Univ. Loughborough - Dept. Chemical Engineering

Name : University of Technology of Loughborough
Dept. of Chemical Engineering
Address : UK - Loughborough Leicestershire LE11 3TU
Head : Dr. G.M. Hall
Tel. : 509.222517 **Fax:** 0509.231746 **Telex:** 34319 G

Type of institute
University Department/Higher Education Institute

Research activities
Upgrading of fishery products
Novel processing methods for by-products, discards, and process wastes

Facilities
Electron microscope, specialized analytical apparatus
Fish-Processing Laboratory
Fermentation modelling, enzyme pH-stat, membrane
filtration/concentration equipment from 1-100 litres, pilot-scale spray drier (computer controlled). Laboratory equipment includes FPLC, GC, HPLC, electrophoresis, amino acid analysis, functionality tests, Kjeldahl, Soxhlet and microbiology. Support available in electron microscopy, particle sizing, rheology etc.

Staff: Scientists: 3 Others: 1

Research budget (ECU/year): 56.900

Description
Research activities:
Projects on the use of enzymic hydrolysis for recovery of fish protein from process wastes including also membrane concentration, spray drying and assessment of the functional properties of the products. Work has also been started on the use of lactic acid bacteria for the recovery of chitin, astaxanthin and protein from shellfish wastes and the preservation of fish and fish products.

90. Univ. Manchester - Agricultural Economics Dept.

Name : University of Manchester
Department of Agricultural Economics
Faculty of Economic and Social Studies
Address : UK - Manchester M13 9PL
Head : Trevor Young
Tel. 061.275 4794 **Fax:** 061.275 4929 **Telex:** 666517 Uniman

Type of institute
University Department/Higher Education Institute

Research activities
Consumer response and demand studies
Economics, econometrics

Staff: Scientists: 2

Description
Analysis of Market Shares for fishery products over time. Study of household demand for fish. Analysis of changing consumer tastes and preferences.

91. Univ. Marine Biological Station Millport

Name : University Marine Biological Station Millport
Address : Isle of Cumbrae
UK - Scotland KA28 OEG
Head : John Davenport(Director)
Tel. : 0475.530581 **Fax:** 0475.530601

Type of institute
University Department/Higher Education Institute

Research activities
Fisheries management
- Stock assessment
- Multi-species biological models
- Environmental problems and their impact on resources

Fishing techniques
- Hydrodynamics of gears

Aquaculture
- Reproduction
- Early stages of reared species

Facilities
Research vessel
Flume tank
Rearing tank
Data bases, extensive libraries
Specialised computer hardware
Specialised software
Field site, electron microscope, specialized analytical apparatus

Staff: Scientists: 9 Others: 17

Research budget (**ECU/year**): 874.600

Description

Main research activities:
Effect on growth of commercial species of flat fish.
Novel feed stuffs. Reproductive success of herring.
Population studies and behaviour studies on scampi.
Red tides.
Growth, nutrition, feed composition, live food production for aquaculture. Disease, immunology, stress.
Aquaculture/environment interaction.
Novel processing methods (biotechnological, physical, chemical) for food and by-products, including recovery of edible protein and other components from fish and shellfish and processes and products to use discards, by-catch and process wastes.

92. Univ. Newcastle upon Tyne - Dept. Biology - Dove Marine Lab.

Name : University of Newcastle-upon-Tyne
Address : Cullercoats
UK - North Shields NE30 4PZ
Head : Dr. S. Evans
Tel. : 091.2524850 **Fax:** 091.2611182 **Telex:** 53654 G

Type of institute
Research centre
University Department/Higher Education Institute

Research activities
Fisheries management
Environmental problems and their impact on resources
Aquaculture
Physiology
Early stages of reared species
Growth, nutrition, feed composition, live food production for aquaculture
Disease, immunology, stress
Upgrading of fishery products
Consumer response and demand studies
Long term monitoring

Facilities
Research vessel
Rearing tank
Controlled environmental chamber
Data bases, extensive libraries
Field site. Full range of analytical equipment and electron microscope on main university campus.

Staff: Scientists: 5 Others: 5

Research budget (ECU/year): 154.700

Description
The main research activities deal with:
Nephrops norvegicus assessment, effects of over-fishing.
Disease and condition in estuarine Platichthys flesus in relation to pollution.
Endocrinology of polychaete ragworm, Nereis. Rearing of Nereis and dover sole Solea.
Energetics of Nereis and Solea. Disease in Nereis.
Biology of Nereids and potential as aquaculture feedstuff, pollution and disease in estuaries. Population dynamics of Nephrops. Bioassay Community (shore) ecology.

93. Univ. Oxford - Dept. Engineering Science

Name : University of Oxford
Department of Engineering Science
Address : Parks Road
UK - Oxford OX1 3PJ
Head : J.M. Brady
Tel. : 0865.73000 **Fax:** 0865.73010

Type of institute
University Department/Higher Education Institute

Research activities
Fishing techniques
- Electronics/acoustics
- Hydrodynamics of gears
- Other improvements in gears and/or fishing vessels

Upgrading of fishery products
- Methods for grading and quality assessment of fish
- Consumer response and demand studies

Staff: Scientists: 12 Others: 2

Description

Examples of projects carried out:

Control methodology applied to national fishing strategies. Signal and Knowledge integration with Directional control for multisensory systems.

Sensor-guided autonomous vehicles.

Detection of fatigue cracks in gears by vibration analysis. Research on "Shape Representation".

94. Univ. Oxford - Dept. Zoology

Name : University of Oxford
Department of Zoology
Address : South Parks Road
UK - Oxford 0X1 3PS
Head : R. Anderson
Tel. : 0865.271234 **Fax:** 0865.310447

Type of institute

University Department/Higher Education Institute

Research activities

Fisheries management
  Multi-species biological models
Aquaculture
  Reproduction
  Genetics and conservation of gametes

Facilities

Coldwater and tropical aquaria

Staff: Scientists: 7 Others: 1

Description

Main research activities concern:

Multispecies biological models and the application of chaos theory in fishery management.

Investigation of schooling behaviour.

Studies of sexual selection with particular reference to the role of parasite and disease.

Genetic diversity and evolution of intraspecific variation in tropical freshwater fish species.

Conservation of fish stocks.

95. Univ. Reading - Animal and Microbial Sciences

Name : University of Reading
School of Animal and Microbial Sciences
Address : AMS Building, Whiteknights
P.O. Box 228
UK - Reading RG6 2AJ
Head : A.R. Jones
Tel. : 0734.875123 **Fax:** 0734.310180 **Telex:** 847813

Type of institute
University Department/Higher Education Institute

Research activities
Fisheries management
  Environmental problems and their impact on resources
Aquaculture
  Physiology
  Reproduction
  Early stages of reared species
  Growth, nutrition
  Aquaculture/environmental interactions
  Ecotoxicology
  Genetics

Facilities
Field site, specialized analytical equipment, coldwater and tropical aquaria.

Staff: Scientists: 4 Others: 1

Description
Projects include gynogenesis using heterologous sperm. Genetic and hormonal sex manipulation. Effects of toxicants on aquatic species. Aquatic resource management in developing countries. Feeding strategies of crayfish.

96. Univ. Salford - Dept. Biological Sciences

Name : University of Salford
Department of Biological Sciences
Address : UK - Salford M5 4WT
Head : Dr. R. Lawson

Type of institute
University

Facilities
Facilities are available for electron microscopy, GLC, HPLC, FPLC, GCMS and for many aspects of fluorescence spectroscopy

Staff: Scientists: 21 Others: 17

Research budget (ECU/year): 809.800

Description
The research group which is involved with fish is concerned with the molecular taxonomy of elasmobranchs and this work seeks to build taxonomies on the basis of comparative immunology and molecular biology.

97. Univ. Southampton - Dept. Biology

Name : Department of Biology
University of Southampton
Address : Medical & Biological Sciences Building
Basset Crescent East
UK - Southampton Hampshire SO9 3TU
Head : Prof. Norman Maclean (Genetics)
Tel. : 0703.594403/594256 **Fax:** 0703 594269 **Telex:** 47661

Type of institute
University Department/Higher Education Institute

Research activities
Aquaculture
Fish genetics
Gene cloning and biotechnology

Facilities
Rearing tanks
Hot and cold water aquaria
Data bases
Electron microscope
Molecular biology laboratory
Well equipped for gene cloning, gene manipulation, and transgenic induction.

Staff: Scientists: 6 Others: 10

Research budget (ECU/year): 75.000

Description
University facility consisting of purpose built hot and cold aquaria for rearing of tilapia salmonids, and zebra fish. Trangenic induction in fish by microinjection and electroporation. Gene cloning, manipulation and sequencing. DNA fingerprinting technology. Fish tissue culture and cytogenetics. Fish genetics, reproduction, and endocrinology.

98. Univ. Southampton - Dept. Oceanography

Name : University of Southampton
Department of Oceanography
Address : UK - Southampton S09 5NH
Head : Dr. I.S. Robinson
(Physical Oceanography)
Head BIOL. GROUP: Prof. A.P.M. Lockwood (Marine Biology)
Tel. : 0703.595000 **Fax:** 0703.593059 **Telex:** 47661

Type of institute
University Department/Higher Education Institute

Research activities
Fisheries management
Multi-species biological models
Environmental problems and their impact on resources
Aquaculture
Crustacean fishery enhancement
Invertebrate immunological responses
Fish genetics

Facilities
Research vessel (38' launch)
Flume tank
Simulated tidal tank
Rearing tank
Artificial reef
Controlled environmental chamber
Data bases, extensive libraries
Field station, specialized analytical equipment
Specialised computer hardware and software

Staff: Scientists: 50 (Biology Group)

Description
The Biology Group of this multidisciplinary department has special interests in the fields of phytoplankton, shelf and deep ocean benthic production, zooplankton ecophysiology and population dynamics, invertebrate physiology, behaviour and reproduction, the genetics of groupers, the physiology and biochemistry of stress assessment, the ecology of lagoons, lobster behaviour and the ecology of artificial reefs. Biologically related aspects of other groups include, the modelling of viral destruction of phytoplankton blooms and satellite studies of phytoplankton and of the coral reefs in Fiji.
The department is expected to occupy a new building with the Institute of Ocean Sciences, Deacon Laboratory during 1995.

99. Univ. St Andrews - Gatty Marine Lab. - Biological and Preclin. Medicine

Name : University St Andrews/Gatty Marine Laboratory
Dept. of Biology and Preclinical Medicine
Address : St Andrews
UK - Fife KY16 8LB
Head : I.A. Johnston
Tel. : 0334.76161/7104 **Fax:** 0334/78299

Type of institute
University Department/Higher Education Institute

Research activities
Aquaculture
  Early stages of reared species
Upgrading of fishery products
  Consumer response and demand studies

Facilities
Research vessel
Flume tank
Rearing tank
Controlled environmental chamber
Field site, electron microscope, specialized analytical apparatus

Staff: Scientists: 20 Others: 12

Research budget (ECU/year): 1.300

Description
The main focus of the research related to fisheries concerns the developmental biology of fish larvae and physiological responses to environmental stress, particularly temperature. Other research groups are concerned with muscle physiology, growth and energy budgets, and osmoregulation. A particular feature of the laboratory is the excellent experimental marine aquaria and fish rearing facilities.

100. Univ. Stirling - Institute of Aquaculture

Name : University of Stirling
Institute of Aquaculture
Address : UK - Stirling FK9 4LA
Head : R. Roberts/J.H. Brown/R.H. Richards
Tel. : 0786.73171 **Fax:** 0786.51462 **Telex:** 777759 STUNIV G

Type of institute
University Department/Higher Education Institute
Fish/Crustacean/mollusc farming
Engineering/consulting firm
Research centre

Research activities
Fisheries management
  Stock assessment
  Environmental problems and their impact on resources
Aquaculture

Facilities
Rearing tank
Data bases, extensive libraries
Specialised software
Field site, electron microscope, specialized analytical apparatus

Staff: Scientists: 57 Others: 43

Research budget (ECU/year): 2.074.400

Description

Research activities:

The research activities have been grouped into four thrust areas: disease, genetics and reproduction, environmental impact and nutrition. In addition, there is an aquaculture systems group which acts as an applied focus for the four research groups.

There are six major continuing research programmes:

Prawn culture, genetics, disease and nutrition. Temperate and tropical fish diseases. Environmental impact of aquaculture. Applied nutrition. Fish and shellfish immunology. fish genetics and reproduction.

Facilities:

Laboratories dedicated to genetics, biotechnology, reproductive physiology, nutrition, environmental studies, virology, micro-biology, parasitology and vaccine development. Gas chromatography equipment.

Electron microscope suite. Extensive aquarium facilities for both tropical and cold water species.

101. Univ. Strathclyde - Dept. Statistics and Modelling Science

Name	:	University of Strathclyde
Address	:	UK - Glasgow GI IXH
Head	:	Pr. W. Gurney
Tel.	:	41.5324400 **Telex:** 77472

Type of institute

University Department/Higher Education Institute

Research activities

Fisheries management

- Stock assessment
- Multi-species biological models
- Environmental problems and their impact on resources

Facilities

Specialised computer hardware

Specialised software

Staff: Scientists: 15 Others: 2

102. Univ. Strathclyde - Ship and Marine Technology

Name	:	University of Strathclyde Ship & Marine Technology Dept.
Address	:	Marine Technology Centre 100 Montrose St UK - Glasgow G4 OLZ
Head	:	Prof. C. Kuo
Tel.	:	041.5524400 **Fax:** 41.5520775 **Telex:** 77472

Type of institute
Research centre
University Department/Higher Education Institute

Research activities
Fishing techniques
  Electronics/acoustics
  Other improvements in gears and/or fishing vessels
Aquaculture
  Aquaculture/environment interaction
  Aquaculture structures, systems, techniques

Facilities
Flume tank
Specialised software

Staff: Scientists: 5 Others: 2

Research budget (ECU/year): 26.900

103. Univ. Surrey - School Biological Sciences - Food Safety Research Group

Name	:	University of Surrey School of Biological Sciences Food Safety Research Group
Address	:	Guilford UK - Surrey GU2 5XH
Head	:	Professor L.J. King/Dr. M.N. Clifford

Type of institute
University

Research activities
Food safety, mechanisms of toxicity

Staff: Scientists: 6 Others: 20

Research budget (ECU/year): 1.000.000

Description
Toxicant burden-occurrence; identification; origin; avoidance.
Interpreting toxicity data-interspecies differences; diet.
Anutrients-dietary phenols; tennins; metabolites.
Protein chemistry-chemical reactions; ligand binding; acceptability.
Foodborne pathogens-physiology; control; detection.

104. Univ. Ulster - Freshwater Lab.

Name	:	University of Ulster Freshwater Laboratory
Address	:	Traad Point-Ballyronan UK - Co Londonderry, Northern Ireland BT45 6LR
Head	:	Prof. R.B. Wood

Type of institute
Research/teaching

Research activities
All aspects of freshwater research
Particularly water chemistry, aspects of pollution and water management, and algal, benthic invertebrate, zooplankton and fish ecology

Facilities
8 m boat and equipment for sampling water and organisms,
Sophisticated equipment for microscopy and water chemistry (e.g. AAS, mass spectroscopy, gas & HPLC).
Aquarium, microbiology and isotope rooms. Well equipped for most aspects of aquatic ecology and computing.

Staff: Scientists: 8 Others: 8

Research budget (ECU/year): 300.000

Description
The laboratory is situated on the shores of Lough Neagh, the largest freshwater lake in Britain and Ireland. There is accomodation for up to 20 people and self catering facilities.

105. Univ. Wales - School of Pure and Applied Biology

Name	:	University of Wales College of Cardiff School of Pure and Applied Biology
Address	:	P.O. Box 915 UK - Cardiff CF1 3TL
Head	:	Prof. M.F. Claridge MA DPhil (oxon) FIBiol

Type of institute
University

Research activity
Biology of fish and aquatic ecosystems

Facilities
Fresh water and marine Aquaria;
Field centre, well placed and equipped for limnological and riverine research;
Genetic fingerprinting and electrophoretic laboratories;
Electron microscopy and microanalysis;
Toxicology unit.

Staff: Scientists: 9 Others: var.

Research budget (ECU/year): 5.000.000

Description
Recent and ongoing research includes:
Fish population dynamics and life history strategy; genetic and morphological definition of salmonids; effects of heavy metals and other pollutants on freshwater fish; biology of freshwater and marine microbes, algae and macroinvertebrates; crayfish distribution and the spread of crayfish plague; population and ecosystem modelling.

106. Univ. Wales - Swansea

Name : University College of Swansea
Address : Singleton Park-Wales
UK - Swansea SA2 8PP
Head : Prof. J.A. Beardmore
Tel. : 792 205678 **Fax:** 0792 295447 **Telex:** 48149 UTCS G

Type of institute
University Department/Higher Education Institute

Research activities
Fisheries management
  Stock assessment
Aquaculture
  Reproduction
  Genetics and conservation of gametes
  Disease, immunology, stress

Facilities
Controlled environmental chamber
Field site, electron microscope, specialized analytical apparatus
Research vessel (10 m)
Wet fish room; large deep freeze
Large warm water aquarium. Sea water aquarium

Staff: Scientists: 8 Others: 25

Description
Research in the fisheries addresses mainly in the following:
1. Reproductive and pre-recruit biology and population genetic aspects of DICENTRARCHUS LABRAX (Prof. J.S. Ryland, Dr. S.E. Shackley, Dr. G.R. Carvalho).
2. Growth, reproduction and population biology of marine fish, both pelagic and demersal, currently including ALOSA FALLAX, CLUPEA HARENGUS, LOPHIUS spp., and SPRATTUS SPRATTUS. (Prof. J.S. Ryland, Dr. S.E. Shackley).
3. Systematics and population genetics of commercially exploited squid (Dr. G.R. Carvalho).
4. Ecotoxicology and pollution effects on fish (Dr. S.E. Shackley).
5. Population analysis of molluscan and crustacean shellfish (Dr. J. Moyse, Prof. J.S. Ryland).
6. Stock improvement in TILIPIA spp. through the application of genetic methods (Prof. J.A. Beardmore, Dr. D.O.F. Skibinski).
7. Population genetics of supreme bivalves (Dr. D.O.F. Skibinski).
8. Genetic factors in reproductive success (Prof. J.A. Beardmore).

9. Stock discrimination in relation to the assessment of Adriatic anchovy and sardine fisheries (Dr. G.R. Carvalho).
10. Genetic diversity and stock structure of Lake Tanganyika kapenta Limnothrissa miodon, Clupeidae) introduced to African lakes (Dr. G.R. Carvalho).
11. Genetic fingerprinting and strain identification of *Tilapia* (Dr. G.R. Carvalho).

107. Univ. East Anglia - School of Biological Sciences

Name : University of East Anglia
School of Biological Sciences
Address : UK - Norwich NR4 7TJ
Head : Dr. I. Gibson
Tel. : 0603.56161 **Fax:** 0603.592250

Type of institute
University Department

Research activities
Fisheries management
Molecular genetics of fish stocks
Diseases
Parasites of flatfish
Conservation
Evolution and ecology of fishes in relation to environmental disturbance

Facilities
Molecular genetic methodologies
Recirculating seawater system, freshwater aquarium rooms
Population dynamics modelling/computing
MAFF Fisheries Laboratory, Lowestoft is Associate Institute with research and teaching collaboration including tanks, labs, library and vessels.

Staff: Scientists: 3 Others: 2

Research budget(ECU/year): 84.700

Description
Research activities:
Identification of molecular genetic variation in marine organisms, concentrating on stock discrimination in Gadoid fish (G.M. Hewitt, C. Rico).
Biology of monogenean parasites of fishes (G.C. Kearn).
Evolution of mating systems and life histories of coastal fish populations in relation to pollution and other forms of disturbance (J.D. Reynolds).

108. Virus Reference Laboratory

Name : Virus Reference Laboratory
Address : Central Public Health Laboratory
61 Colindale Avenu
UK - London NW9 5HT
Head : Dr. P.P. Mortimer

Type of institute
Provides National Public Health
Virology Reference Services

Research activities
Broad based research in public health virology, wide interest in the diagnostics of a viral agent associated with gastroenteritis.
A specific interest in the role of shellfish in human
gastroenteritis and methods of inactivating viruses in shellfish.

Staff: Scientists: 60 Others: 22

109. Water Research Centre - Medmenham Lab.

Name : WRC plc
Address : Henley Road-PO Box 16, Medmenham
UK - Marlow SL7 2HD, Bucks
Head : Dr. J. Moss/Dr. A. Dobbs
Tel. : 0491.571531 **Fax:** 0491.579094

Type of institute
Research centre

Research activities
Fisheries management
- Stock assessment
- Multi-disciplinary approach to fisheries management
- Environmental problems and their impact on resources

Aquaculture
- Aquaculture/environment interaction

Facilities
Controlled environmental chamber
Data bases, extensive libraries, toxicity databases
System of 8 mesocosms (4 x 4 x 1 m)
Electric-fishing equipment
Specialised analytical equipment

Staff: Scientists: 10 Others: 2

Description
Research projects:
Salmon Tracking in Ribble Estuary. - Stock policies for put and take fisheries. - Review of UK Fish Stock Management Techniques. - Protein-binding of heavy metals. - Water quality requirements of non-salmonid fish and setting of Environmental Quality Standards. - Effect of Chrysodine on carp. - EIFAC report on aluminium effects on fish. - Status of freshwater fish in relation to water quality. - Setting standards for storm sewage discharges. - Antibiotics in fish farm effluents. - Effects of sludge disposal. - Development of an intake protection system using fish responses. - Environmental impact of fish farms. - Farm pollution and impact on rivers. - Effect of land-use change on salmonid fisheries. - Development of fishery classification schemes. - Development of salmonid stream habitat evaluation system. - Development of fish stock assessment methodologies and methods. - Assessment of best stocking practices. - Survival and dispersal of stocked coarse fish. - Review of implementation of EC freshwater fish directive.

Index

MIX
Papier aus verantwortungsvollen Quellen
Paper from responsible sources
FSC® C105338

If you have any concerns about our products,
you can contact us on
ProductSafety@springernature.com

In case Publisher is established outside the EU,
the EU authorized representative is:
Springer Nature Customer Service Center GmbH
Europaplatz 3, 69115 Heidelberg, Germany

Printed by Libri Plureos GmbH
in Hamburg, Germany